The Existential Toolkit for Climate Justice Educators

The Existential Toolkit for Climate Justice Educators

How to Teach in a Burning World

Edited by   JENNIFER ATKINSON and SARAH JAQUETTE RAY

UNIVERSITY OF CALIFORNIA PRESS

University of California Press
Oakland, California

© 2024 by The Regents of the University of California

Library of Congress Cataloging-in-Publication Data

Names: Atkinson, Jennifer Wren, editor. | Ray, Sarah Jaquette, editor.
Title: The existential toolkit for climate justice educators : how to teach in a burning world / edited by Jennifer Atkinson and Sarah Jaquette Ray.
Description: Oakland, California : University of California Press, [2024] | Includes bibliographical references and index.
Identifiers: LCCN 2023044871 (print) | LCCN 2023044872 (ebook) | ISBN 9780520397118 (hardback) | ISBN 9780520397125 (paperback) | ISBN 9780520397132 (ebook)
Subjects: LCSH: Climate justice—Study and teaching. | Social justice and education.
Classification: LCC GE220 .E95 2024 (print) | LCC GE220 (ebook) | DDC 304.2/8071—dc23/eng/20231108
LC record available at https://lccn.loc.gov/2023044871
LC ebook record available at https://lccn.loc.gov/2023044872

Manufactured in the United States of America
33 32 31 30 29 28 27 26 25 24
10 9 8 7 6 5 4 3 2 1

CONTENTS

Acknowledgments and Gratitude ix

Introduction. Climate, Justice, and Emotions in the Classroom: Why a Toolkit? 1
Sarah Jaquette Ray and Jennifer Atkinson

PART I Getting Started with Emotions in the Climate Justice Classroom

1. A Pedagogy for Emotional Climate Justice 17
   Blanche Verlie

2. Balancing Feelings and Action: Four Steps for Working with Climate-Related Emotions and Helping Each Student Find Their Calling 22
   Andrew Bryant

3. Transformative Psychological Approaches to Climate Education 31
   Leslie Davenport

4. From Existential Crisis to Action Planning: Building Individual and Community Resilience 42
   Jessica D. Pratt

5. Empathy and Care: Activities for Feeling Climate Change 49
   Sara Karn

6. The Emotional Impact Statement 55
   Christie M. Manning

7. The Politics of Hope 59
   Daniel Chiu Suarez, Sophie Chalfin-Jacobs, Hannah Gokaslan, Sidra Pierson, and Annaliese Terlesky

8. Unfucking the World 74
   Leif Taranta

PART II Justice as Affective Pedagogy

9 Preparing Students to Navigate a Harrowing Educational Landscape: Accessibility and Inclusion for the Climate Justice Classroom   81
Ashley E. Reis

10 Photovoice for the Climate Justice Classroom: Inviting Students' Affective and Sociopolitical Engagement   89
Carlie D. Trott

11 Leveraging Affect for Climate Justice   99
Michelle Garvey

12 Infrastructure Affects: Registering Impressions of Mega-Dams   108
Richard Watts

13 From Principles to Praxis: Exploring the Roots and Ramifications of the Environmental Justice Movement   116
Shane D. Hall

PART III Embodied Pedagogies

14 Working with Ecological Emotions: Mind Map and Spectrum Line   123
Panu Pihkala

15 Building Somatic Awareness to Respond to Climate-Related Trauma   133
Emily (Em) Wright

16 Using Poetry to Resist Alienation in the Climate Change Classroom   141
Magdalena Mączyńska

17 Prompts for Feeling-Thinking-Doing: Somatic Speculation for Climate Justice   147
Sarah Kanouse

PART IV Futurity, Narrative, and the Imagination: Visualizing What We Desire

18 The Tool of Imagination   155
Doreen Stabinsky and Katrine Oesterby

19 Overcoming the Tragic   162
Peter Friederici

20 Practicing Speculative Futures   170
April Anson

21 Cultivating Radical Imagination through Storytelling   177
Summer Gray

PART V Unsettling Pedagogies: Discomfort and Difficult Knowledge

22 Critical Journalism, Creative Activism, and a Pedagogy of Discomfort   187
Kimberly Skye Richards

23 Why Worry? The Utility of Fear for Climate Justice   193
    Jennifer Ladino

24 The Social Ecology of Responsibility: Navigating the Epistemic and Affective Dimensions of the Climate Crisis   201
    Audrey Bryan

25 Beyond the Accountability Paradox: Climate Guilt and the Systemic Drivers of Climate Change   210
    Marek Oziewicz

PART VI   Joy and Resilience as Resistance

26 Joyful Climate Work: The Power of Play in a Time of Worry and Fear   221
    Casey Meehan

27 Finding Hope in the Influence and Efficacy of Native/Indigenous Rights   228
    Kate Reavey

28 Teaching Climate Change Resilience through Play   234
    Jessica Creane

29 Building Capacity for Resilience in the Face of Environmental Shocks   240
    Abosede Omowumi Babatunde

30 Releasing Growth   247
    Terry Harpold

31 Ecotopia versus Zombie Apocalypse: Collaborative Writing Games for Existential Regeneration   254
    Marna Hauk

PART VII   Community, Collaboration, and Kinship

32 Facilitating "R&R": Student-Led Climate Resilience and Resistance   267
    Jessica Holmes

33 Climate Justice and Civic Engagement Across the Curriculum: Empowering Action and Fostering Well-Being   272
    Sonya Remington Doucette and Heather U. Price

34 Come for Climate, Stay for Community: Acting, Emoting, and Staying Together through the Climate Crisis   279
    Alissa Frame, Charlotte Graf, Lydia O'Connor, Jillian Scannell, Amy Seidl, and Emma Wardell

35 The Climate Imaginary: Reading Fiction to Make Sense of the Climate Crisis   285
    Benjamin Bowman, Chloé Germaine, Pooja Kishinani, and Charlie Balchin

PART VIII **These Skills Are Needed in the World: Career Planning for the Climate Generation**

36   How Will Climate Change Affect My Career?   295
     Debra J. Rosenthal, Jeffrey Johansen, and Ruth Jacob

37   Fostering Student Agency for Climate Justice through Vocational Exploration   299
     Rachel F. Brummel

Appendix: Chapters Sorted by Themes   307

List of Contributors   313

Index   317

ACKNOWLEDGMENTS AND GRATITUDE

To the students, who are our teachers, whose love for the world and for climate justice pushes us to do better.

To the authors in this collection, and all the climate educators, activists, and leaders working to create a vibrant and livable future.

To the Rachel Carson Center for supporting the Existential Toolkit conference that launched this book project and to Elin Kelsey, who helped facilitate that gathering.

To all the participants in the Existential Toolkit for Climate Justice Educators discussion series.

To the Mind & Life Institute, especially Vivian Valentin and Kevin Gallagher, for hosting a symposium on climate emotions and education in 2021.

To Trang Dang for developing our website and to Sydney Verga for keeping us organized through the initial phase of this project.

To Dan Suarez, Jade Sasser, Blanche Verlie, Maria Vamvalis, Panu Pihkala, and the myriad educators who are so passionate about having these conversations with us.

To the funders who made this work possible through grants and other support: the University of Washington Royalty Research Fund and the UW Bothell Scholarship, Research, and Creative Practice Seed Grant Program.

To our institutions, flawed and fabulous, University of Washington Bothell and Cal Poly Humboldt.

Thank you.

INTRODUCTION

Climate, Justice, and Emotions in the Classroom

Why a Toolkit?

SARAH JAQUETTE RAY and JENNIFER ATKINSON

The climate predicament is demanding that higher education radically change. If we are in a womb, not a tomb, as the poet Valarie Kaur stated in a speech in 2016,[1] how are educators to help midwife what's to come?

It is no longer morally or strategically appropriate to berate students with dire forecasts and reports of the sixth great extinction, to teach case studies of slow violence and ecological destruction, or present problems so deep and entrenched that solutions appear elusive. With students coming into the classroom already aware of how bad things are, the old model of scaring them into caring is no longer working.

Even worse, the doom and gloom model is backfiring, as shown by a growing body of scholarship on the role of emotions in climate action. In *Living in Denial,* Kari Norgaard demonstrates that the "information deficit" assumption of most environmental messages, which maintains that people and systems will change when enough climate facts are heard, is a myth. Instead, apathy is the most likely result of these doom-focused change messages, as climate psychologists like Renee Lertzman have demonstrated. Even if pro-social, pro-environmental, and activist behavior are the initial responses to the fear and urgency peddled in most climate messages, the longer-term costs (to the planet and to our students) of climate anxiety are well documented.[2] Indeed, Haltinner and Sarachandra have shown that skepticism is actually a reasonable way to cope with being overwhelmed by such messages. Environmental melancholia, eco-phobia, degradation desensitization,[3] and other emotion-focused terms are emerging to explain how the apocalyptic approach to climate messaging simply doesn't work. If psychologists and climate communication experts know that piling on the doom makes us feel powerless, despairing, skeptical, and apathetic, why are so many of us still teaching this way?

One fundamental premise of this collection arises from the insight that many educators have come to: emotions shape not only students' capacity for learning,[4] but also their ability to respond meaningfully and effectively to what they're learning. Most environmental educators want students to feel inspired to engage in sustained, lifelong action in service of climate justice, environmental protection, and both personal and societal healing. And most educators are aware that we are at a turning point in education, requiring enhanced skills to respond to these times.[5] But few are aware of what psychologists, affect theorists, narrative experts, and students of most wisdom traditions and social movements all know—that the key to these outcomes is not logic, information, data, or facts, but (as much, and relatedly) *emotion*. This book seeks to bring the most up-to-date research on climate emotions to bear on college-level pedagogy.

Further, learning how to best teach "difficult knowledge"[6] has long been necessary in fields such as gender, sexuality, and women's studies, critical race theory, ethnic studies, Native American and Indigenous studies, history, social work, child development, and other fields where trauma and violence characterize the central human experience, both in the content and for the students in the room. Learning trauma-informed pedagogy to facilitate healing conversations with students is essential for educators in these fields.

Yet climate educators have mostly viewed their subject matter as technological, ecological, physical, atmospheric, or at best, only distantly related to human systems, much less to trauma, injustice, and oppression. This approach reflects the relative insulation of most faculty and institutions of industrialized nations from the worst effects of climate change. The pedagogy of climate studies has primarily emerged from a Western, Eurocentric, positivist epistemology, often failing to engage in pedagogies of power and injustice—much less offering support to traumatized students.

As climate education has become enshrined as a topic of the sciences, some experts exert their privilege in part by remaining oblivious to the ways climate change is changing higher education, how student demographics are shifting, and the fact that climate change is directly and indirectly harming students and their communities. Yet it is no longer effective or ethical to teach climate topics without some training in trauma-informed pedagogy and without an analysis of white supremacy, heteronormative patriarchy, and Indigenous epistemology—lenses which are the domain of fields typically not in the sciences. In this collection, we aim to move climate pedagogy in these directions, bringing other frameworks to bear, in addition to the insights of psychology, on our awareness of how climate information works on and through students.

Another argument for centering emotions in climate pedagogy is that educators themselves are exhausted, overwhelmed, and often wrestling with climate emotions like despair, anger, apathy, and hopelessness. The coronavirus pandemic has made matters worse. It revealed in stark relief that the working conditions of educators, including at the college level, are precarious and exploited. This was worse for female faculty, and more so for female faculty of color, faculty with disabilities, and any faculty who were also relied upon for caregiving of children or elders. The absence of a social safety net, the conditions for social reproduction, and access to technology and medical support profoundly affected educators and their students, turning spaces of education into spaces of survival where basic needs for food, shelter, and medicine took precedence over academic learning, much less content acquisition, career advancement, and skill development. Many environmental studies courses already highlight the larger failures of capitalist society, but the pandemic further forced this analysis to the fore. More than ever, students are ready to imagine, desire, and build a different world. They don't want to "go back to normal," and they often see how "normal" was already bad for people and the environment. Abolition, mutual aid, intentional communities, new definitions of kinship, public banking, and taking governance, education, and infrastructure into their own hands—these are the skills students desire, since they have lost faith that existing systems will fix what ails us.

So what is the role of the climate educator in this context? How, in good faith, can we train students for careers in a world where the current economic system is unraveling? We are teaching in an existentially pivotal moment. What does pedagogy for a just transition look like? What is the task of higher education in what the famous eco-Buddhist social activist Joanna Macy calls "the Great Turning"? As we see in ongoing global youth activism, and as many of us feel in our classrooms daily, students barely have the patience to go through the motions of getting an education as a means to an individualist, career-oriented end. They are waking up to the fact that their time on this planet is limited, and that what they—and we—do now will significantly shape the future of life for all beings on this planet. Students want something different from their education than what their professors studied. How will we as educators—often exhausted, burned out, and despairing, too—rise to this moment?

CENTERING JUSTICE IN CLIMATE EDUCATION: IT GETS EMOTIONAL

The climate movement has only recently started to learn from and integrate insights from transnational environmental justice movements. For the most part, climate educators have been teaching in the same way—from a dominant, positivist, settler-colonial set of assumptions about how social change works. With political changes bringing together the climate and justice movements, a powerful new moment—often dubbed *climate justice*—is shaping a whole generation. Social movements such as Idle No More, Standing Rock, and Black Lives Matter all demonstrate that the climate movement is pivoting toward a justice orientation. In this historical context, many of our college-aged students, who are coming to consciousness and coming of age, are bringing this lens to our classes. Moreover, in many higher-ed contexts, the "traditional" student is changing. What does this mean for intergenerational conversation, and how can we stage it effectively in our classrooms?

Out of these shifting forces, the climate movement—and sometimes education—is finally focusing on the disproportionate costs and benefits of climate change: the ways that disposability, racism, patriarchy, and extraction are at the root of both social injustice and environmental degradation. It is suspicious of climate policy that doesn't also address inequality. Following the work of thinkers like Julie Sze and Lindsay Dhillon, David Pellow, Rob Nixon, Olúfẹ́mi O. Táíwò, adrienne maree brown, and Mary Annaise Heglar, for example, the climate justice movement connects the *ability to breathe* to both the spectacular violence of police brutality and the less visible, long-term pollution of living in neighborhoods targeted for toxic siting. These systems-level analyses locating the root of the problem not in emissions but in injustice have become mainstream. Climate *justice,* not just the climate movement, has arrived.

Yet many educators struggle to teach from this insight, to meet students where they are. Some of the challenges educators report in their efforts to bring climate justice to the classroom are:

Challenge 1: Cancel culture and the politicization of climate justice. In our classes, the culture wars do play out and we, as well as our students from all political parties and backgrounds, and the institution of higher education itself, are often getting "canceled." Many conservative students feel their very identities are under attack when the word "justice" is added to conversations about climate change (climate topics seem "safer" than conversations about race). Meanwhile, students coming to climate change with a robust justice lens express frustration and harm by repeatedly hearing climate and the environment uncritically discussed in privileged ways. Without skill, educators risk reinforcing the same dynamic of the culture wars at large.

Identity politics have everything to do with climate emotions and they will dictate where we go from here. This signals the shift from climate *change* as a central focus to climate *justice* and is a necessary part of teaching climate change in this moment. Student positionality—background, race, assumptions about nature and science, their relative position to power, etc.—all shape how they will respond to both the projects of climate change and social justice. They also shape which emotions are available to different students, and how different students will feel in relationship to different themes.[7] These conversations are not "academic" nor are they just about "identity politics" as has been co-opted by dominant politics;[8] they touch students' intimate lives. We shouldn't want it any other way—students learn best when they feel "relevance" between the topic and their lives, and when they have an emotional connection to the material they're studying.[9]

An older problem with teaching climate was to make it relevant, touchable, imminent, because it was framed in dominant climate spaces as distant, abstract, and uncertain; this is no longer the case—a majority of students feel climate change is relevant to their lives right now.[10] This is both a cause for concern (climate change is happening to us, *now*) and celebration (such intimacy engages students in the subject much more readily). Thus, pedagogies that take identity as their primary concern (in fields such as ethnic studies, queer studies, disability studies, gender/sexuality studies) have much to teach climate educators. Climate concern is emotionally potent not just for scientific reasons (reports from the UN's Intergovernmental Panel on Climate Change, etc.); it is also emotional because it is complexly overlaid with politics, identity, power, race, gender, geography, privilege, and proximity to climate effects, for example.

Moreover, we contend, students' emotional assumptions about identity, climate, and power ought to be a subject of analysis to begin with. These differences otherwise are likely to involve significant conflict in the climate classroom, which will in turn affect students' emotional capacity to engage in the content. How can climate justice educators create conditions in their classrooms for these difficult conversations? How can these conversations become the generative force that helps students build new worlds? How can we harness our classes as labs for precisely the social healing that is required to move where we need to on climate justice, at the scales of both the interpersonal and the political?

It is naive for educators to assume that all students come into their classrooms with neutral or no emotional connection to any of these issues, and that they will all develop them together, uniformly, over the course of the class, with the educator directing this arc. Emotions, trauma, and violent communication patterns are absorbed from the culture we all marinate in, and are in the classroom before we even sit down to pen a syllabus.

The challenge for educators is this: How can we create the conditions in the classroom for these conversations to be productive, and not just repeat the culture wars we see everywhere in mainstream media and that threaten to undermine the very democratic system we need to leverage to mitigate climate change? Might climate justice educators need training in nonviolent communication, deliberative dialogue, trauma-informed pedagogy, and compassionate facilitation?

Challenge 2: Climate and justice sometimes don't get along. It's easy to teach case studies where the solutions to climate change are the solutions to social justice as well. It's easy to show that climate change is a matter of social justice. We can see the ways climate change is a poverty-multiplier everywhere—in our medical systems, in extending colonial-capitalist extraction, in geographical marginalization . . . the list goes on. But what about when the frames and solutions of climate and justice don't agree with each other? What if a particular climate mitigation strategy (such as carbon taxes or wind farms) disproportionately impacts an already marginalized community? Or if a major triumph for social justice comes at the expense of faster climate mitigation? We often hear students say "we can't have social justice if we don't have a planet" or claim they're happy to dispense with democracy and inclusion if it means we can move faster, more unilaterally, and even violently in favor of climate protection.

On the other hand, some students "choose" social justice over climate, because they worry their limited resources for mobilization will be diluted if "shared" with the climate cause. And, tired of being sidelined in the dominant political discourse for white, privileged environmental causes, they understandably hesitate to build coalitions or join mainstream environmental spaces. Even with all of the systems-thinking, intersectional environmental analyses that many students learn before they get to college, a battle of single-issue politics often plays out in the classroom, and it can get heavy, heated, and hurtful.

Are we prepared to navigate the messiness of some students thinking that justice can be sidestepped to do what they think is needed to mitigate climate, or vice versa? How do we facilitate those conversations? Can we bind together the means and the ends for our students, that is, are we prepared to teach climate change as rooted in inequality, and are we prepared to do the *slow work* of cultivating relationships and equitable processes required to address inequality on the way to addressing climate change? Solidarity is aspirational and easy to theorize, but we can feel its challenges play out in our classes and institutions. For climate education to become climate *justice* education, educators will need to move beyond the politics of representation and integrate pedagogies much more familiar to those who center justice. Among climate educators trained in colonial systems of education, this requires humility and collaboration, which arguably are also the core of our climate justice solutions.

Challenge 3: The content may be triggering. We no longer can teach about climate change as if it were an abstraction, distant in time or space. The closer our students are to the frontlines of environmental injustice, the more climate change is affecting them—and there is nowhere that climate change hasn't touched. Do we know what to do when these climate traumas are triggered in our classrooms? While it has never been considered required training in climate classes to teach what is termed by pedagogy experts as "difficult knowledge," that is changing. The relative privilege of the climate movement and climate classrooms is crumbling, and therefore, we contend, climate educators must become trauma-informed.

Challenge 4: Climate content is often taught as depressing and debilitating. Doom and gloom in general make people shut down, as discussed above. Moreover, weaving a story for students in which the apocalypse is inevitable creates a self-fulfilling prophecy where the "pseudo inefficacy effect" predisposes students to give up rather than try to fix this flurry of problems. Andrew Bryant describes this phenomenon of doomism and the pseudo-inefficacy effect in his essay in this collection:

> When we experience fears or anxieties, or see something in the world that needs changing, a common impulse is to jump into a new activity, hoping to ease our difficult emotions by trying to make a difference. If we haven't processed our underlying emotions, we often choose actions that are not in alignment with our strengths, capacities, and resources, and we can end up feeling disillusioned, deficient, or burned out. In other cases, we have trouble identifying any action that feels worthwhile. Nothing seems like enough, so we throw up our hands, feeling despondent or deficient. We blame ourselves, or decide that no action is worth doing because it won't be enough.

If we want students to do their part in a collective effort to reduce harm in the world, we have to help them live in a story where that is a possible, likely, and desirable outcome. We also could examine the doom-and-gloom narrative from a justice lens: For whom is that narrative compelling and persuasive? How does privilege shape the stories we live in? In what ways is despair a luxury or hope a survival strategy?

Challenge 5: Educators are burned out and not trained as therapists. One of the primary push-backs we hear from educators is that they don't want to become therapists. As Jessica Pratt notes in her essay in this collection, "science faculty, including me, are not trained to confront psychological distress among their students, let alone support them through it." Most educators are not trauma experts, and they do not want to do harm to students by inviting emotions into the classroom.

While we are not suggesting that all educators must take care of their students' emotional lives (and this is of particular relevance to gendered and racial analyses of cultural taxation and emotional labor for females, and female faculty, of color), our hope is to convince skeptics that students are *already* bringing their emotional lives into college, and that it may in fact be *more* ethical, effective, and generative for educators themselves, too, to collectively address students' (and to some extent, our own) emotional experiences in the classroom in more explicit ways, so that these responses aren't stigmatized, shamed, shunted, or suppressed. In this latter case, educators are more likely to be asked for their time and energy *outside* of the classroom to support students anyway, often in individualized ways that just reiterate the atomism at the root of both our climate and mental health crises. At a minimum, as most contemporary pedagogical research shows, we should pay attention to emotions in our classrooms because emotions are essential to learning, knowledge retention, and decision-making.

We offer the possibility that gaining tools for centering emotions in the classroom supports both students *and* educators, especially for those who have long been expected to serve as emotional supports to an increasingly despairing population. In treating students as whole people, enabling them to find support in each other (not just from their instructors) as an explicit part of our work during class time, and leveraging current psychological research, we can meaningfully transform our classrooms to respond to this moment.

At the very least, as Ashley Reis reflects in this collection, even if she herself cannot (and should not) become an overnight therapist, she *is* "equipped to establish an accessible and inclusive classroom, wherein students feel secure and valued to the extent that they are prepared to face and interrogate the emotions that will inevitably arise as we navigate the affective landscape of ecological degradation and social injustice in the age of climate disruption."

We see these as the central challenges and opportunities for integrating emotions, climate, and justice in higher education. To summarize, this project takes up the following questions, even if our collection itself is only an initial response:

- How can climate justice educators create conditions in the classroom for these conversations to be productive, and not just repeat the culture wars we see everywhere in mainstream media these days?

- Are we prepared to do the slow work of cultivating relationships and equitable processes required to address inequality on the way to (and as fundamental to) addressing climate change?
- What would it take to help climate educators become trauma-informed?
- How can educators help students imagine thriving in a climate-changed world? How can we help students desire, rather than fear, their future?
- How can educators do this work without doing further harm to themselves or to others? Or even better, how might this work enhance educators' well-being, and increase planetary flourishing?

BACKGROUND OF THE PROJECT

While this collection hardly offers a comprehensive or definitive toolkit for all of climate education, it points in some useful directions and gives a flavor of how different educators are reinventing their climate pedagogy for these times. This is a moving field, with new research and new pedagogical approaches pushing the edges at every turn.

In 2020 we hosted a three-day online workshop supported by the Rachel Carson Center in Munich on the topic that was to evolve into this book, "An Existential Toolkit for Climate Justice Educators." Many participants hailed from our networks in the environmental humanities and the fields of climate emotions and psychology. That gathering ultimately launched a website and a two-year-long discussion series with educators from at least nine different countries.

As the network expanded in the years that followed, we learned that participants were keenly interested in a book project that curated existential tools for the climate classroom. The resulting book represents a snapshot of the kinds of experiments in teaching that people in our network have been exploring. It reflects conversations that emerged between clinical practitioners and climate educators. And instead of putting this collection forward as a "canon" of best practices, our intention is to draw attention to the necessity of bringing these themes of climate, emotion, and justice together in the classroom and elsewhere.

There is so much work happening at the interface of human healing, climate injustice, and social movements, and that work is increasingly shaping climate education. The social movement backdrop of our own engagement with these ideas is Joanna Macy's the Work That Reconnects, adrienne maree brown's *Emergent Strategy,* and ideas of pleasure activism, Movement Generation, Project Drawdown, and the Just Transition, the pedagogical tools from the All We Can Save project, the Practical Handbook for Climate Educators and Community (titled "Climate Doom to Messy Hope: Climate Healing and Resilience") published by Meghan Wise for the University of British Columbia's Climate Hub, the teaching tools of Movement Generation, and, as ever, the liberatory pedagogy of figures such as Paulo Freire and bell hooks. In the realm of climate pedagogy scholarship, we are thinking in relationship with Jo Hamilton's "emotional methodologies," Panu Pihkala's "taxonomy of ecological emotions," and their implications for environmental education, such as Audrey Bryan's "difficult knowledge" and Daniel Suarez's ongoing multimodal project, "A Clear and Present Pedagogy." Others we have learned about in this work are Education Ecologies Collective, the Bard College's World-Wide Teach-in for Climate Justice Education, and the mindfulness approach to climate justice pedagogy emerging from the Garrison Institute and Mind & Life Institute. And there is certainly more, as the field of climate pedagogy is evolving faster than the time it takes to produce a book.

ORGANIZATION OF THE BOOK AND HOW TO USE IT

In our call for submissions, we encouraged all kinds of tools—short ones, long ones, easy ones, difficult ones, quick ones, deep ones. We wanted to include tools that could be used in non-higher educational settings, and ones that could be used across disciplines. We wanted the collection to be like a cookbook: a variety of options for people looking for different things, to inspire the chef when they're feeling unmotivated, or to help them dip their toes in the water of a new recipe. We also wanted to hear the voices of students and educators who are bringing emotions into their climate work.

We got a wide range of tools and voices from a variety of spaces. We organized the book primarily by theme, and within each theme, you will find an assortment of types of pieces—longer theoretical essays, shorter assignments, quick modules, and testimonies.

Part One offers an introduction to "Getting Started with Emotions in the Climate Classroom." It includes two essays by mental health providers, other pieces demonstrating why climate emotions matter in higher education, and a few assignments for those looking to dip their toes into this realm, which we call "sample assignments for getting started."

Part Two, "Justice as Affective Pedagogy," collates pieces that focus on how we teach as itself a practice of justice and liberation. For Ashley Reis, inclusion and healing can happen at the level of what may otherwise seem to be the most prosaic domains of our teaching: syllabus policies. Content gets all the attention, but the policies communicate as much or more. The theoretical essays and assignments in this section put environmental justice at the center of their arguments, explore what it would mean to decolonize climate education, and center participatory action research as a justice-based methodology. Michelle Garvey's piece, which offers a checklist for such justice-centered methods, explains that such methods are decolonial to the extent that they are "relevant to stakeholders on the frontlines of climate injustice; inclusive of community ways of knowing, including cultural and spiritual considerations; participatory at every stage of development; transparent about researcher bias; accountable to the community within which knowledge is acquired; and inclusive of more-than-human needs."

Part Three, "Embodied Pedagogies," challenges the dominant Western European education model that divides mind from body and asks that students bring only the former into our work together. Similarly, educators tend to devalue students' embodied knowledge and rarely if ever consider what an overemphasis on the cerebral teaches students about the *corporeal* wisdom, traumas, and knowledges they already bring or might need to cultivate for the turbulence we are living through. How can we begin to bring the body, and the robust scholarship on embodied knowledge, into our teaching? Might an embodied approach to the climate crisis also be a more justice-oriented one? How can we balance intelligences of heart, body, and mind?

Part Four explores "Futurity, Narrative, and the Imagination," because the call for submissions on emotions and climate justice resulted in many contributions focused on the emotion of hope and its temporal partner—the future. If it is true, following Fredric Jameson and others, that it is easier for most people to imagine the end of the world than it is for them to imagine a post-capitalist or post-fossil fuel society, then the lack of imagination about what we desire surely will become a self-fulfilling prophecy. The contributions in this section support educators who want to facilitate students' radical imagination for a just transition toward climate justice. For April Anson, for example, cultivating a radical imagination is a "practice" that has to be taught. Only the fossil fuel industry benefits from all of us living in a story of apocalypse,[11] and so cultivating muscles of storytelling that move us in other directions is a kind of resistance.

Part Five, "Unsettling Pedagogies: Discomfort and Difficult Knowledge," intervenes in the conversation about the classroom as a "safe" space and the more justice-oriented move to consider classrooms as "unsettling." This is a play on words that illustrates how necessary *discomfort* is for the work of racial justice and also brings the often "invisible" power of settler-colonial, white supremacist privilege into the light. What is the role of emotional discomfort in the classroom, and how might it be harnessed for climate justice? When is discomfort generative, and when is it triggering? How do we navigate these various "stretch zones" for different students?

Part Six, "Joy and Resilience as Resistance," complicates the assumption that all climate education and all climate emotions are dreadful, negative, and unpleasant. On the contrary, as contributions in this section show, the work of climate education often involves more pleasant emotions like joy and humor. The neuroscience of resilience shows that our brains are pleasure-seeking machines, and so we will keep coming back when the work is pleasurable on some level. And justice movements inform this move toward what adrienne maree brown calls "pleasure activism" as well; fiercely protecting joy and vitality has long been understood as a methodology of resistance, while uncritical despair can serve as complicity in the status quo.

Casey Meehan's essay poses the question this way: "Where in our work and teaching do we build space for surprise, humor, celebration, laughter, and the time to connect with each other? This isn't time wasted: it is a recognition that we must nourish our whole selves when engaging with the emotionally fraught topics we teach." In this sense, joy and pleasure are strategies for keeping engaged and resilient, and forms of resistance in themselves. Social movements and wisdom traditions show that the emotions of social change are seemingly paradoxical—we feel anger because we feel love, and vice versa, and we feel joy because we feel grief, and vice versa.

The emotions of climate justice education are similar: How can we create space for "both/and" in our classrooms? How can we more explicitly leverage the pleasures of participating in collective liberation for climate justice, making it both a pedagogy in our classes and a practice in our lives?

Part Seven, "Community, Collaboration, and Kinship," theorizes that the foundation of climate justice is solidarity and training in skills for collectivity and collaboration, and offers case studies of what happens when several educators prioritize community over content in (and outside) their classrooms. If we are going to get to agreement on climate justice within democratic societies, it will happen at the level of interpersonal relationships, one conversation and one bridge built at a time. Following two of adrienne maree brown's principles of emergent strategy, "less critical mass, more critical connection" and "move at the speed of trust," this section experiments with skill-building around community. Spaces of education must follow what are referred to in activist circles as "movement spaces." This ought to be a no-brainer, yet higher education is designed to prioritize knowledge acquisition, content delivery, and career-building at the *individual* level, often at the expense of interdependence and relationality. Care-giving, mutual aid, humility, and nonviolent communication have typically been the domain of movement politics, not pedagogy. Yet, as Magdalena Mączyńska writes in her essay in Part Three, quoting from Cavanagh's *Spark of Learning*, "Community and affective engagement are—as great educators have always understood, and the neuroscience of learning confirms—fundamental to successful learning."[12]

Finally, our last section is about the prickly issue of "career building" in the apocalypse: "These Skills Are Needed in the World." Yes, as the Department of Labor can tell you, there will be many careers in sustainability and climate over the next few decades. And in a historical moment where neoliberalism has turned higher education into a commodity rather than a social good, it is now arguably the job of higher education to help students ensure that they are trained for those vocations.

And yet, much of what environmental studies teaches is that the existing, dominant systems of governance and economics are not working, and that graduates of this area of study will be part of the movement to dismantle those systems and build better ones. Thus, if "we are in a womb, not a tomb," what does that mean for vocationalizing this content? Can we in good conscience train students to get jobs in a dying system? The emperor has no clothes. Should we still be teaching as if it were otherwise, with our heads in the sand, modeling our own form of denial and possessive investment in the status quo? What would it look like to train students for jobs in a world that is being born but not yet here?

Although, admittedly, the pieces in this section do not adequately answer these pressing lines of inquiry, we want to raise such concerns in the introduction, and note that more discussion is worth having, if not in this book, then in all of the spaces where educators are working with students to build the world they desire.

We have sought to organize contributions according to the larger argument we are advancing that underlies the book's purpose—that a new pedagogy is needed for climate justice, and that these themes are some of the central ways to invite that new pedagogy into being.

However, readers may come to this book with other aims, and so we offer alternative groupings for readers looking for different categories, as well as for those who might like to conceive of the contributions in a more expansive organization. The appendix at the back of the book sorts chapters by the following themes, if you want to explore this book through other lenses:

- **Discipline-specific** (for readers looking to filter for authors or assignments that are in a particular discipline)
- Pieces that are about **students leading** the direction of the course/material (for readers looking to center student problem-solving, wisdom, or existing assets as a kind of climate justice pedagogy)
- **Experiential** (for readers looking to integrate field, embodied, or open-ended pedagogies)
- **Type of text** used in the assignment (for readers looking to integrate a particular genre or type of text, such as poetry, cli-fi, games, news media, etc.)
- **Type of affect** (for readers looking to focus on a particular emotion)

Our intention for including these alternative groupings is to make the book accessible to a wide variety of readers and to facilitate their diverse uses of these tools. Like a cookbook, we seek in this organization to invite creativity and to try to avoid being dogmatic about how to use them. Many of the chapters could go in other categories, and there may even be other, more useful groupings for users we haven't yet imagined. In this sense, there is some level of arbitrariness and assumptions we are making about you, dear reader, and we appreciate your grace in allowing us to guide you through the materials in the (admittedly limited) ways our own minds have organized them.

NOTES

1. Kaur, "Sikh Prayer for America."
2. See, for example, Ogunbode et al.'s study of climate anxiety in 32 countries, which concluded that fear and anxiety were associated with action, but inversely with wellbeing and longer-term engagement.
3. Degradation desensitization is defined as the loss of sensitivity to difficulties, due to consistent exposure to that stimulus. See Alhadeff, "Numb to the World."
4. See, for example, Cavanagh's *Spark of Learning*.
5. A review of scholarship on pedagogy for uncertain times is beyond the scope of this introduction, but a few examples are hooks's *Teaching to Transgress* and *Teaching Community,* Maniates's "Teaching for Turbulence," and Macrine's *Critical Pedagogy in Uncertain Times*. Many syllabi have been created to support educators integrating recent historical events in their classes, from Standing Rock to Black Lives Matter to the COVID pandemic. Similar efforts have been made around the climate crisis.
6. The idea of difficult knowledge in pedagogy we refer to here comes from Britzman's *Lost Subjects, Contested Objects*.
7. See Ray, "Who Feels Climate Anxiety?"

8. This is the subject of Táíwò's book *Elite Capture*, which argues that an effective identity politics can in fact be deeply coalitional and organize across differences toward shared interests, such as climate justice (which he later takes up in *Reconsidering Reparations*). We use "identity politics" here in the same way he advances it, not as it has been captured by political parties to divide rather than forge those possible coalitions.

9. See, for example, Duncan-Andrade's "Note to Educators."

10. See, for example, Hickman et al.'s 2021 report, "Climate Anxiety in Children and Young People."

11. See Nosak's research on the legal strategy of the fossil fuel industry, as summarized in Britt Wray's newsletter *Gen Dread*.

12. Cavanagh, *Spark of Learning*.

REFERENCES

Agee, Christopher. *The Streets of San Francisco: Policing and the Creation of a Cosmopolitan Liberal Politics, 1950–1972*. Chicago: University of Chicago Press, 2014.

Alhadeff, Alexandra C. "Numb to the World: Degradation Desensitization and Environmentally Responsible Behavior." *Tropical Resources: The Bulletin of the Yale Tropical Resources Institute* 34 (2015).

Britzman, Deborah P. *Lost Subjects, Contested Objects: Toward a Psychoanalytic Inquiry of Learning*. Albany: State University of New York Press, 1998.

brown, adrienne maree. *Emergent Strategy: Shaping Change, Changing Worlds*. Chico, CA: AK Press, 2017.

———. *Pleasure Activism: The Politics of Feeling Good*. Chico, CA: AK Press, 2019.

Bryan, Audrey. "Affective Pedagogies: Foregrounding Emotion in Climate Change Education." *Policy and Practice: A Development Education Review* 30, no. 1 (2020): 8–30. https://www.developmenteducationreview.com/issue/issue-30/affective-pedagogies-foregrounding-emotion-climate-change-education.

———. "Pedagogy of the Implicated: Advancing a Social Ecology of Responsibility Framework to Promote Deeper Understanding of the Climate Crisis." *Pedagogy Culture and Society* 30, no. 3 (2022): 329–48. https://doi.org/10.1080/14681366.2021.1977979.

Cavanagh, Sarah Rose. *The Spark of Learning: Energizing the College Classroom with the Science of Emotions*. Morgantown: West Virginia University Press, 2016.

Dillon, Lindsey, and Julie Sze. "Police Power and Particulate Matters: Environmental Justice and the Spatialities of In/Securities in U.S. Cities." *English Language Notes* 54 (2016). https://doi.org/10.1215/00138282-54.2.13.

Doppelt, Bob. *Transformational Resilience: How Building Human Resilience to Climate Disruption Can Safeguard Society and Increase Wellbeing*. London: Routledge, 2016.

Duncan-Andrade, Jeff. "Note to Educators: Hope Required to Grow Roses in Concrete." *Harvard Educational Review* 79, no. 2 (2009).

"An Existential Toolkit for Climate Justice Educators." https://www.existentialtoolkit.com/.

Freire, Paulo. *Pedagogy of the Oppressed*. New York: Continuum, 2000.

Haltinner, Kristin, and Dilshani Sarachandra. "Climate Change Skepticism as a Psychological Coping Strategy." *Sociology Compass* 12, no. 6 (2018). https://doi.org/10.1111/soc4.12586.

Hamilton, Jo. "Emotional Methodologies for Climate Change Engagement." PhD dissertation, University of Reading, 2020.

Heglar, Mary Annaise. "Climate Change Isn't the First Existential Threat." *Medium*, 2019. https://zora.medium.com/sorry-yall-but-climate-change-ain-t-the-first-existential-threat-b3c999267aa0.

Hickman, Caroline, et al. "Climate Anxiety in Children and Young People and Their Beliefs About Government Responses to Climate Change: A Global Survey." *The Lancet Planetary Health* 5, no. 12 (2021). https://doi.org/10.1016/S2542-5196(21)00278-3.

hooks, bell. *Teaching Community: A Pedagogy of Hope*. New York: Routledge, 2003.

———. *Teaching to Transgress: Education as the Practice of Freedom*. London: Routledge, 1994.

Kaur, Valarie. "A Sikh Prayer for America on November 9, 2016." Accessed September 24, 2022. https://valariekaur.com/2016/11/a-sikh-prayer-for-america-on-november-9th-2016.

Lertzman, Renee. "The Myth of Apathy." *The Ecologist*, June 19, 2008. Accessed October 21, 2022. https://theecologist.org/2008/jun/19/myth-apathy.

Macrine, Sheila, ed. *Critical Pedagogy in Uncertain Times*. London: Palgrave Macmillan, 2020.

Maniates, Michael. "Teaching for Turbulence." In *State of the World 2013*, 255–68. Washington, DC: Island Press, 2013.

Nixon, Rob. *Slow Violence and the Environmentalism of the Poor*. Cambridge, MA: Harvard University Press, 2013.

Norgaard, Kari Marie. *Living in Denial: Climate Change, Emotions, and Everyday Life*. Cambridge, MA: MIT Press, 2011.

Nosack, Grace. "How the Fossil Fuel Industry Seeds Doomism to Protect Continued Extraction." *Gen Dread Newsletter*, September 23, 2022.

Ogunbode, Charles, et al. "Climate anxiety, Wellbeing and Pro-environmental Action: Correlates of Negative Emotional Responses to Climate Change in 32 countries." *Journal of Environmental Psychology* 84 (2022). https://doi.org/10.1016/j.jenvp.2022.101887.

Pellow, David Naguib. *What Is Critical Environmental Justice?* Cambridge: Polity Press, 2018.

Pihkala, Panu. "Toward a Taxonomy of Climate Emotions." *Frontiers in Climate* 3 (2022). https://doi.org/10.3389/fclim.2021.738154.

Ray, Sarah Jaquette. "Who Feels Climate Anxiety?" *Cairo Review of Global Affairs*, Fall 2021.

Suarez, Daniel Chiu. "A Clear and Present Pedagogy." *Antipode* (forthcoming).

Táíwò, Olúfẹ́mi O. *Elite Capture: How the Powerful Took Over Identity Politics (And Everything Else)*. London: Pluto Press, 2008.

———. *Reconsidering Reparations*. New York: Oxford University Press, 2022.

PART ONE

GETTING STARTED WITH EMOTIONS IN THE CLIMATE JUSTICE CLASSROOM

ONE

A Pedagogy for Emotional Climate Justice

BLANCHE VERLIE

This theoretical piece applies principles of environmental justice (including those focused on distribution, recognition, participation, capabilities, and relationships) to the issue of environmental emotions, and how emotional climate justice can be embedded in and practiced in education.

The climate crisis is often said to be a problem of science, politics, or technology. It is all these things but, as my anxious-overwhelmed-cynical-hopeful-frustrated-bitter-devastated-joyful-appalled students have taught me, it is also a matter of emotion.[1] As one student stated in an end-of-semester reflection, climate change "is such an emotional challenge to deal with, especially when you accept the fact that you and the society we live in today are to blame."[2] The experience of being both vulnerable and complicit, manifesting as both fear and guilt, leads many to refer to the experience of learning about climate change as "a roller coaster of emotions." For many of my students, while apathy and hope, inspiration and disgust may be the ups and downs the carriage visits along the roller coaster, anxiety seems to be what the tracks are made of; it is our constant companion,[3] taking us to the highs and lows but never going away.

I believe the vast majority of people in the developed world are anxious about climate change. Some of us—like my students—are anxious about the impacts of climate change on our kids, on biodiversity, or on other people around the world. Others are anxious about losing jobs, livelihoods, and identities if and when carbon-reduction policies transform industries, such as fossil fuel extraction. Still others are anxious about the sense of personal inefficacy that can arise from tackling such a big problem; it's the ego that feels threatened.

Many politicians, fossil fuel executives, and right-wing elites are also anxious, I believe, about the threat climate *action* poses to their taken-for-granted cultural, economic, and political power. This political ecology of (often unexamined) climate anxiety leads to an outsourcing of the emotional burden of climate change onto those who have less responsibility and power, such as young people. Those who are most responsible for causing climate change deflect the job of exploring, interrogating, and addressing their anxiety to those least responsible for greenhouse gas emissions, specifically to those who most bear the emotional costs of the ecological, economic, and social impacts of climate change. Yes, someone should be worrying about climate change, but it should be our elected leaders and those who have benefited the most from extraction and who have the most power to make change.

> This unfair distribution of not just the ecological but also the emotional burdens of climate change makes climate change an issue of emotional injustice.

This unfair distribution of not just the ecological but also the *emotional* burdens of climate change makes climate change an issue of *emotional injustice*. Research shows that both specific climatic disasters and knowledge of climate change can lead to post- and pre-traumatic stress disorder,[4] and that this is most intensely experienced by those most exposed to the impacts of climate change.[5] However, rather than offering compassion and support, in Australia where I am from, former Prime Minister Scott Morrison denigrated those who are distressed by climate change. While refusing to develop any nationally effective policy for reducing emissions, he told youth climate activists that their concerns were "needless anxiety"[6] and dismissed the emotional pain of those who lost their homes, families, and communities in the devastating 2019–2020 bushfires by claiming they were just "feeling raw."[7] In such a context of greenhouse gaslighting,[8] understanding the political dimensions of climate anxiety, as well as the emotional dimensions of climate politics, is imperative. These issues play out in school and university classrooms, and for this reason, climate educators (of all kinds) can benefit from an understanding of emotional climate in/justice, which I attempt to begin outlining here.

An account of emotional climate in/justice can build on the frameworks developed for conceptualizing climate injustice. Climate injustice is not just about the unequal *distribution* of the harms of climate change. Scholars and activists have argued that climate injustice is also about who is, and who is not, *recognized* as valuable and enabled to *participate* in discussions and responses to climate change.[9] Some have suggested that aiming for relatively equal *capabilities* of different people to, for example, meet their basic needs and adapt to climate change, is another useful way of thinking about climate justice.[10] These are arguably Western approaches to justice. Indigenous peoples conceptualize climate justice as more about the maintenance of healthy, respectful *relationships*.[11] Drawing holistically on these approaches, a conceptualization of emotional climate in/justice helps us analyze who is (not) feeling what in relation to climate change, how that both empowers and disempowers them, and what kinds of relationships we need to establish in order to cultivate climate justice.

Emotional climate justice would require the mainstreaming and embedding of *recognition* of the emotional dimensions of climate change in all walks of life. This would, for example, mean that public events, communication strategies, government policies, and formal education recognize people's intense yet diverse emotional experiences of climate change to be valid and important; deserving of respect, engagement, and response; and fundamental to any and all efforts to address climate change. Recognition of this would demand rethinking, reorganizing, and regenerating *relationships* between people, and between people and the planet, towards ones that systematically cultivate care, respect, trust, and accountability.[12]

Mapping the unequal *distributions* of climatic trauma would emphasize the need for privileged people to learn to live with climate change.[13] Research is finding that the everyday denial of people who are relatively insulated from climate change is often a result not of apathy, but of their inability to manage worry and guilt about climate change.[14] That is, our inaction is often a result of our inability to engage with painful emotions. Thus, privileged people—like me—need to build our capacity to engage with, interrogate, grapple with, and transform our distress about climate change. We need to do this emotional work in order to prevent further harms to others, human and not, now and in the future.

Nevertheless, climate change is already here, and is unavoidably getting worse. So emotional climate justice would also require supporting those who are most vulnerable to climate change to enhance their (already admirable) emotional *capabilities* to continue living with it. Marginalized people have long histories and strong cultures of survivance and are often experts in supporting each other through trauma.[15] However, they often do not have access to resources that would enhance their survival abilities—because of systemic exclusion from access to land, money, education, and/or professional services. To work within an emotional climate justice context would not mean trying to impose Western/bourgeois/medicalized mental health practices onto such communities, but instead would involve ensuring people have the skills, capacities, and resources they want so that they can support their communities in their own ways.

> While any kind of effort towards justice requires some kind of discussion about the injustice, forcing people to engage with or recount their trauma can be just as harmful, if not more so, than the original trauma.

Achieving all of this would require discussions and deliberations between different groups of people, which brings us to the issue of *participation*. I once had a student write on a reflective assessment task that she was "often too anxious to speak" in class because of the intensity of her emotions about climate change. While any kind of effort towards justice requires some kind of discussion about the injustice, forcing people to engage with or recount their trauma can be just as harmful, if not more so, than the original trauma. Learning from advocates in the arenas of sexual violence and other interpersonal injustices, we must appreciate that talking about our experiences of climate change can sometimes be liberatory and empowering, but at other times can be retraumatizing and harmful. Careful consideration of the kinds of structures, practices, formalities, and protocols required in each specific context is needed to ensure that participating in discussions about climate change and emotions are indeed contributing to justice, rather than perpetuating violence (see also chapters by Garvey, Gray, and Anson in this volume).

For climate justice educators, this task has specific relevance. We might find ourselves teaching *about* emotional climate in/justice. For example, I used to teach undergraduate urban planning students who, if all goes well for them, will graduate into jobs where they conduct public consultation as representatives of governments or businesses. For them, appreciating the high emotional stakes of the consultations they might end up running about climate change and learning how to structure and manage such conversations are crucial skills if they are to contribute to emotional climate justice. Other degrees taught at the tertiary level—such as, but not limited to, degrees in education—might also benefit from learning about these issues and the practical skills students will need in their professional lives.

Perhaps more important, we need to embed practices of emotional climate justice into our pedagogies. While students in formal education are not always children and young people, this is most often the case, and we know that today's young people are already deeply distressed about climate change because they are facing the consequences, at this point in time, of a 3-degree hotter world by the end of the twentieth-first century (although this will hopefully be avoided by drastically reducing fossil fuel emissions, but at the time of writing, the world has yet to take these actions). Emotional climate justice in the classroom will require careful attention to which students, in which circumstances, would benefit from engaging with their emotional experiences of climate change, and for whom this would be too traumatic. It requires a whole suite of tools to be able to do this carefully; this book offers many valuable activities, ideas, and strategies for doing this.

NOTES

1. Verlie, *Learning to Live with Climate Change*, 48.
2. Ibid.
3. Head, *Hope and Grief in the Anthropocene*.
4. Hayes et al., "Climate Change and Mental Health"; Pihkala, "Anxiety and the Ecological Crisis."
5. Manning and Clayton, "Threats to Mental Health and Wellbeing"; Hickman et al., "Climate Anxiety in Children and Young People."
6. Murphy, "Morrison Responds to Greta Thunberg."
7. Remeikis, "Scott Morrison Brushes Off Angry Bushfire Reception."
8. Verlie, "Greenhouse Gaslighting."
9. Agyeman et al., "Trends and Directions in Environmental Justice."
10. Schlosberg, "Justice, Ecological Integrity, and Climate Change."
11. Whyte, "Too Late for Indigenous Climate Justice."

12. Ibid.
13. Verlie, *Learning to Live with Climate Change*.
14. Norgaard, *Living in Denial*; Haltinner and Sarathchandra, "Climate Change Skepticism."
15. Vizenor, *Survivance*.

REFERENCES

Agyeman, Julian, David Schlosberg, Luke Craven, and Caitlin Matthews. "Trends and Directions in Environmental Justice: From Inequity to Everyday Life, Community, and Just Sustainabilities." *Annual Review of Environment and Resources* 41, no. 1 (2016): 321–40. https://doi.org/10.1146/annurev-environ-110615-090052.

Haltinner, Kristin, and Dilshani Sarathchandra. "Climate Change Skepticism as a Psychological Coping Strategy." *Sociology Compass* 12, no. 6 (2018): 1–10. https://doi.org/10.1111/soc4.12586.

Hayes, Katie, G. Blashki, J. Wiseman, S. Burke, and L. Reifels. "Climate Change and Mental Health: Risks, Impacts and Priority Actions." *International Journal of Mental Health Systems* 12, no. 1 (2018): 28–40. https://doi.org/10.1186/s13033-018-0210-6.

Head, Lesley. *Hope and Grief in the Anthropocene: Re-Conceptualising Human–Nature Relations*. Milton Park, UK: Taylor & Francis, 2016.

Hickman, C., E. Marks, P. Pihkala, S. Clayton, E. R. Lewandowski, E. E. Mayall, B. Wray, C. Mellor, and L. van Susteren. "Climate Anxiety in Children and Young People and Their Beliefs About Government Responses to Climate Change: A Global Survey." *The Lancet Planetary Health* 5, no. 12 (2021): e863-73. https://doi.org/10.1016/S2542-5196(21)00278-3.

Manning, Christie, and Susan Clayton. "Threats to Mental Health and Wellbeing Associated with Climate Change." In *Psychology and Climate Change*, edited by Susan Clayton and Christie Manning, 217–44. Cambridge, MA: Academic Press, 2018.

Murphy, Katherine. "Morrison Responds to Greta Thunberg by Warning Children against 'Needless' Climate Anxiety." *The Guardian*, September 25, 2019. https://www.theguardian.com/australia-news/2019/sep/25/morrison-responds-to-greta-thunberg-speech-by-warning-children-against-needless-climate-anxiety.

Norgaard, Kari Marie. *Living in Denial: Climate Change, Emotions, and Everyday Life*. Cambridge, MA: MIT Press, 2011.

Pihkala, Panu. "Anxiety and the Ecological Crisis: An Analysis of Eco-Anxiety and Climate Anxiety." *Sustainability* 12, no. 19 (2020). https://doi.org/10.3390/su12197836.

Remeikis, Amy. "Scott Morrison Brushes Off Angry Bushfire Reception, Saying He Doesn't Take It Personally." *The Guardian*, January 3, 2020. https://www.theguardian.com/australia-news/2020/jan/03/scott-morrison-brushes-off-angry-bushfire-reception-personally.

Schlosberg, D. "Justice, Ecological Integrity, and Climate Change." In *Ethical Adaptation to Climate Change: Human Virtues of the Future*, edited by Allen Thompson and Jeremy Bendik-Keymer, 165–83. Cambridge, MA: MIT Press, 2012.

Verlie, Blanche. "Greenhouse Gaslighting: Scott Morrison's Emotional Manipulation from Climate Apathy to Fake Empathy." *Sydney Environment Institute*, January 14, 2020. https://sei.sydney.edu.au/opinion/greenhouse-gaslighting-scott-morrisons-emotional-manipulation-from-climate-apathy-to-fake-empathy/.

———. *Learning to Live with Climate Change: From Anxiety to Transformation*. London: Routledge, 2022.

Vizenor, G. R. *Survivance: Narratives of Native Presence*. Lincoln: University of Nebraska Press, 2008.

Whyte, Kyle Powys. "Too Late for Indigenous Climate Justice: Ecological and Relational Tipping Points." *WIREs Climate Change* 11, no. 1 (2020): e603. https://doi.org/10.1002/wcc.603.

TWO

Balancing Feelings and Action

Four Steps for Working with Climate-Related Emotions and Helping Each Student Find Their Calling

ANDREW BRYANT

The future of this world is so much in question that each person needs to be considered a potential subject of a genuine "calling" to serve in some meaningful way. . . . Everyone has some gift to give if they learn to give from their essential nature.

MICHAEL MEADE, *The Genius Myth*

This chapter presents a four-step process for supporting students to explore and express their feelings about ecological crisis, to unite with others, and to take steps towards meaningful and sustainable action.

As a therapist focused on ecopsychology who talks with people about climate-related emotions, I've sat with and listened to many clients, young and old, who were struggling to find a sense of purpose in the face of anger, despair, grief, or confusion about the state of the world and the state of our ecosystem. And I've struggled with those feelings myself: fear for my children's futures, sadness at the suffering to come for the most vulnerable, grief over lost or damaged ecosystems, rage at those in power, and anger at myself for not doing more.

These feelings can lead to flailing and self-doubt about whether what I'm doing really matters in the face of the enormity of the climate crisis, and I've watched many of my clients go through similar stretches of confusion and doubt. How to integrate all of these challenging feelings? How to engage with the world in a way that provides a sense of meaning and purpose while contributing to a just and healthy future?

Those of us who are involved in communication and support around climate issues, as educators, activists, counselors, parents, caregivers, and mentors, are faced with a difficult balancing act: First, we must find a way to convey the facts of the situation we are in, in a way that's honest and accurate. Second, we must attend to the emotional impact of this information and provide a space for students to process their feelings. Third, we must support students in taking action that is meaningful, fulfilling, and sustainable.

On top of these challenges, there is a fourth: managing our own reactivity, terror, denial, and pain. If we don't understand and attend to our own emotions, we may unconsciously act them out in the classroom, with students.

The purpose of this chapter is to provide a model for understanding and working with the relationship between feelings and action, so that we can effectively apply the specific tools, exercises, assignments, and activities in this book. I present a simple, four-step process:

Step 1: Feel
Step 2: Talk
Step 3: Unite
Step 4: Act

Before I go into detail about each of these steps, I will describe one of my own attempts at activism, back in 1999 when I was 19 years old, and the lesson I learned about what happens when we don't process our feelings and talk them through.

BURNOUT AND RECOVERY

I arrived in Seattle in the fall of 1999 to start my first year at the University of Washington. Soon I learned that a big event was coming to town in November: a conference of the World Trade Organization. Activists were organizing a protest to highlight the negative impacts of globalization on labor, human rights, and the environment. I didn't know much about globalization; I didn't have any models in my life for how to get involved in activism; and I didn't have any idea what I, personally, could contribute. But I knew the protest would be important, and I wanted to be a part of it, so I joined a student organizing group at the university, and began attending meetings.

Unsure of myself, I stepped into a support position in the group, following those who knew more about the WTO and who were (through age, education, experience, or bluster) more confident, adamant, and certain of themselves. I attended trainings and helped with preparations for what would eventually be known as the "Battle of Seattle." It was a crash course in street protest organization. We sent out emails, posted flyers, chalked sidewalks, painted signs, and concocted street theater pieces. But over the course of those three months, my sense of uncertain excitement, unity, and purpose turned into frustration, confusion, and insecurity. As it turned out, I didn't really like organizing street protests. I struggled to find my voice in the group. I didn't like chants, and I never really felt that I had much that was meaningful and personal to contribute. The actual days of protest in the streets were chaotic and exhilarating, but I walked away from the "Battle of Seattle" feeling burnt out, confused, cynical, and full of self-doubt. Comparing myself to activists who were more energized, more idealistic, and more confident, I concluded that I was simply not a very good activist.

In itself, this could have been an excellent lesson. Everyone needs to experiment, to try new forms of engagement in order to home in on what's best for them. There's no shame in trying something new and moving on, and if I'd had the opportunity to sit down with someone I trusted and talk over my doubts, worries, frustrations, and self-judgment, I might have learned something about the type of activism I was better suited for, and how to deal with my emotions when I came up against people or situations that frustrated me. Instead, I simply gave up. It was only fifteen years later that I reconsidered what had happened and started thinking about activism in a new way. Two experiences woke me up from this fog of doubt and helped me get unstuck.

The first was in a therapy session, working with a client struggling with climate change anxiety. His partner wanted to have a baby, but he was preoccupied with visions of the hellish, apocalyptic future he felt sure his child would experience. I knew enough about the science to know that his anxieties were reasonable, if not guaranteed, and his struggles brought up for me my own feelings of fear, guilt, and self-doubt about my decision to become a parent.

The second experience was a few years later, when a cloud of wildfire smoke spread over the Pacific Northwest of the United States and covered Seattle in an eerie haze that scratched at your throat and stung your eyes. The smoke had just settled in when, one morning, tired and distracted, I decided to take my kids out to a lake to swim. Spreading out our picnic blanket at the beach, I looked across the water and realized that the smoke was so thick that I could barely see the other side. The air around us was a weird pinkish-orange, and the few walkers nearby were wearing masks. I realized at that moment how privileged I'd been to live all my life in an environment that was safe, and I got a taste of what many on the front lines of climate change already knew: the anxiety and helplessness of living in an environment that is unsafe for those I loved. I knew that I, and many others, would face more days, and more summers, like this, and that this would affect us all in powerful and complex ways.

None of my training as a therapist had prepared me for these types of existential fears. I realized that I had to learn more about the psychological impact of climate change if I were to manage my own worries, let alone support others with theirs.

FEEL-TALK-UNITE-ACT (AND REPEAT)

We all struggle, in our own ways, to find a balance between the need to be still, to look within, to acknowledge our often challenging, confusing, and contradictory emotions; and the need to take action in the world, both to address the environmental injustice and devastation that we are witnessing, and to find a sense of agency and meaning in our lives.

The four-step process I call "Feel-Talk-Unite-Act" is meant to support students, clients, and others to act from their own hearts, with an awareness of their own feelings, and contribute to the joint human effort to stave off the climate crisis.

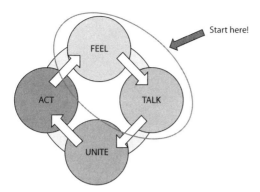

FIGURE 2.1 Feel-Talk-Unite-Act Process Model.

The most important part of this process is where we begin. When we experience fears or anxieties, or see something in the world that needs changing, a common impulse is to jump into a new activity, hoping to ease our difficult emotions by trying to make a difference. If we haven't processed our underlying emotions, we often choose actions that are not in alignment with our strengths, capacities, and resources, and we can end up feeling disillusioned, deficient, or burned out. In other cases, we have trouble identifying any action that feels worthwhile. Nothing seems like enough, so we throw up our hands, feeling despondent or deficient. We blame ourselves, or decide that no action is worth doing because it won't be enough.

As educators, our goal should be to help students find actions that are sustainable, impactful, and personally rewarding; this will maximize the likelihood that their efforts will benefit the issues they care most about. To do this, we must help students start with their feelings, stay with their feelings, and talk about their feelings with others, before jumping to action. And it is essential that we, as educators, work through this process on our own so that we can be as clear and attuned to our students as possible, and avoid letting our unconscious, unprocessed feelings interfere with our students' own, distinct expressions.

Step 1: Feel

The most important part of this whole process is for us, and our students, to recognize and attend to what we feel in the face of unchecked climate breakdown. This doesn't mean we have to stay at this initial stage forever, but we need to slow down and let ourselves feel what we're feeling in the moment, however muddled, contradictory, or distressing. Here is a short sampling of the types of feelings that I've heard clients express:

- *Sadness* about the future they are expecting
- *Grief* for species, ecosystems, and human communities being harmed
- *Anger* at those in power

- *Shame* at their own complicity in climate change
- *Guilt* about their consumption patterns or their lack of involvement
- *Hatred* towards humanity or towards a specific group of people
- *Hopelessness* about the possibility for beneficial change
- *Denial* about the true impact and extent of the crisis

Staying with these feelings is often the hardest part of this process, because we feel vulnerable and exposed to existential fears and potential losses that we'd rather not think about. But if we allow ourselves to feel what is real in our experience, however scary, it tends to transform into something new.

The feelings I listed above are generally thought of as difficult or negative. It's worth remembering that other, more positive feelings might also be present in a student's experience:

- *Excitement* at being part of a cause that is bigger than oneself
- *Hope* about creating a better world
- *Love* for our planet, our ecosystem, and our fellow humans
- *Joy* when we see others joining together, united in a cause
- *Compassion* for the suffering of others or for our own suffering
- *Trust* in the natural system to heal itself

The task we have as educators is to attune to our own fluctuating emotional world, while supporting our students to stay with and identify the complex ways that their hearts respond. Staying with and being conscious of our emotions doesn't mean getting stuck in feelings of sadness or despair. It simply means recognizing the feelings that are there, without judgment and without the need to act them out or push them away. Only then can we use the energy in these emotions to find purpose and meaning, both in ourselves and out in the world.

Some, including many in the activist community, insist that paying attention to climate-related emotions is a waste of time: self-indulgent, unproductive, or distracting from the crisis. The implication is that time spent attending to feelings will *compete* with time spent taking action. But this is not, in fact, the case. Ignoring our feelings tends to lead to misdirected, panicked, or stunted action, as it did for me in my early days of activism. The step of *feeling* is actually essential to identifying which actions we are best suited for, and which will have the greatest impact.

Step 2: Talk

The second step is to support the student in expressing their feelings to another person. Many of us feel a sense of shame, guilt, or aloneness in our climate emotions, or feel confused about how to handle them. By expressing our feelings aloud and having them reflected back by a trusted friend or mentor, we feel more empowered and less alone. Talking also helps us make sense of and manage complex emotions.

In an educational context, this means providing space in the classroom for students to share their feelings aloud and to hear what others are experiencing. Understandably, many educators don't feel equipped to process emotions as part of their curriculum. But we don't need to be trained therapists or counselors to incorporate psychological themes into the classroom. Simply acknowledging that climate-related content may have an emotional impact, and naming possible feelings that might arise, can have a powerful impact. Here are some other ideas:

- Leaving time at the end of each class for students to voice their feelings in a word or phrase
- Devoting a portion of one class to a group discussion about the emotional impact of climate crisis
- Inviting a guest speaker to talk on themes of ecological grief and anxiety
- Providing or assigning relevant resources (books, articles, websites) about climate mental health
- Being prepared to refer students to school counselors or other resources if they need support beyond what can be provided in the classroom

So far, we have focused on attending to what might be called "anticipatory climate distress." But, as climate impacts increase, we will encounter more students who have directly experienced traumatic events related to climate change, such as forced migration; permanent or temporary displacement due to wildfires or floods; and other extreme, frightening weather events like hurricanes, droughts, and heat waves. The severity of these impacts will be experienced disproportionately by students from already vulnerable and marginalized groups. These students' emotional experiences are likely to differ from those of students more removed from the direct impacts of climate change and may require additional attention, including opportunities for discussion tailored to their particular experiences and, at times, support from mental health professionals outside the classroom.

Whatever the context of the climate-related emotion, the goal is to let students know that their feelings are understandable and that talking about them is welcome. If the feelings expressed go beyond what the classroom can contain, educators can encourage students to talk with each other outside of class or with others they trust, such as friends, relatives, therapists, or neighbors.

Step 3: Unite

Humans are social animals. We thrive in community and gain strength in unity. On the flip-side, we struggle in isolation. We also tend to think that we, as solitary individuals, must find a solution on our own, and if we don't, there's something wrong with us. Finding and uniting with others who share our feelings and concerns counters isolation, helplessness, and depression. So the next step, after feeling our feelings and talking about them, is *uniting* by joining with others who share a common experience or purpose. This can come in many forms and should emerge from each person's exploration during the first two steps.

Our classroom can be the first place where students unite, through the process of learning, feeling, and talking together. Beyond the classroom setting, examples of groups my clients have joined or formed include:

- Climate activist organizations
- Trail maintenance crews or community gardens
- Mountaineering or outdoors groups
- Online support groups for climate grief
- Climate book clubs
- Identity-based organizations supporting communities on the frontlines of climate injustice
- Professional development organizations aligned with students' career aspirations in the climate field

As educators, we can integrate this step into our classes by encouraging students to connect with each other around these issues, inside or outside class; by providing examples of possible ways to unite, such as those listed above; or by assigning students to seek out, investigate, or form groups as part of the curriculum itself.

The key here is not to find the one group that seems to be the most effective, most outspoken, or most "activist" but rather, the one that will be the most supportive of each student's particular process of discovery. Spending time with others who share similar concerns about the climate, or similar appreciation for the Earth, enriches us and energizes us to stay engaged.

Step 4: Act

Clients I have worked with who slowed down and worked through steps 1–3, exploring their inner experiences along the way, were more likely to engage in sustainable, rewarding actions. During this fourth step, educators should avoid imposing their own preferred actions on students, but rather see what emerges organically. We don't know what is best for each person, and we may unconsciously impose our own ideas of "right action," based on our own biases and anxieties. I am always surprised by the ideas that arise when I let clients explore on their own. Some are predictable, like joining a protest movement, volunteering for an advocacy organization, or donating money. But others I could never have predicted. One client who worked in the tech sector decided to develop a video game to communicate, in image and sound, the possible consequences of unchecked climate change. Another client, who spent much of her time online reading about climate and politics, decided to buy a bike. She realized that she was spending all her energy worrying about the planet without giving herself a chance to experience what she loved about the natural world. Biking was her way to connect back with that love and to feel reinspired. I am sure this was not the final step in her journey, but it was clearly something she needed in order to remain engaged. Small changes like this, made in alignment with what we value and what we need, have unexpected and expanding ramifications.

Repeat

The exciting part of this process is that it never needs to end. Each time we move through a cycle of the four steps, from feeling to action, we learn something new about ourselves. As we engage in new actions, new feelings emerge: deeper feelings we were not aware of, and new emotions, emerging from new experiences and engagement with the issue. This sets the stage for more sharing, uniting, and yet-to-be imagined ways of engaging. In some cases, this can lead us to change course. Rather than stalling (like I did after the WTO protest), we can return to our feelings and get guidance from them, shifting towards actions that are in greater alignment with what is needed, what we know about ourselves, and what we are capable of contributing.

The process is recursive and potentially endless. The outcome is unknown because each new experience leads to new realizations and new, unknown actions. And this is just what we need as a species, because a problem so intractable as climate change requires new, never-before-seen solutions, ones that have not yet been developed or imagined.

FINDING OUR INDIVIDUAL CALLING

The forces of climate crisis, environmental destruction, and ecological injustice are so huge, so multifaceted, and so potentially overwhelming that no one solution, strategy, or style of engagement is sufficient to save us. We need each person to engage their particular unique strengths, capacities, and resources in defense of our ecosystem and our future, rather than trying to emulate others, compare themselves, or expect that they need to "save the world" on their own.

The purpose of this chapter is to present a model for supporting students in finding their calling by looking within. We can guide them, we can inform them, but we cannot tell them what to do. The best we can do as teachers and guides is to create a safe environment for those solutions to emerge in an organic and creative way. Essential to that process is slowing down and attending to complex climate emotions. The steps described here can help students find their own calling around the ecological crisis. Each student's action expresses their particular "genius,"[1] discovered in the process of working through feelings, in connection with others. Those solutions will emerge from the heart, when we look within, stay with our emotions, share them aloud, and unite in common cause with others.

NOTES

Epigraph: Michael Meade, *The Genius Myth* (Seattle: Greenfire Press, 2016), 4.
   1. Ibid.

THREE

Transformative Psychological Approaches to Climate Education

LESLIE DAVENPORT

This chapter explores how unconscious biases fuel climate change and offers ways to shift to a more eco-wise orientation. It includes guidelines for facilitating a climate emotions group, and two emotional resiliency practices that can be used in classrooms from high school though graduate school.

Climate change is often explained through references to carbon emission reports and warnings about rising seas. However, little attention is paid to the emotional impact and root cause of this traumatic reality and the domain where the solutions will ultimately be found: the human psyche. Though climate change is now widely acknowledged to be the result of human behavior, our climate efforts don't skillfully attend to the deep-seated psychological factors that drive our unsustainable lifestyles and unjust cultural norms, which in turn shape the economic and political systems that condition our perceptions and options for action.

There are three key concepts that can be incorporated in climate education in order to clarify and shift the psychological causes of climate change. They're accompanied by practices drawn from clinical experience and research in the behavioral sciences. These practices don't just help people cope with the high level of existential distress generated by climate change; they also support deeper systemic healing. The three concepts are:

1. Hidden Drivers of Climate Change: We can recognize how our unconscious emotions influence our beliefs and behavior, and contribute to climate change, and learn how to transform these patterns to support eco-conscious living.

2. Synergistic Thinking: We can discover ways to reawaken dormant forms of human perception that are undervalued in the dominant culture—including intuitive, creative, somatic (body-based), relational, and emotional intelligence—so that we can develop a more holistic and expansive understanding of our humanity and enable our eco-wise transformation. This process decolonizes our mind and reveals a more organic route to embodying climate justice perspectives.
3. Emotional Resiliency in a Changing World: We can learn a new definition of resiliency in the face of the climate crisis, and ways to cultivate different forms of resiliency.

HIDDEN DRIVERS OF CLIMATE CHANGE

Our human psychology has evolved to protect us from strong emotional reactions to real or imagined threats that could overwhelm or paralyze us. The human brain and nervous system have built-in "defense mechanisms" meant to help us function in a complex world by blocking, diminishing, or deflecting overpowering emotions. These inner defenses are active in all of us to varying degrees, but when it comes to the climate crisis, they come with a high price, shielding us from the recognition of difficult truths and hindering our much-needed engagement in our changing world.

It's important to recognize that defense mechanisms are largely unconscious. Our brain automatically activates them when significant threats are perceived, and we generally aren't aware that they're operating. This is distinctly different from emotional resiliency, where we consciously cultivate self-care and find ways to provide emotional respite and regulation without losing our connection to the realities of climate change. A few common examples of psychological defenses include:

- Repression: Climate change is real, but the harsh realities are removed from our conscious awareness so that our daily life remains in the realm of "business-as-usual." We develop subtle internal narratives, such as "Climate change isn't going to affect my family."
- Doomism: Being aware of all the unknowns of climate change feel unbearable, so the global situation is framed as "already doomed" or "It's too late, nothing I do will matter." These aren't reasoned statements that predict the outcome of climate escalation. Rather, doomism is a reactive position resulting from cognitive and emotional overwhelm. Paradoxically, it provides emotional relief because the mind rests on a conclusion, easing the tension of the unknown—but this position can catalyze self-destructive and reckless acts.
- Splitting: The world is divided into an all-or-nothing view of "people who care for the environment and those who destroy it." There's no tolerance for the

challenge and ambivalence of navigating a world where our daily actions are often entwined with fossil fuel use: for example, the everyday dilemma of buying produce at the grocery store that's unsustainably produced and transported. There's often a lack of empathy for our complex predicament and a dismissal of less-than-perfect efforts toward solutions.

As the gap increases between our environmental values and systemic life-threatening actions by governments and corporations, so does our distress. Psychological defenses may appear in the form of climate apathy or a teen's surly "I don't care" attitude, which are common disguises for underlying anxiety or overwhelm. We're just not skilled at managing the complex feelings that arise when the severity of climate change truly registers.

A valuable first step in freeing ourselves from the powerful and sometimes destructive pull of unconscious emotional suppression is to create a safe place to express feelings. There's a pervasive myth that acknowledging and expressing our feelings about climate change is unproductive, a waste of valuable time, and that we'll end up falling into an abyss of pain and sorrow with no way out. But just the opposite is true: it's our unacknowledged and disenfranchised emotions that keep us stuck. We can dispel the taboo of expressing feelings as part of climate work by learning and promoting a more accurate understanding of our emotional landscape and creating the space and time to talk about emotions in a group setting.

> Leading a climate circle doesn't require a mental health degree, and this kind of activity is well suited for classroom and co-curricular settings in higher education.

You'll find a reference in the Resources section at the end of this chapter to a free workbook that contains group exercises. Leading a climate circle doesn't require a mental health degree, and this kind of activity is well suited for classroom and co-curricular settings in higher education. Climate circles can be led by instructors or students, online or in person. In all of these cases it may be useful to draw on the following guidelines developed by therapists for facilitating group experiences and conversations. Feel free to adapt them to fit the exercise and setting.

- Confidentiality: In order to create a safe space and treat each other's experiences with respect, we need to agree to refrain from sharing what occurs in the group with others, unless the entire group chooses otherwise.
- Self-care: Participation is encouraged but not required. Participants are supported in caring for their emotional well-being and comfort level. This may

include keeping the eyes open during a mindfulness practice, opting not to share, or briefly stepping out of the group if feeling overwhelmed or triggered.
- Expression of feelings: The aim is to create a safe space for the fears, sorrows, and aspirations of each person in the group. Being witnessed and received in a nonjudgmental atmosphere is one of the foundations of healthy growth, and it builds trust and community. Speak from your own experience of eco-grief or anxiety rather than giving advice, interrupting, trying to fix another group member, or entering problem-solving mode.
- Share the air: Everyone who wishes to share has an opportunity to do so. Create space for one person to speak at a time, with no side conversations. It can be helpful to use a timer, bell, or "talking piece" to clearly define whose turn it is to speak.
- Provide resources: Sharing feelings can reveal a desire for emotional support that goes beyond the purpose and capacity of the meeting. It is useful to have resources available for referring group members to professional support. See the directory of climate-aware therapists in the Resources section at the end of this chapter.

A note on trauma: I often hear from educators that because they are not mental health professionals, they are afraid of centering emotions in their classrooms. There are some best practices for doing so that do not require you to be a mental health practitioner, and the benefits of validating students' inner lives and their emotional relationship to the material you're teaching are invaluable. If you feel that the conversation could trigger a participant's own experience of trauma, or even retraumatize someone, the following guidelines are helpful:

- In an intake questionnaire in the first days of class, it is common practice for educators to get to know students (pronouns, nicknames, where they're from, concerns or questions they have, and so on). This is a good practice, and if you use it, you can add a question about students' level of willingness to engage in discussions that may get heavy, and what kinds of support resources (if any) they already have in place or would like more information about. If you don't use this practice, now is a great time to start! Such a practice ensures that there is a contact person in the worst-case scenario that the student gets severely dysregulated. You might suggest in the questionnaire that each student identify a volunteer "buddy" to accompany them if they decide to leave the room. It can be very disconcerting for all if someone is upset and chooses self-care by exiting, without knowing what's happening and if they have what they need. And for the exiting student themselves, this is when they need support the most, so it's good to have an explicit plan in place of which everybody is aware. This signals to all students that they will be supported, whether or not they choose to participate in the discussion, and that they can choose the forms of self-care they find most helpful.

- Learn the signs of dysregulation. All classrooms ought to be trauma-informed at this point in history, given the myriad forms of trauma we all carry with us and which epigenetics are proving, from intergenerational and historic trauma to sexualized violence and abuse, and now, collective trauma (pandemics, politics, climate change). Sadly, however, such training is not required of most educators. Trauma is in your classroom, whether you are prepared for it or not, and will increasingly be so. And again, educators need not be professionals or trauma experts to facilitate healing experiences, or at a minimum, not do further harm. Edutopia lists common signs of dysregulation in the classroom. Consider what you've assumed about your students when they have acted in dysregulated ways and what it would take to transform these "challenges" into transformational learning experiences for all.

Signs of Dysregulation in the Classroom

- Disproportionate reactions to setbacks and unexpected changes
- Trouble managing strong emotions (extreme anger, excessive crying, etc.)
- Extreme shyness and difficulty engaging with others
- Clinginess
- Difficulty transitioning from one activity to the next
- Spacing out or chronic forgetfulness
- Frequent complaints of feeling sick, especially triggered by difficult conversations
- Lack of safety awareness
- Poor academic performance, such as missed deadlines, lack of effort
- Extreme perfectionist tendencies
- Physical and/or verbal aggression towards oneself or others
- Consistently conflictual interpersonal relationships
- Excessive substance use

Be explicit about the possibility of the conversation being uncomfortable from the beginning and let students know that discomfort is okay—it's not a sign that something is terribly wrong. That said, assure students that only they know when the discomfort transitions to dysregulation and that should they feel they are becoming dysregulated, they are very welcome to do what their body and heart need—leave the room, express the response, or ask for support. Assure them that they have tools, and you will support them. Assure them that the conversation is not graded, and that there is no academic evaluation occurring. Second, it is helpful for you to learn to identify when a student is experiencing dysregulation or retraumatization. Learn the signs, so you can be responsive in the moment as well. If you notice signs, slow down, shift course, or take a pause to check in. Third, it is critical that you offer resources that you know are in fact reliable. It is recommended to develop a relationship with a professional on your campus, so that you can trust you are offering students a safe resource. This relationship can also help you build confidence facilitating emotional conversations. As a facilitator, it isn't required that you be an expert on healing trauma, just that you can identify it, that you can offer many options for participation that are not required, shamed, or formally assessed, and that you are prepared to offer information about how students can access mental health resources.

It can be helpful to have a facilitator or witness who serves as a timekeeper and reminds the group of the guidelines. This role can rotate in ongoing meetings. These are the best practices for creating a safe space for discussing difficult emotions in a group setting, whether in a classroom or other forum.

Grounding Practice to Follow Climate Emotions Exploration

Each session, whether individual or group-based, can end with a grounding exercise which brings closure to deep emotional dialogue and supports a transition to whatever activities lay ahead. One person slowly reads the following exercise aloud for the group to experience:

> *Sit with your feet on the floor/ground. Close your eyes or give them a soft downward focus. Find a position where your upper body is balanced over your hips and your back and shoulders are in a comfortable but alert posture. Notice and enjoy all the places that your body is being supported by the chair and the floor. Take three slow, full breaths at a comfortable pace. Notice your breath filling up your body with oxygen and then emptying it out. (Continue your focus on this breath cycle for about one minute.) Recognize the gentle pull of gravity on your body. Feel yourself present in your body, being aware of your head . . . neck . . . shoulders . . . upper and lower torso . . . hips . . . legs . . . feet. Imagine the soles of your feet sending out roots into the ground, growing down through your socks, shoes, and the floor. Let these roots stretch down into rich soil, connecting, stabilizing, and grounding you. As you breathe in, image vitality entering your body through the soles of your feet. As you exhale, send down anything that wants to be released into the center of the Earth. (Continue your visualization of this breath cycle for about one minute.) Remain connected to the feeling of your body and the Earth as you reorient to the room around you—and when you're ready, open your eyes.*

SYNERGISTIC THINKING

The introduction of functional magnetic resonance imagery (fMRI) has provided unparalleled access to the activities of our brain and patterns of cognitive and emotional expression. We've learned that various areas of the brain perceive and filter information in very different ways. Our dominant Western culture places a high premium on a rational, analytical orientation, which is driven by areas of the brain that use logic to understand ourselves and the world by breaking things apart and examining them in separate pieces. Information is processed in a linear, sequential, atomized fashion. We've grown accustomed to relying on words and logic, which are essential—but they're only a tiny fragment of the myriad ways our brain is designed to understand our experiences.

As we learn to think, our mental habits establish neural pathways in the brain, and these neurological grooves continue to reinforce our cognitive patterns. Our thoughts tend to run the same limited course over and over, like the proverbial horse that always returns to the barn on a well-worn path. We can activate other parts of the brain, but this requires awareness, motivation, and practice. When it comes to climate education, it is essential—and well worth the effort—to cultivate synergetic thinking.

The dynamic neural networks that are active in synergetic thinking help us understand life by putting the pieces together, recognizing patterns, and connecting the dots; diverse inputs like images, sounds, spatial relationships, kinesthetic impressions, and emotions build up a holistic "big picture." This form of processing recognizes how seemingly disparate elements are connected to the whole. It allows us to put ideas together in new ways, experience flashes of insight, and discover creative solutions to old problems. It supports relational forms of thinking that are cooperative and empathetic. These neural dynamics are naturally active in children but gradually become less accessible from lack of use as we adapt to the prevailing cultural values and educational emphases.

If we group these primary types of brain processing into two broad categories—analytic and synergistic—we can think of them as two halves of a whole. If our analytical half is over-developed and "muscle-bound," and our synergistic half is atrophied, we're navigating life at half-power. When we're lost and walking through an unfamiliar landscape with anxious, unbalanced steps, we come to realize that we've been walking in a circle. This is an apt metaphor for how we continue to create the same problems by relying on internal perceptions that are incomplete and out of balance.

> Without using the synergetic parts of the brain to grasp the big picture, we can't recognize the interconnected nature of life and our place in the biosphere.

Without using the synergetic parts of the brain to grasp the big picture, we can't recognize the interconnected nature of life and our place in the biosphere. Perception is powerful, and meaningful behavioral shifts naturally occur when we cultivate synergistic, interconnected forms of awareness. We can use these two types of learning to create climate education guidelines that rely on whole-brain functioning, which Sherri Mitchell/Weh'na Ha'mu' Kwasset (She Who Brings the Light) of the Penobscot Nation attributes to the destructive

Western mindset, or "colonized mind." The colonized mind commodifies life and builds a power-over orientation into social structures. Mitchell describes some of the values rooted in the colonized mind as individual versus communal, competition versus cooperation, aggression versus patience, conquest versus harmony, arrogance versus humility, exclusiveness versus inclusiveness, and fragmentation versus wholeness. She explains that a decolonized mind approaches life relationally, with a quality of mindful responsibility for every living being. If you recall the features of synergetic thinking described earlier, which include interdependence, empathy, and cooperation, it's easy to see how synergistic thinking helps decolonize the mind by dismantling the myths of separation, othering, and the objectification of living systems.

Experiencing Connection, a Synergistic Thinking Practice

One person in the group reads the following script aloud slowly, allowing pauses of about five seconds at the end of each paragraph. Optionally, the exercise can be followed by small-group sharing or time to write or draw.

> *Sit upright comfortably in a way that allows your breath to flow easily. Let your shoulders open up and place your feet firmly on the floor. Either close your eyes or give them a soft downward focus.*
>
> *Consider that this moment right now has never been lived before. It arises fresh and uncluttered. And to the extent that we keep it free of thoughts about the past or future, it can be a spacious and peaceful moment.*
>
> *As you sense the feelings of spaciousness and peacefulness, connect these qualities to your breath, as though you could breathe those feelings in throughout your body. While keeping your breath in a rhythm that's comfortable for you, you might want to welcome in those qualities even more by taking a few breaths that are deeper and slower.*
>
> *Notice how the in-breath is vitalizing and fills you with fresh oxygen, and how the exhalation has a natural "letting-go" feeling. Filling up as you breathe in, and emptying out as you exhale. Notice how this reflects a very ancient rhythm: the rising and setting of the sun, the ebb and flow of the tides, the waxing and waning of the moon, the turning of the seasons.*
>
> *You may notice that a full and conscious in-breath and a full and conscious out-breath create a feeling of balance inside.*
>
> *Now bring your awareness even deeper inside yourself, below the fluctuation of sensations, behind the coming and going of thoughts, underneath the movement of your emotions, to the very vitality that animates your body, mind, and emotions—the same spark that animates and sustains all of life, the quality of aliveness.*
>
> *Can you sense where this aliveness originates, and where it ends?*
>
> *Imagine and sense the Earth below you—through your socks, shoes, floor, and building foundation—and how it's supporting you in this moment. Sense the spacious-*

ness of the sky above you—through the ceiling and roof—and get in touch with the edge of the biosphere, with deep space.

Every in-breath brings the sky into your body; with every out-breath, you mingle with the atmosphere. Sense how the food you've eaten has become your bones that are strengthened by minerals from the soil, and how the rivers, streams, and oceans literally course through your veins from the water you drink. Feel how life is nourishing and breathing you—and rest for a moment in its interconnected web that's like an energetic hammock.

Perhaps you can also sense our energetic connections to one another—more threads in life's supportive web.

Sense the circular flow of the web of life that moves through you. Now see if you can discover a quality of reciprocity—we're receiving, but how are we contributing to the whole? Open to this question without jumping up into your thoughts. Simply notice if there's an image, a phrase, an impression, or a sound that arises in response through your awareness.

As we prepare to close, notice any more shifts that have occurred in your awareness and any changes to your body, emotions, and mind. Let anything of value truly register and create an imprint that is easy to recall and integrate into your daily life.

Begin to assimilate your inner experiences while also reconnecting to the room around you—and when you're ready, gently open your eyes.

EMOTIONAL RESILIENCY IN A CHANGING WORLD

Synergistic thinking comes with an additional benefit. It helps build emotional resiliency by connecting us to our internal experiences of vitality and wholeness, rather than what may be a habitual focus on problems or distressing feelings. The conventional definition of emotional resiliency is the capacity to emotionally "bounce back" from the upset of a stressful event to our previous state. But this definition needs to be updated. There's no "going back" to the way things were before, to life-destroying systems and their depletion of environmental and emotional resources.

I define emotional resiliency as the practice of cultivating the ability to remain present, open-minded, and empathetic in the face of increasing stress. Building resiliency is building the capacity to bear witness to and be present with more intensity without lashing out or "checking out." Emotional resiliency can be learned, and there are now a wide variety of valuable tools and practices emerging from neuroscience research and trauma studies. They're based on a growing understanding of how our nervous system works, and they offer direct and effective ways to self-regulate, soothe, and emotionally replenish ourselves.

A good way to build emotional resiliency over time is to apply a concept called "toggling." We need to be truth-tellers when it comes to climate education, but we also need to use the tools to work with our natural emotional reactions. Toggling refers to intentionally moving back and forth between difficult climate truths and emotional self-regulation. Practicing our ability to move between the two grows and strengthens our emotional resiliency. Examples include doing a mindful breathing practice after reading an Intergovernmental Panel of Climate Change (ICPP) report; ending a climate feelings circle with a grounding exercise; calling a like-minded friend after watching an eco-documentary (to experience the calming hormones of co-regulation); or journaling our feelings after an advocacy event. Each of these self-regulating exercises activates parts of the brain that help regulate the nervous system. Co-regulation in particular offers benefits that ripple out, as emotions are contagious. Polyvagal theory, as espoused by Deb Dana and Stephen Porges, describes co-regulation as that which occurs when two (or more) nervous systems send and receive signals of safety by biologically adapting to and regulating each other. Porges believes that co-regulation lies at the heart of all human relationships, and that it explains interpersonal connection as a part of a vital evolutionary survival system. If we improve our emotional self-regulation, we can nourish and help others to regulate as well. And if we increase our social connections, we help strengthen our community's resiliency.

Befriending the Nervous System with Creativity

Neurobiological research on trauma confirms the ways that creative imagination can facilitate both self- and co-regulation. By including nonverbal forms of exploration, we tap into a deeper emotional integration than words alone can provide. Creative expression has often been at the heart of social change, action, and justice, and supporting each person's natural creative instinct is important for building culturally informed social justice perspectives that include diverse ways of depicting our experiences.

For this exercise, each person will need paper and something to write with.

[To be read aloud by a member of the group.] *Take five minutes (inside and/or outside) to find a simple object like a leaf, button, or stone. Let the object "find you": Notice something that calls your curiosity or attention to it and bring it back to the group.*

[Once everyone is back, continue with these instructions and then with the questions below.] *Sit quietly with your object, getting to know it as though it joined you for a reason today. Imagine that it's communicating with you through impressions, inner words, or a felt-sense.* [Pause] *Write down the answers you receive from your object after asking it one or more of the following questions:*

[Read the questions aloud, allowing about one minute of silence after each one.]

- *What's important for me to know about you?*
- *What message do you have for me today that relates to my climate justice work?*
- *What do you want or need from me?*
- *How are we meant to be in relationship?*
- *Is there anything else you want me to know?*

[Following the exercise, group members may be asked to share their experiences in pairs or small groups.]

RESOURCES

See the sources below for other reflective questions for the exploration of climate feelings.

- Edutopia.org has great resources on trauma-informed pedagogy.
- Decolonialfutures.net has wonderful inquiry-based card decks.
- More group resiliency practices are included in this free PDF: https://secureservercdn.net/198.71.233.33/vjx.f43.myftpupload.com/wp-content/uploads/2019/10/Emotional-Resilience-Toolkit-for-Climate-Work-v1.5-04Oct19-2.pdf.
- Referrals for climate-aware therapists: https://www.climatepsychology.us/climate-therapists.

REFERENCES

Dana, Deb, and Stephen W. Porges. *The Polyvagal Theory in Therapy: Engaging the Rhythm of Regulation.* New York: W. W. Norton, 2018.

Davenport, Leslie. *All the Feelings Under the Sun: How to Deal with Climate Change.* Washington, DC: Magination Press, 2021.

———. *Emotional Resiliency in the Era of Climate Change: A Clinician's Guide.* Philadelphia: Jessica Kingsley, 2017.

Federmeier, Kara. *The Psychology of Learning and Motivation.* Cambridge, MA: Academic Press, 2021.

Lertzman, Renee. *Environmental Melancholia: Psychoanalytic Dimensions of Engagement.* Milton Park, UK: Routledge, 2016.

Malchiodi, Cathy. *Trauma and Expressive Arts Therapy: Brain, Body, and Imagination in the Healing Process.* New York: Guilford Press, 2020.

Mitchell, Sherri. *Sacred Instructions: Indigenous Wisdom for Living Spirit-Based Change.* Berkeley: North Atlantic Books, 2018.

Rossman, Martin. *Guided Imagery for Self-Healing.* Novato, CA: New World Library, 2000.

Todd, Roisleen. "Recognizing the Signs of Trauma." Edutopia, October 27, 2021. https://www.edutopia.org/article/recognizing-signs-trauma.

FOUR

From Existential Crisis to Action Planning

Building Individual and Community Resilience

JESSICA D. PRATT

This chapter provides an overview and materials for a class module focused on confronting eco-grief and climate anxiety in the classroom, including pre-reading suggestions, journal prompts, in-class discussion activities, and possible extensions. It aims to address student emotions in a way that builds resilience, fosters community, and promotes a sense of self-efficacy and action.

Conservation and environmental science students who study climate change and other planetary crises often suffer emotional distress and anxiety related to what they learn. To effectively address such eco-grief or climate anxiety in students and empower them to meaningful action, educators must not just teach their disciplinary content, but inspire hope, foster optimism, cultivate coping mechanisms, and promote relational learning and connectedness to community.[1] How can we do this, recognizing that by and large, science faculty, including me, are not trained to confront psychological distress among their students, let alone support them through it? We may also find little support from our own institutions and subdisciplines for bringing emotions into the science classroom in the first place. For me personally, it began with confronting my own discomfort with discussing eco-grief, creating space to talk about it, and being open to listening to what students had to say. Over the past several years, I have increasingly integrated discussions of emotional responses to course content into my classes and studied the impacts of doing so.

Here I share a lesson module with assignments and classroom activities that directly confront the topic of eco-grief with students in the hopes of enabling instructors to address this phenomenon in a way that fosters a sense of purpose and builds resilience in students—both of which are essential to turning knowledge into action and may be important for retaining students in environmental fields.

Regardless of the "stages of grief" model you subscribe to (an internet search will show models of five to twelve stages and everything in between), the same pattern emerges—moving through the experience of grief can bring feelings of anger, denial, guilt, depression, and hopelessness. Effectively moving through these stages of grief to a place of acceptance and hope can instill a sense of personal efficacy and ability to take meaningful action towards addressing the source of that grief. This lesson plan aims to help students move through the stages of grief or across the "affective arc"[2] from a place of emotional overwhelm to one of action, self-efficacy, and resilience. In this lesson module, students will individually consider and write about their emotional response to learning about climate change and other so-called "wicked problems," directly confront those feelings in community via a structured and supportive class discussion, and make commitments to meaningful action on issues they care about.

Student Learning Objectives

1. Describe links between knowledge/learning about wicked global problems like climate change and mental and emotional well-being.
2. Define eco-grief and climate anxiety.
3. Consider your own psychological and emotional responses to various course content.
4. Discuss eco-grief and climate anxiety with classmates and devise strategies to turn anger and despair into power, hope, and action.
5. Make a commitment to action by completing a Personal Change Plan from the Social Transformation Project.

LESSON PLAN

Pre-Class Reading and Reflective Journal or Discussion Post

Overview: Students are assigned to complete the following pre-class reading and informal writing assignment. This works well as either a reflective journal entry read only by the instructor or as a class discussion post, where students can read and comment on each other's submissions. After trying both options, I find that the public class discussion posts create space for affirmations, supportive comments, and camaraderie. These are valuable in promoting discussion and participation in class as students come in knowing that their feelings are shared by their peers. When providing the following writing prompt to students, you can embed the article links from the resources list directly into the prompt.

Writing Prompt
In environmental education, there has been increased attention on student's emotional responses to learning—with growing awareness of feelings of "eco-grief" and "climate

anxiety." Additionally, research on climate change is beginning to document the mental health impacts of both learning about and experiencing the impacts of climate change. Simultaneously, there is recent work connecting values of sustainability and eco-friendly behaviors with increased personal well-being and happiness.

Much of what we face today in thinking about environmental change can be quite depressing. For example, see Calvert's "So What If We're Doomed? Climate Chaos, Mass Extinction, the Collapse of Civilization" or Butera's "Flooding, Heat Waves, and Destabilized Ecosystems." There is SO MUCH bad news.

Initial prompts:

As a student who is learning about and studying these problems day in and day out (and sometimes for years), how do you feel?

What are your emotional and psychological responses to these crises?

Peruse the readings below (cited in full in References section at end of this chapter) and address the subsequent questions in your submission:

Ray, "Do You Suffer from Eco-Despair? Seek Critical Thinking Treatment Right Away." Blog post on eco-grief and the student experience.

Clayton et al., *Mental Health and Our Changing Climate*. Report on the mental health impacts of climate change (look at pgs. 7, 15–17).

Barrington-Leigh, "Sustainability and Well-Being: A Happy Synergy." Highlights positive relationship between sustainability and happiness.

Questions to consider:

Do you experience eco-grief or climate anxiety?

How do you reconcile these feelings with hopefulness and empowerment to work towards a better future?

What can instructors do to effectively confront these feelings in the classroom?

What are your general thoughts on any of the articles you skimmed? Did anything surprise you?

Do you find any of the course content to be particularly troubling in terms of your emotional response?

In-Class Discussion Activity and Action Planning

Overview: This in-class activity was designed for an 80-minute class session. It can be adapted for a shorter session by eliminating the Sunrise Movement case study or by introducing a different case study example more relevant to your local context. This class activity has been implemented with undergraduate and graduate students but could also work well for a high school environmental science course.

Interactive Mini-Lecture on Eco-Grief and Climate Anxiety (10 minutes)
Description: This short lecture introduces learning goals and provides an overview of research on eco-grief and climate anxiety.[3]

Sunrise Movement Case Study and Video (15 minutes)
Description: Embedded within the interactive mini-lecture, this case study may be introduced by asking students the following questions:

What does it take for individuals or groups (like our class) to turn their psychological and emotional response to crises into meaningful action or even "hope"?

Have you heard of Sunrise Movement or other youth-empowered activist organizations dedicated to taking action on environmental or climate crises?

Following this reflection, Sunrise Movement is introduced with an 11-minute video by NBC News, *Inside the Sunrise Movement: How Climate Activists Put the Green New Deal on the Map*.[4] Before playing the video, ask students to note the various emotions they witness in the video. Upon the video's completion, consider sharing information for your local Sunrise Hub, if relevant. A list of active hubs can be found at sunrisemovement.org. One could select a different youth-empowered environmental or climate-action organization to highlight in place of this particular case study.

Small Group Discussion and Reporting (20–30 minutes)
Description: Students are placed in groups of four to six to share reflections and build community support. This can take place in online breakout rooms, in the classroom, or in an outside space on campus. When teaching this activity in person, I suggest letting students head outside, if possible, for some extra privacy and nature connection. I even bring a stack of picnic blankets into the classroom. Optionally, each of the discussion questions below can be placed on a Google Jamboard or a whiteboard in the classroom so students may report out their thoughts.

Instructions: In the discussion groups, ask each student to share a two-minute reflection from their eco-grief writing prompt and discuss their emotional and/or psychological responses to learning about climate change and other so-called "wicked" problems. Have one student in each group serve as the note-taker and add the discussion notes from their group to the Jamboard.

After each student shares, focus your discussion on the following:
How can you contribute to meaningful change in the world? For instance, how can you harness your fear, anger, despair, etc. and turn it into action, hope, and change?

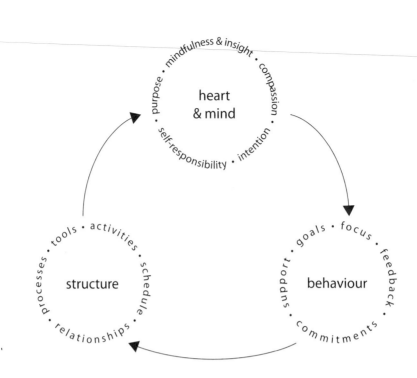

FIGURE 4.1 Heart and Mind, Structure, Behavior Diagram. © Tools for Transformation, https://atctools.org/tools-for-transformation.

TABLE 4.1 Worksheet from the Tools for Transformation Project

The Change I Want to Make	
P: (Purpose) Why am I working on this?	
O: (Outcome) What outcomes do I want? What will be different?	
P: (Process) How will I go about implementing these changes?	
Heart and Mind What needs to shift in the way I think and feel?	
Behavior What needs to change in the way I act?	
Structure What external changes do I need to make in my world?	
How I will track my progress:	

Note: This worksheet from the Tools for Transformation project is used for individual action planning. The full tool is available online at https://atctools.org/tools-for-transformation; search "Personal Change Plan." © 2015 Robert Gass.

What do you need to believe about yourself, society, and the world to commit to meaningful action and feel hopeful?

What do you need from your community (or professors) to be able to do this?

Action Planning and Changemaker Quiz (25 minutes)

Description: A common result of the class discussion is that students feel hopeful and move past their grief and anxiety by taking action, and this matches the conclusion from climate psychologist Dr. Patrick Kennedy-Williams. "The positive thing from our perspective as psychologists is that we soon realized the cure to climate anxiety is the same as the cure for climate change—action. It is about getting out and doing something that helps."[5] Thus, after the group discussion, students individually complete an action planning tool from the Tools for Transformation project (Figure 4.1, Table 4.1). This guides students towards making one commitment to personal change or action and walks them through the purpose, outcome, and process for making that change. Students then share their commitments to action with the class, and it can be easily seen how these individual commitments add up to substantial and meaningful actions. I want students to feel a sense of self-efficacy in making meaningful change in their communities and the world around them and understand that there are many changemaker roles to fill. I end class by having students take the Changemaker Personality Quiz from the Story of Stuff Project, which asks the question, What kind of changemaker are you? This quiz highlights the extent to which an individual fits into the categories of resister, networker, nurturer, investigator, communicator, or builder. It then provides suggestions for meaningful actions that fit naturally within a student's personal tendencies and emphasizes that there is a role for *everyone* in creating the future we want to live in.

ASSESSMENT OF LESSON PLAN

From 2018 to 2020, Dr. Karissa Lovero and I conducted a study at the University of California, Irvine designed to examine the impacts of eco-grief on student learning and to assess the efficacy of this eco-grief lesson module. We found that while eco-grief doesn't just "go away" as a result of these activities, the majority of students felt that this lesson helped them develop better mechanisms to cope with their emotional responses to climate change, and 67 percent of students felt they could better turn their emotions into action. Finally, 89 percent of students agreed that this lesson was effective and impactful.[6]

NOTES

1. Ray, *Field Guide to Climate Anxiety*.
2. Ibid., 13.
3. Materials available for download at https://faculty.sites.uci.edu/prattecology/teaching-resources.
4. Available on YouTube, https://www.youtube.com/watch?v=N28iaWIzJzg.

5. Check https://www.climatepsychologists.com for references to Patrick Kennedy-Williams's work.

6. Lovero et al., "Addressing Eco-Grief and Climate Anxiety."

REFERENCES

Barrington-Leigh, Christopher. "Sustainability and Well-Being: A Happy Synergy." *Great Transition Initiative,* April 2017. https://greattransition.org/publication/sustainability-and-well-being.

Butera, Candace. "Flooding, Heat Waves, and Destabilized Ecosystems: Here's What the Next 100 Years of Climate Change Could Bring." *Pacific Standard,* March 9, 2018. https://psmag.com/environment/climate-change-timeline.

Calvert, Brian. "So What If We're Doomed? Climate Chaos, Mass Extinction, the Collapse of Civilization: A Guide to Facing the Ecocide." *High Country News,* July 24, 2017. https://www.hcn.org/issues/49.12/essay-climate-change-confronting-despair-in-the-age-of-ecocide.

Clayton, Susan, Christie M. Manning, Kirra Krygsman, and Meighen Speiser. *Mental Health and Our Changing Climate: Impacts, Implications, and Guidance.* Washington, DC: American Psychological Association and ecoAmerica, 2017. https://www.apa.org/news/press/releases/2017/03/climate-mental-health.

Lovero, Karissa G., Amy K. Henry, Kathleen K. Treseder, and Jessica D. Pratt. "Addressing Eco-Grief and Climate Anxiety in the Undergraduate Classroom." Unpublished manuscript, March 1, 2023, typescript.

Ray, Sarah Jaquette. "Do You Suffer from Eco-Despair? Seek Critical Thinking Treatment Right Away." Blog post, March 27, 2018. http://writingattheendoftheworld.blogspot.com/2018/03/do-you-suffer-from-eco-despair-seek.html.

———. *A Field Guide to Climate Anxiety: How to Keep Your Cool on a Warming Planet.* Oakland: University of California Press, 2020.

The Story of Stuff Project. "Changemaker Personality Quiz." Accessed July 14, 2021. http://action.storyofstuff.org/survey/changemaker-quiz/.

FIVE

Empathy and Care

Activities for Feeling Climate Change

SARA KARN

This chapter presents a series of guiding questions and activities that K-12 and postsecondary educators can use to address the affective dimensions of climate change histories, while fostering empathy and care towards humans and more-than-humans in the past and present.

History has an important role to play in helping the climate generation navigate the emotional aspects of ecological degradation and social injustice in the age of climate disruption. While typical approaches to history are centered around content knowledge and historical thinking, learning about the past also involves engagement with affective elements, including empathy and care. According to James Garrett, when engaging with difficult knowledge in social studies, "affect and new ideas come to us in surprising ways. The various crises are not just in the past, they are also in the present and can be anticipated as inevitably coming in the future."[1] This is particularly true when learning about climate crisis, as environmental histories of destruction, suffering, and injustice are closely tied both to the present and to uncertain futures.

Historical empathy is a cognitive-affective process of understanding the thoughts, feelings, experiences, and actions of people in the past within their specific historical contexts. In conceptualizing historical empathy, Keith Barton and Linda Levstik outline four varieties of caring in history education: (1) *caring about* people and events in the past; (2) *caring that* particular events took place; (3) *caring for* people in history who have suffered injustices or oppression; and (4) *caring to* change our beliefs and behaviors in the present.[2] Caring—and the range of emotions and feelings that may flow from it—engages students in wanting to learn about the past and apply their learning beyond the history classroom.

Using this caring framework, I suggest a series of guiding questions and activities that educators can use to address the affective dimensions of climate change histories. The guiding questions can be applied to different historical contexts to generate discussions and support students' emotional connections to the past. The activities encourage students—including those who may be apathetic to climate change—to empathize with human and more-than-human experiences over time, care about diverse perspectives, and express their feelings about environmental injustices. Together they highlight how learning about the past can inform our present and future as we face the uncertainties of climate crisis.

CARING ABOUT HUMANS, MORE-THAN-HUMANS, AND EVENTS IN THE PAST

Guiding Question 1: How might the perspectives, successes, and setbacks of environmental activists in the past offer us stories of hope and resilience for today?

- Activity 1: Have students research environmental activists from different times and places (e.g., Rachel Carson, David Suzuki, Wangari Maathai). Present hopeful stories or "lessons" for today based on what they learned. Record these stories and lessons, along with drawings and QR codes linking to more information, on mural paper and display them on campus.
- Activity 2: Teach students about art activism and show examples by environmental activists from different decades. Consider how the artwork and the issues they represent have changed over time. Have students create their own artwork that captures emotions and represents environmental issues they care about today.

Guiding Question 2: How have animals, plants, and natural environments been treated by humans in the past?

- Activity 1: Look up information and resources on the internet about what might be gleaned from an Elder about Indigenous knowledges and ways of being on the land, if you yourself do not hold this knowledge. Invite an Elder from your community to speak with your class (being sure to offer compensation for their expertise and labor) or watch a recorded discussion online that is locally situated. Facilitate a group discussion or talking circle with students to reflect on what they learned, being sure to uplift (without tokenizing) students who identify as Indigenous or who come to the class bearing a lot of knowledge about and from the land already.
- Activity 2: Have students research how the environment in your area has been exploited by setter-colonialists over time. Students select a local natural resource (e.g., limestone, timber) and learn about the ecological and cultural consequences of extracting that resource.

Guiding Question 3: How have our families and communities been impacted by climate change over time?

- Activity 1: Students conduct oral history interviews with Elders or grandparents about the environmental changes they have witnessed over time in their communities. Provide students time to share these stories with their classmates.
- Activity 2: Encourage students to spend time exploring their community. Have them take a photograph of something (e.g., forest, glacier, park, neighborhood) they do not want to see impacted by climate change. Students present their photographs to the class, explaining how it is likely to be impacted by climate change. Students use a digital platform to edit the image to show what it may look like in 25, 50, or 100 years. Create a collage for a campus display or blog.

CARING THAT PARTICULAR EVENTS TOOK PLACE AND CONTINUE TO IMPACT THE ENVIRONMENT TODAY

Guiding Question 1: What can we learn from environmental policies implemented in the twentieth and twenty-first centuries by governments, intergovernmental organizations, and industries?

- Activity 1: Provide students with background on the 1972 United Nations Conference on the Human Environment in Stockholm. Students select one of the Stockholm Declaration's 26 principles and research its impacts on policies in different countries, or specific to where you live. Students should identify shortcomings and suggest ways to address the principle moving forward.
- Activity 2: Engage students in learning about the oil industry's history of climate denial by creating interactive timelines, which highlight key events and policy decisions. Identify and discuss the techniques used by corporations to promote doubt in climate science and avoid developing environmental policies. Discuss how knowing these strategies can help combat present-day climate denial.

Guiding Question 2: How do historical instances of anthropogenic environmental degradation continue to impact environments today?

- Activity 1: Students research human-caused environmental disasters (e.g., Dust Bowl, Chernobyl, Fukushima) that are related to course topics and/or personally relevant. Read about the disaster, listen to oral histories, and develop a report about ongoing impacts from human and more-than-human perspectives.
- Activity 2: Locate a nearby military base or former conflict zone. Students conduct research on the environmental pollution and destruction that occurred and

continues to occur due to military presence and warfare. Examine photographs of the site from different years or decades and discuss changes in the landscapes over time.

Guiding Question 3: How do examples of environmental decisions and actions in the past make us feel in the present?

- Activity 1: Encourage students to identify words they associate with how they feel about environmental policies and decisions discussed in class. Display the words. Have students write short journal entries to reflect on their feelings, which may or may not be shared/submitted.
- Activity 2: Students research the history of different renewable energies and sustainable technologies, and then present their findings to the class, with a particular focus on how each technology addresses problems that contribute to climate change. Facilitate a discussion about the power of technology to assuage anxieties about the future.

CARING FOR HUMANS WHO HAVE SUFFERED INJUSTICES OR OPPRESSION

Guiding Question 1: How does learning about historical examples of environmental racism make us feel in the present?

- Activity 1: Teach students about the history of NIMBY (Not In My Back Yard) and the implications for particular neighborhoods (e.g., low-SES, minority groups) to bear the burdens of environmental degradation. Read accounts from those who are vulnerable to toxic waste in your community. Discuss how learning about environmental racism makes students feel and identify ways to bring greater awareness to the issue. Note that many students have themselves experienced environmental racism; explicitly facilitating a conversation between students who are just learning about their own complicity in harm and those on the receiving end of that harm is critical here. Assume students in the class experience environmental trauma of one form or another, so the emotions may be significant.
- Activity 2: Have students research a historical example of a community that did not have access to clean drinking water (e.g., Indigenous people living on reservations). Consider how their situation has changed or remained the same into the present, and how climate change will impact water supplies moving forward. Students can create a poster to represent the issue of rights to clean drinking water and share it on campus or social media.

Guiding Question 2: What emotions do we experience when thinking about environmental degradation in the past, present, and future?

- Activity 1: Students research an extinct or endangered plant or animal species in the region where they live or attend school. Present the history of its destruction (e.g., clearcutting, overhunting) from the perspective of that plant or animal.
- Activity 2: Engage students in an imagining activity: invite them to close their eyes and think about a place (somewhere outdoors) they have a long history with and vivid memories of. What do they see? Hear? Smell? Next, have them imagine that place without any trees, plants, animals, or flowers. The water has dried up and the glaciers disappeared. How would these changes make them feel?

Guiding Question 3: What emotions do people experience when learning about human and more-than-human suffering as a result of environmental degradation and injustice?

- Activity 1: Students construct a profile of an imaginary person who is different from themselves in every way (e.g., gender, age, race, religion, politics). Throughout the course, students reflect on their own feelings about the topics of study and also consider each topic from the perspective of their imaginary person (who likely represents a real individual). How does this other person's background shape their responses to environmental degradation and injustice?
- Activity 2: Facilitate a group discussion or talking circle for students to express their emotional responses to learning about human and more-than-human suffering. Use your discretion to determine whether students are comfortable sharing in this way. Begin by establishing some guidelines (e.g., the right to pass, active listening) and open with your own responses before turning it over to students.

CARING TO CHANGE OUR VALUES, BELIEFS, AND BEHAVIORS IN THE PRESENT AND FUTURE

Guiding Question 1: How have we personally contributed to the climate crisis and what steps can we take to improve our relationships with environments moving forward?

- Activity 1: Engage students in creating short "videos for change" that could be shared on YouTube or other social media platforms. Students should begin by explaining the historical roots of a climate-related issue and suggest ways that individuals, companies, and governments can take action to effect change.
- Activity 2: Encourage students to identify three ways that their values and beliefs changed because of their learning in the course. Develop a personal action plan for improving their relationships with the environment in ways that address their changing values and beliefs.

Guiding Question 2: How might we change our relationships with more-than-humans in meaningful ways?

- Activity 1: Watch a documentary or video clip (e.g., *Cowspiracy, Kiss the Ground*) that depicts the history of animals and ecosystems suffering as a result of factory farming and agricultural exploitation. Visit a local farm that supports humane and sustainable sources of food. Have students develop a list of other farms and organizations in their community they could support.
- Activity 2: Engage students in planting a garden or maintaining a walking trail on campus. Use an app (e.g., iNaturalist) to record plant and animal species in the area. To make the garden or trail interactive, have students develop signs (physical or QR codes) that explain the history of human and more-than-human relations in the community, suggest better ways to interact with more-than-humans moving forward, and offer reflection questions for visitors. Note that some students may have deep knowledge already; be sure to create conditions for that knowledge to be validated and shared, so the students can learn from each other.

Guiding Question 3: What can we do to address environmental injustices within our communities?

- Activity 1: Partner students with community groups that are working to address environmental injustices. Encourage students to develop relationships with the community throughout the semester and discuss appropriate ways to get involved. Have students write reflections on the experience and present what they learned to the class.
- Activity 2: Students create protest signs, poems, or songs that address an issue related to environmental racism that is meaningful for them. They should identify the components that make their messages effective (e.g., drawings, memorable words). Encourage students to consider attending a community protest to see these messages in action.

NOTES

1. Garrett, *Learning to Be in the World with Others*, 69.
2. Barton and Levstik, *Teaching History for the Common Good*, 229.

REFERENCES

Barton, Keith C., and Linda S. Levstik. *Teaching History for the Common Good*. Mahwah, NJ: Lawrence Erlbaum Associates, 2004.

Garrett, H. James. *Learning to Be in the World with Others: Difficult Knowledge and Social Studies Education*. New York: Peter Lang, 2017.

SIX

The Emotional Impact Statement

CHRISTIE M. MANNING

This chapter introduces a reflective writing assignment that invites students to express, share, and process feelings in response to environmental degradation and injustice. It therefore presents an alternative to the analytic impact statements required for industrial developments, which are neutral and detached in describing damage to natural and human communities.

A student at a midwestern college returned home for fall break one year to discover that the patch of woods he'd played in as a child had been cut down to make way for a huge self-storage building. He was distraught and irate. How could this have been allowed to happen? The student's research revealed that the developer had, in fact, gone through a specific legal and regulatory process to gain approval for the removal of the trees: they had done an environmental impact statement (EIS). "Apparently," the student shared with his peers in class, "the environmental impact statement didn't take into consideration the emotional impact of losing those trees." And thus the emotional impact statement assignment came to be.

Most industrial developments in the United States—like the building of a self-storage building, a strip mall, a pipeline, a new road, or a power plant, or anything that might "affect the quality of the human environment"—require an environmental EIS before it can begin.[1] In analytic language, the EIS describes the potential environmental impacts of the planned development. Despite the requirement of an EIS, however, numerous industrial development projects, like the self-storage unit on the student's home ground, have destroyed fragile ecosystems, polluted waterways, and severely impacted the health of the nearby human and more-than-human communities. This takes a toll on the people whose lives are touched, directly or indirectly.

One way to process that toll is to write a different kind of EIS, an *emotional* impact statement. In contrast to the analytic environmental impact statement, which is neutral and detached in describing damage to natural and human communities, the emotional impact statement invites feelings to be expressed, shared, and processed.

The emotional impact statement assignment is meant for college students, activists, and anyone learning about and bearing witness to the destruction of the natural systems upon which all life depends and the unjust burdens this destruction imposes. Writing the emotional impact statement is powerful for nearly everyone, and particularly so for those who have at some point directly experienced a cherished natural space being harmed or destroyed, be it a wild forest or a small city park. Mainstream U.S. society offers few opportunities to express, or even slow down and feel, the emotional weight of environmental destruction and injustice. Writing an emotional impact statement allows people to come to terms with their feelings rather than avoid them, to express their sadness, anger, grief, frustration, and other complex and sometimes conflicting reactions. For some, writing the emotional impact statement is a cathartic relief. For others, this reflection exercise helps them identify and move beyond less-than-ideal coping mechanisms they use to avoid thinking too deeply about searing environmental issues.

Students in our classes who have completed the emotional impact statement assignment have told us that it was clarifying for them. It helped them realize the deep well of care they hold for nature and for humanity, and in many cases it spurred them toward greater action and activism. Many reported feeling less alone in their grief, anger, and anxiety after hearing the emotional impact statements of their peers.

Of course, a personal reflection activity like this one is not always easy to complete. In our experience, however, very few students have been unwilling to engage in this writing exercise. We do encourage students to share some or all of their writing with their peers. Sharing works best when the class has established a sense of community and people feel safe with being vulnerable, so we typically have students do this assignment in the second half of the semester, after students have gotten to know one another and we have established norms of respect, listening, and support for one another in the classroom. We have found it most effective to have students first share in small groups (two or three) for 10–15 minutes before inviting people to read parts of their writing, or share some of their learning, with a large group. Typically, we let students volunteer to read their emotional impact statement out loud to the whole class. The readings are nearly always very powerful.

The basic instructions for the emotional impact statement assignment are described below.

Gathering resources: To begin, the instructor or facilitator chooses a reading or a film/YouTube video that summarizes some of the ways that industrial pollution and climate change have altered ecosystems and harmed people. For example, in recent years we have successfully used the essay "The Uninhabitable Earth" by David Wallace-Wells.[2] Choose something that is relevant to your class materials and that has touched you emotionally. Another option we have used effectively is to share an actual EIS for an in-progress or completed project. For example, the 2018 federal EIS is available online for the now-completed Line 3 pipeline in Minnesota, a highly contested oil pipeline replacement and expansion project carrying tar sands crude oil from Alberta, Canada to a terminal facility on the shores of Lake Superior, in Superior, Wisconsin. The pipeline was opposed by an ongoing movement led by Indigenous water protectors. The FEIS is available at https://mn.gov/eera/web/legacy/34776/. We have shared small segments of Chapter 9 (Tribal Resources) and Chapter 11 (Environmental Justice) from https://mn.gov/eera/web/legacy/34776/. Alternatively, we have also assigned the emotional impact statement without a specific reading, but after spending following several weeks reading materials that are heavy and depressing.

Writing prompts: We use the following prompts: "Consider your thoughts, emotions, and behavior as you read about the various ecological and/or environmental justice issues created by humanity. What were your reactions, and what did you learn about yourself as a result? Write one to two pages of reflection with the following questions to guide you."

1. Thoughts: Did you skim the material, thinking "I already know all this?" Did you question or doubt the content in the reading? Was your curiosity aroused: "Really? I need to look into that . . . "
2. Emotions: What emotions were evoked by the reading? Did you feel angry? Sad? Overwhelmed? Did you feel numb and paralyzed? Or did the reading make you feel determined and energized? Did you feel excited by the potential for creating a more sustainable and justice-oriented community or world?
3. Behaviors: What did you do with your reactions? Distract yourself with the internet, social media, TV, or your text messages? Did you crawl into a bag of potato chips or crack open a beer? Did you call a friend to talk about the thoughts and feelings you were having about the climate crisis? Did you investigate advocacy or activist groups you could join?
4. Optional: Draw an image that captures your response to the reading, the emotional dimension in particular.

One final note: A personal reflection exercise is a tough one to grade, if you choose to use it as a classroom assignment. In our experience, this kind of assignment works best as a "pass/fail" (and everyone gets a "pass" if they complete it) or as an automatic B+ for completion, with grades of A- or A easily available with revision based on kind feedback.

NOTES
1. Environmental Protection Agency, "National Environmental Policy Act Review Process."
2. Wallace-Wells, "Uninhabitable Earth."

REFERENCES

Environmental Protection Agency. "National Environmental Policy Act Review Process." Accessed July 15, 2021, https://www.epa.gov/nepa/national-environmental-policy-act-review-process.

Wallace-Wells, David. "The Uninhabitable Earth." *New York Magazine,* July 17, 2017. https://nymag.com/intelligencer/2017/07/climate-change-earth-too-hot-for-humans.html.

SEVEN

The Politics of Hope

DANIEL CHIU SUAREZ, SOPHIE CHALFIN-JACOBS, HANNAH GOKASLAN, SIDRA PIERSON, AND ANNALIESE TERLESKY

We are frequently told we must "never give up hope." But what is at stake in hoping? In this course, we will interrogate this ubiquitous injunction to hope. We will analyze contemporary debates about the possibility of hope in the face of uncertain planetary futures to consider the politics of how, in what ways, toward what ends, and why we hope. At what point does hope become misplaced, turning into a "cruel optimism"? How is hope mobilized politically? How are different futurities—optimistic and pessimistic, utopian and dystopian, redemptive and apocalyptic—distributed among different groups? And what might happen if we let go of commonly held yet narrowly conceived hopes and tried imagining something different?

Course description for DAN SUAREZ'S course, "The Politics of Hope"

This chapter explores what emerged from a class on the "politics of hope." It includes reflections from the instructor and short essays by four undergraduate students coming to terms with how they were relating to their futures (in other words, how to hope) in the face of escalating planetary crises.

The following student reflections were written during an intensive one-month course I designed for Middlebury College called "The Politics of Hope," which I built around the course description above. For their final papers, I invited my students to respond to the question, *How should we hope?* The prompt asked students to reflect on the variety of hopes and ways of hoping we had engaged over the semester and to subject these hopes to critical scrutiny. Their task was to try to reconstruct for themselves something that felt personally compelling, analytically rigorous, and politically generative: something that was honest and unflinching in its reckoning yet something that also amounted to more than just paralysis or despair.

The course was premised on the idea that hoping is political and therefore an inherently social process: it is collectively derived and relationally experienced. To help reconcile our more theoretical concepts *about* hope with the more dynamic, contradictory, and everyday lived experiences people have *with* hope, I thus had my students go out and talk to some people. They conducted interviews with their professors, older family members, and peers (their age or younger), which they contextualized with discussions and texts from class and with their own continuing efforts to discover where they stood among the big questions animating the course.

It goes without saying that learning about the enormous stakes and dire implications of contemporary ecological crises can be deeply troubling. Yet for this class I made a conscious effort to let those sensations in rather than keeping them at bay. Admittedly, designing a class around a sustained confrontation with acute planetary devastation was a risk. Although I had come up with what I thought were careful plans for addressing the range of powerful affects inherent to the subject matter, I could not be certain of what the results would be. This lack of guarantees felt both exciting and dangerous.

Many notable things started to happen when we ditched the careful avoidance and forced optimisms that had defined many of the environmental classes I myself had taken as a student. Insofar as those treatments did acknowledge the systemic character and profound severity of current crises—confounding hopes in efficient light bulbs, reusable straws, and smart thermostats—those earlier instructors would often conclude the class by finally arriving at such realizations (right before releasing students and wishing us a happy holiday). In contrast, we began with this observation as a bewildering premise, and a defining challenge for the course, and we spent most of our time together essentially picking up the pieces and trying to recover something coherent (and potentially even rigorously "hopeful") in the aftermath.

As we proceeded, I came to realize how this starting point for the class aligned with the starting points of many of my incoming students and the sensibilities they were bringing to questions of planetary transformation, upheaval, and crisis. Annaliese, for instance, described a "painful, mostly terrifying" reality, while Sidra lamented a predicament often compelling her to feel "sad, scared, and angry." Whether acknowledged or not, these feelings are already out there roiling our classrooms. In this context, students seemed to appreciate the willingness of the course to drop the circumspection and level with them honestly about the unsettling implications of what was presently unfolding. In perusing their writing, you can see them not just stress-testing logical propositions and trying out different arguments derived from the assorted literatures I'd curated for them: they were undergoing a discernibly and sometimes powerfully affecting process, demanding not just analytical rigor from their instructor but deliberate scaffolding responsive to the intense emotions that came to saturate our pedagogical engagements with the topic.

In this regard, I found myself grappling with a dilemma each time I taught the course. What if I exposed my students to these materials and they began to doubt, or worse, lose hope? And yet, expecting a preordained conclusion from them—instructing them that they should not doubt but instead force themselves to believe—seemed cheap and manipulative. Their final paper was intended to let them confront their own doubts and hesitations and grapple with whatever arose from the encounter. Rather than presuming that students must or necessarily would hope by the end of class, or furnishing them with whatever personal reassurances I could offer that things would work out, I let them do it themselves: to face the sources of their dread, to come to terms with what they believed was actually happening, and to reforge their own orientations toward the future (and past and present) tempered by a hard look at things. I was fairly candid with my students about not having answers to the momentous questions I was posing to them and we proceeded with the understanding that we would simply have to do our best to muster a satisfying response, both collectively as a group and individually for ourselves.

After hearing from Hannah, Sidra, Annaliese, and Sophie reflecting in their own words on the politics of hope—each of them, in the process, struggling to come to grips with an outrageously disconcerting situation—I will conclude with some final reflections highlighting the generative potential of such an approach to teaching about (and with) the climate crisis.

HANNAH GOKASLAN

In a high school literature class, at the impressionable age of seventeen, I read Dante's *Divine Comedy*, an epic poem written centuries before I was born. The story details the protagonist's journey through the nine circles of Hell, up the slope of Mt. Purgatory, and, finally, into the splendor of Heaven. My teacher was a soft-spoken man named Dan Christian who had the reputation of being one of the "life-changing" teachers at my small, Baltimore high school. He was trained as a theologian and spoke in a medley of song lyrics and quotes from his favorite authors, sprinkling in his own meditations between the words he memorized.

To my surprise, I grew to love the class and often find myself thinking about the conversations we had. Hope is at the center of Dante's story. The opening lines, arguably the most famous from the text, mention the absence of it explicitly, reading "Midway upon the journey of our life / I found myself within a forest dark / For the straightforward pathway had been lost."[1] The narrator has lost hope and spends the next ninety-nine cantos of the poem on a mission to reorient himself. He feels the helplessness and weight of an uncertain future, symptoms associated with modern understandings of clinical depression. Dante's conception of hope, which he defines in one moment as "sure expectancy of future bliss," is what I see as a traditional understanding of hope.[2] It centers a future that is better than the reality of the current day. It is this sort of hope that lends itself to a reimagining, especially in the context of the climate crisis.

Even so, the poem does address questions similar to the ones that we encounter in our discussions today. In Dante's vision, Hell is a stagnant world of despair and the complete absence of hope; the very entrance gates read "Abandon hope all ye who enter here."[3] Meanwhile, Purgatory manifests as a mountain rising towards heaven populated by the toiling souls who work, seemingly against all odds, to climb towards a paradise that they cannot even imagine. The hope that motivates these souls is one that requires action embedded with the belief that what they are working towards will be better than their reality. Once in Heaven, time ceases to matter. There is one eternal present. No longer a future to strive for, Dante's Heaven becomes the embodied accumulation of those acts and the manifestation of that hope.

Although the story exists within these Judeo-Christian frameworks of the afterlife, it provides us with insight into culturally traditional understandings of hope and in doing so, can help guide our reimagining of Hope. The seemingly unexpected connection between these worlds, one of hope in the face of climate change and the other from this archaic, religious story, is what compelled me to reach out to my high school teacher in the context of this class.

When I probed him about hope in the context of climate change more specifically, he paused for a minute before responding "I don't feel hopeful, but I am hopeful." This very distinction between feeling and being hopeful, and the ability to hold the two in tandem, is central to how I believe we will exist in the coming decades.

Our future may require a reimagining of hope from a primarily emotional experience, something relatively passive and situational, to a verb that describes an approach—to hope rather than to feel hopeful, or perhaps in addition to that emotional response. The latter conception of hope is also the one that we tend to mistake for a sort of reckless optimism, which Nietzsche so directly criticizes. Hope as action has been central to abolition and radical social movements across the world for centuries. Climate educators, students, and activists can look to these examples to see how reimagining hope as an action allows it to become a form of resistance rather than the form of interpretive denial that Jem Bendell rejects in "Deep Adaptation."[4]

Rebecca Solnit articulates a vision of hope as an active pursuit, beautifully writing: "To hope is to gamble. It's to bet on the future, on your desires, on the possibility that an open heart and uncertainty is better than gloom and safety."[5] Solnit's work not only highlights the distinction between feeling and being hopeful but also addresses a secondary, crucial aspect of hoping today: the uncertainty of what is to come. In order to act with a hopeful orientation, there has to be a confrontation with the unpredictability of the future and an understanding that it may be hard to foresee the impact of our actions.

In a similar vein to Solnit, Laurie Penny emphasizes the active part of hope when she writes "It's not a mood. It's an action. It's behaving as if there might be a future even when that seems patently ridiculous."[6] We do not have to feel hopeful. We can (and responsibly should) acknowledge the harm and inequities associated with the climate crisis. We can wholeheartedly believe that we are doomed and that no amount of intervention now will save us, and nevertheless act in ways informed by hope. I often find myself navigating this space. My hope becomes less about the emotional response, although that is visceral and important, but rather an orientation of action. If we can emphasize this contradiction, this new existence of hope allows us to experience emotion, to acknowledge the sadness and frustration, and to still act. It does not ignore the sadness, the pain, and the fear associated with this work, but it does give us space to reduce harm, to think creatively, and to actively care for our communities.

While conducting my interviews, I identified two threads of this more traditional understanding of hope in the responses: it was external, meaning that the hope was located in other (often younger) people, and there was a faith in the existence of a certain future. My mother immigrated to the United States when she was my age. Newly married and unfamiliar with her new home, she and my father built the life to which I was born. Their transition was eased in many ways (they came to the U.S. with advanced degrees, they spoke English well, and they were white), but they still faced periods of deep uncertainty. She and my father found work in separate cities and spent the first years of their marriage apart. When I asked her about navigating this period of time, she placed her hope on her future family; her hope was based on the belief that my brother and I would live full and consequential lives. While it was a touching sentiment to hear as her daughter, my mother's hope built on my future existence during some of her hardest moments does not translate to the sort of hope I envision in the face of the climate crisis. It is too specific and isolated. We need a hope that extends beyond the future of individuals. We must hope for a collective future that we take an active role in constructing.

Mr. Christian also talked about future generations when I asked him about where he finds hope. Although he does not have children of his own, his 40-year career as a high school English teacher means that he has spent a great deal of time engaging closely with the minds of young people. He said, "I am hopeful because of young people—because people like you exist and are studying and thinking about these kinds of questions." Although I know his faith in our abilities to enact change is well-intentioned (and not entirely unfounded), there is also an unfair weight that comes with this sort of hope, one that young activists have critiqued before. An externally-based hope tends to absolve older generations from their own responsibility and can justify inaction, therefore becoming a form of denial.

In both cases, their hope carries a weight of certainty. My mother was motivated by the belief that she both could and would provide a better life for me and my brother, and her hope was contingent on the possibility of that future existing. Mr. Christian's hope in his students' ability to change the future is based on an expectation that the younger generation will find some sort of solution. What I am imagining is a hope that shifts from a basis in certainty to one that not only acknowledges the unknown but emphasizes it, allowing it to become a catalyst for creativity.

The context of the climate crisis is uncertain on an entirely new scale, so our hope will have to be different from what we believe today. We can balance being realistic without falling victim to a sort of reckless optimism or ineffective denial. In acknowledging these framings and tendencies in these traditional conceptions of hope, I am working to bring attention to the ways in which they can be limiting and by doing so, call to reimagine a hope that resists them. Our future will be different than today, that we are almost sure of. So, it only makes sense that our conception of hope should adapt along with us.

SIDRA PIERSON

The more my awareness of climate change grew, the more I came to see hope as senseless; anyone holding hope was practicing a form of denial, refusing to face both the facts of climate science and the reality of global governance. I recognized that we still have control over just how bad things are and will be, but I shared David Wallace-Wells' belief that the world is "surely not alarmed enough."[7] Now I recognize that what I saw as realism has a darker, more dangerous side, and that while some hope is naive and misguided, there are other more nuanced forms of hope that are not only justifiable but essential.

In today's world, it is easy to be sad, scared, and angry. As Rebecca Solnit writes, I "spend a lot of time looking at my country in horror."[8] Despair is not much of a leap, and it can feel like a nobler and more pragmatic stance than an insistence on unjustifiable hope. In fact, despair requires much less emotional work. It is often a premature determination induced by impatience in the face of uncertainty. While hope demands clear-sightedness and imagination, despair "demands less of us, it's more predictable, and in a sad way safer."[9] Despair can make the latest findings and forecasts more tolerable because you cannot be surprised or devastated by confirmation of a belief you already hold. But this resignation eats away at motivation for change-making; as Solnit notes, "activists who deny their own power and possibility likewise choose to shake off their sense of obligation: if they are doomed to lose, they don't have to do very much except situate themselves as beautiful losers or at least virtuous ones."[10]

Despair is therefore dangerous. Defeatists replace "the superstition of progress with the equally vulgar superstition of doom,"[11] providing a new cause for a familiar paralysis. Defeatism supports narratives that characterize climate change as an inevitable result of human nature and that frame continuing down our current emissions path as similarly unavoidable.[12] "[Attributing] tragedy to evil actors" disallows "individual or collective autonomy."[13] It is no wonder then that those in positions of power and privilege "prefer that the giant remain asleep," promoting media depictions of resistance as ridiculous and pointless.[14]

To say that despair is unproductive and dangerous is not to say that the feelings that fuel it should be ignored. The scientific reality of climate change and the lack of existing urgency to address it "[require] that we get very worked up indeed."[15] My peers and I have found that expressing our fear and cynicism is often necessary in order to get to a place of feeling motivated and capable of making change as an individual. There is a fine line between sitting with these feelings and letting them overwhelm you. Finding this balance means clinging tighter to what we cherish, which becomes more vivid in the course of loss.[16] For "to wallow in despair that the natural world is dying is to fail to be aware that it is still, in many ways, very much alive."[17] Fury and sadness are justified, but they must be channeled into passion. This means recognizing that "the fury you feel is the hard outer shell of love: if you are angry it is because something you love is threatened and you want to defend it," and that such anger needs to serve as motivation.[18]

Hoping is terrifying. It is an "anticipatory consciousness [that] involves risks," risks of betrayal and incredible disappointment.[19] These risks exist because hopefulness is "a form of trust, trust in the unknown and the possible even in discontinuity."[20]

But hoping is imperative and powerful, for many of the same reasons. Acknowledgment of uncertainty promotes an acceptance of agency and responsibility.[21] Solnit believes this lays the groundwork for an entirely different kind of hope: "that you possess the power to change the world to some degree or just that the world is going to change again."[22] This hopefulness does not depend on success or happy endings. It is fueled by determination to fight for a less tragic future, with the knowledge that the act of fighting will allow different futurities to become more intuitive.[23] Hope therefore is an act of defiance, "or rather [. . .] the foundation for an ongoing series of acts of defiance."[24]

Hope is further strengthened by the possibility that comes with a future that is still in the dark. The past becomes a torch we can carry forward into that darkness, and hope is born out of the opportunity for action created by "the spaciousness of uncertainty."[25] While we know that it is too late to "solve" or "stop" climate change, we also know that "the nature of that change is still up to us."[26]

Depending on the day or the hour, I may or may not be able to lean into this uncertainty or illuminate it with hope. News alerts, class readings, and personal observations of the world around me threaten to pull me back toward the comfort of despair. What keeps my cynicism at bay is remembering the positive impacts of holding such a difficult hope.

If we understand hoping to be a social process with political consequences, then we know it is not a self-contained act. Sometimes hopefulness is an act of self-care, but sometimes it is a choice to influence politics, ways of knowing, and other people's affects. Hope inspires imagination and actively encourages creative thinking, not only in ourselves but in others. It means being at "peace within shifting terrain" and recognizing that uncertainty presents an opportunity for radically undertaking "a grand project of mutual reinvention."[27] It means avoiding unrealistic utopian visions that place a better existence in temporal isolation without a clear path from now to then, and instead emphasizing process, cultivating "real and imagined strivings for a livable and social existence."[28]

This radicalism and boldness must include practicing freedom daily, which involves doing away with the distinction between activism and daily life. Practicing freedom entails a refusal to cooperate with grief and despair and a refusal to "collaborate by lending energy to that which oppresses you."[29] By defiantly hoping in a sick society, we can withdraw our consent from the systems we wish to change and make ourselves "unavailable for servitude."[30] Welcoming uncertainty and allowing it to form a foundation from which to hope allows us to delegitimize systems and rules that have become entrenched.[31] Rather than making ourselves out to be victims who must endure and suffer through climate change, we can use hope to frame ourselves as subjects with agency.[32]

As we break free, we can also recognize those who have done so before us and who are striving alongside us now. Effective technological, political, and social solutions to the climate crisis exist, and in our eagerness to rebuild, we must not negate the valuable effort and thought that has already been put into changemaking. We must therefore strike a balance between embracing uncertainty and recognizing knowledge, practices, and values that we should take with us into the unknown.

ANNALIESE TERLESKY

"The foundation of nature seems to be hope." In an interview with my mother, she concluded that hope exists all around us—in acorns, in all of nature's pregnancies, in an octopus laying half a million eggs though only a handful will survive. Hope seems to be embedded in nature, so how shall we, a collectivity of humans staring into the face of climate change, hope ourselves? As I interviewed Professor Hatjigeorgiou from Middlebury's Religion department, she brought up the three Christian cardinal virtues announced in 1 Corinthians 13:13: faith, hope, and love. Paul, she explained, cleverly placed hope in the middle because it is a deep biological, transcendental state built into us as human beings, and a "bridge" between the "prerequisite" foundation of faith and the end goal of love. Looking then to the process and practice of hope in the climate crisis, this framing can be reinterpreted and applied: faith is the foundation of first trusting and then understanding the reality of climate science and coming crisis; hope is a bridge we imaginatively, collectively, and actively create to adapt and reach a future we desire; and love is the end goal, a future state of harmony between peoples and respect for life on earth.

The facts are painful, mostly terrifying, but exposure to them is necessary. In order to spring into hope and into action to make changes within ourselves and within harmful broader systems, a foundation of faith must be developed, however unsettling and overwhelming. Faith, in this context, moves beyond common cultural and political connotations of the word; it is neither blind nor necessarily religious. Rather, in exposing ourselves to the truths of climate change, we must interpret this faith in its most basic definition, stripped of overdetermined understandings. It is trust in something, founded upon realism and clarity. This being said, David Wallace-Wells puts it simply: "It is, in fact, worse than you think."[33] There is a very real sense of doom that must be clearly evaluated and faced. As Roy Scranton argues, global climate change is the greatest threat to the world—and we will face an "apocalyptic future: no matter what changes we make."[34] Thus, as Jonathan Franzen writes, if we abandon our "false hope of salvation"—of "stopping" climate change or "saving" the planet—and accept coming disaster, then we can remodel our hope with acceptance.[35] We can then act from a place of reality, not blind optimism. We can do the best we can, and do good—do better—for the sake of good, not for a utopic future or a future akin to the present.

Beyond avoiding blind optimism, we cannot shrink into despair after facing the worsening present right before us along with the kind of future Scranton and Wallace-Wells warn against.[36] However, as Jennifer Atkinson suggests, grief is not only "a healthy and necessary process we have to undergo in order to heal" after a loss, but it is also a sign of "deep attachment and connection"—of love and praise.[37] In the face of such ecological catastrophe, as we lose more and more of the natural world and its wild creatures, our grief can actually remind us to honor earth's remaining lifeforms, "to hold even tighter to our values, and to resist with all we have any act that threatens to extinguish the life that remains." In other words, we have to sit with the negative emotions that come with learning the dooming reality of climate change. However, we cannot stop there: we must learn from our grief, anger, and anxiety, and channel them into both gratitude and loving, transformative action in order to lose as little as we can moving forward. Grief and hope can coexist.

This brings us to the bridge that is hope, and the more important question of how we can hope in the face of a future that is undoubtedly complicated if not catastrophic. Hope, as

Rebecca Solnit explains, "doesn't mean denying" reality, but rather facing it.[38] Perhaps the object of our hope isn't a continuation of how we comfortably live now. Hope is a bridge from faith to love: it is something that requires not only our imagination to build as a collective, so that it can stand the weight of many bodies, but requires that we move our feet and walk over it—hope requires action if we want to achieve our goal of getting to the other side, past the chasm beneath. Without the bridge, our life is a fall into this chasm. If we do not build the bridge of hope, we remain where we are or fall; and if we do not cross the bridge, putting our hope into action, we also stay exactly where we are.

Our hope must be radically *imaginative*. As Sarah Jaquette Ray explains of moving past eco-grief into hope and action, "We will get nowhere if we do not [first] imagine the future we hope to live in."[39] It is for this reason that if we believe we are entirely doomed to extinction and the problem is out of our hands, we have decided on a future we have no part in shaping, and we become entirely immobilized, even nihilistic. Eagerly desiring something different will help us better adapt to a changing world. Jem Bendell proposed just this imaginative adaptation and argues that we cannot continue operating and hoping with the assumption "that we can slow down climate change, or respond to it sufficiently to sustain our civilization."[40] Facing reality, he argues we should prioritize adaptation over mere mitigation of climate crises. Bendell redefines and reimagines hope: the object is not that our current way of life will continue, but rather, with what he calls "radical hope," we can creatively, imaginatively adapt to things falling apart, to everything changing before our eyes.

Our hope must also be *active*: drawn out of the imagination into action. While hope, through the power of our imagination, builds and is the bridge, action gets us across the bridge. Atkinson describes what Joanna Macy has called "active hope." As a practice, it is "something we do rather than have."[41] Hope is not the passive belief that everything will be fine. Rather, hope is powerful in that, informed by reality and embracing uncertainty, it demands that we act against this suffering and destruction. Solnit champions uncertainty's space for action, writing, "Anything could happen, and whether we act or not has everything to do with it."[42] In uncertainty, the space of hope, we can act, taking informed risks. Solnit concludes, perfectly bringing together the relationship between hope and action, writing, "Hope calls for action; action is impossible without hope."[43]

Our hope absolutely must be *collective*. Turning again to Solnit, we understand that "popular power is real enough to overturn regimes and rewrite the social contract. And it often has."[44] The unimaginable has been made possible in the past, and the question of how has always been a matter of active hope championed by more and more active individuals. Solnit describes the public as the "sleeping giant," for when it finally wakes up, having gathered faith and clarity, it is not just the "public" but civil society, forcefully capable of creating change.[45] Crossing our bridge of hope, toward the future we have imagined, "we write history with our feet and with our presence and our collective voice and vision."[46] Her words raise the question: what if we were all to resist the current status quo, which is so clearly wreaking havoc on the planet and inducing vast suffering for most human beings and prosperity for a select few? It could not survive in the presence of such active, immense resistance. We cannot and do not do anything alone, and thus we must come together with our grief, channel it into loving hope, and act. We must build the bridge that is hope as a collective, centering joy and love in our activism to sustain us, and we must cross this bridge as a collective in order to create and experience a changed world.

At the other side of the bridge is *love* and the pursuit of a collectively created future of equity and compassion. Love directs attention outward from the self, and moves us to act on behalf of others, whether that be a person or a planet. Maybe doomsayers are correct in saying it's too late to do anything, but I really cannot see how this helps anyone, or moves anyone to act from compassion and agency. There are so many other maybes, and perhaps the one we choose to fight for is acting as compassionately and ethically as we can. If a future of widespread joy and love is what is on the other side of the bridge, we can, starting today, live in accordance with this end. Emily Dickinson wrote, "Hope is the thing with feathers / That perches in the soul." Hope has wings. It waits, "perches" deep within us, ready to take flight if we choose. As Professor Hatjigeorgiou explains, once we, the "vessels" of such hope, rise, there is a lift from the inside and we can overcome our circumstances, no matter how bleak. In our hope, if we ourselves in the reality of the crises likely to come, imagine the best, most equitable future, if we act, and if we form a collective, our hope can work. The daunting, exciting part is that it is a choice.

SOPHIE CHALFIN-JACOBS

When I think of the question, "How should we hope?" my first thought is who is "we"? There is much consensus that the climate crisis was neither caused by nor are its effects equally distributed across a collective "we." As Meehan Crist puts it, "We know that climate risk and the worst effects of ecological disaster are unevenly distributed across race, class and gender, and among industrialised and developing countries—for many people, conditions tantamount to the end of the world have already arrived."[47] However, theorists from Roy Scranton to Steven Pinker invoke this "we." I believe these thinkers who appeal directly to intellectuals in academic arenas such as TedTalks or research journals are not really speaking to all of us but rather those of us untouched by the horrors of climate change or, as Rob Nixon describes, the "comfortable minority in the boat [who] ponder how many of the drowning masses they can afford to take on board."[48] The way that we each approach hope, as my professor Spring Ulmer states, is "dependent on context."

I would also like to consider the nonhuman species excluded from this "we." Max Chalfin-Jacobs, my younger brother, is a birder. He notes the changes he has seen in bird populations within his lifetime, reaching a conclusion that "maybe there's a chance for people, maybe there's a chance that we will create inventions that clean the air for us, but for birds and for other animals, I don't see any hope." While humans have outlasted many species, and many species will likely outlast humans, I think it is important that the nonhuman be a part of the conversation. As Jennifer Atkinson asserts, "The ability to mourn for the loss of other species is an expression of our sense of participation in and responsibility for the whole fabric of life."[49] "We" are a part of nature and should not forget that.

At the heart of these exclusions lies a certain amnesia. For Mary Annaise Heglar, white supremacy amplifies this "existential exceptionalism" (the belief among those with privilege that climate change is the first or most significant threat to survival) and perpetuates a view that is "not only inaccurate, shortsighted, and arrogant—it's also dangerous. It serves only to divorce the environmental movement from a much bigger arc of history."[50] She urges us to remember that communities of color "have even more to teach you about building movements, about courage, about survival."[51] Rebecca Solnit also speaks of the dangers of forgetting, arguing that "amnesia leads to despair in many ways. The status quo would like you to believe it is immutable, inevitable, and invulnerable, and lack of memory of a dynamically changing world reinforces this view."[52] She connects memory with hope and history, saying, "Things don't always change for the better, but they change, and we can play a role in that change if we act. Which is where hope comes in, and memory, the collective memory we call history."[53] These remarks point to the same conclusion: Forgetting the violence and inequality embedded into our history, or forgetting the small victories in our struggles for justice, can lead to cruel optimism or blind pessimism about the path forward, and forgetting the successes of social movements that have come before us closes the door to powerful generational knowledge about creating change. Just as we should not forget the implications of a collective "we" or the nonhuman cohabitors of this earth, we should not forget our complex history.

While hopelessness need not be debilitating, it is exceedingly common these days to hear that, without hope, people lack the motivation and vision to put in the work of building a better world. Speaking on her own feelings of hopelessness, Laurie Penny reminds us that "the same muscles that are required to survive an episode of depression are the muscles that are required for what is nebulously called 'resistance' to this current dark tide."[54] Perhaps hopelessness itself presents an opportunity for growth and perseverance. I, too, am not convinced that hopelessness itself is inherently dangerous, although there are circumstances where it can be. I asked Ulmer about hopelessness in the classroom. She responded, "The hopelessness that I've encountered is apathy . . . I think apathy is the biggest evil in some way. Or like flat-out denial. Apathy seems even worse because at least denial has passion attached to it." This reminds me of Lisa Duggan's postulation that "hope and hopelessness exist in a dialectical rather than oppositional relation, and that the opposite of hope is complacency—a form of happiness that will not risk the consequences of its own suppressed hostility and pain."[55] In other words, denial (while still problematic) and "bad" sentiments like cynicism and depression have strong emotion behind them, while apathy and complacency imply a lack of caring altogether and an aversion to reexamining the status quo. Another takeaway from this notion that hopelessness need not be debilitating is that despite the multitudes of feelings we experience facing the climate crisis, we tend to move through them in dynamic ways. Inheriting a climate emergency, I have felt my fair share of hopelessness, optimism, and a seemingly apathetic numbness. Yet each time I find myself in these numb periods, they ultimately pass, as I pull myself forward or am jolted by a call to action.

The question of how we should hope still remains. Essentially all of the authors we have read are in agreement that major political, economic, and cultural transformations are necessary if we want any chance at combating the climate crisis. Scranton proposes we "learn to die," Duggan and Muñoz propose "educated hope" and "concrete utopianism," Stephanie Wakefield proposes we explore the possibilities of "the back loop," Sarah Jaquette Ray proposes resilience that is "bound up with resistance." For Ulmer, my professor, hope means "ongoing life, planetary and otherwise." While I find myself averse to the death metaphor and partial to analogies of perseverance and transformation, different frameworks will reach different people. I still find myself left with questions I strive to answer, like "Why am I unable to envision a world without capitalism?" and "How can I build this imagination?" So, once we confront our complex histories and move past apathy and complacency, the question for me becomes not how we hope but how we move forward, because the world keeps turning nonetheless.

DAN (AGAIN)

This course was, and remains, an experiment. While I initially reeled at the intense feelings it dredged up, the courage my students showed when approached in this manner—when given the chance and without too many guardrails—offered glimpses into a rich reservoir of experience, wisdom, and insight that I have since learned not to condescend to or to underestimate. Interestingly, the process of streamlining these reflections in preparation for this volume—while helping to sharpen claims and clarify arguments—has also had the noticeable effect of smoothing over the more visceral sensations of bewilderment, existential self-questioning, and creaturely messiness I remember defining our experiences together in the course. Indeed, many students emphasized this emotional core of the class as one of its most lasting and important features. Crucially, as I came to appreciate, opening the door to such feelings appeared not only to court "negative" emotions like grief and anxiety but to also create openings for curiosity and excitement, fellowship and solidarity, and even humor and joy, as we struggled (professor included) to make sense of the fateful questions set forth by the class.

While harrowing, at times, I have learned to see these emotional dynamics as not only compatible but welcome and even necessary aspects of a more critical environmental pedagogy. Although critical educators working in other settings have long sought ways of teaching aimed at undermining complacency in the face of suffering and injustice, I have observed the manifestly radical implications of the climate crisis—when tackled directly—prompting potent reckonings among my students and serving to interrupt (if not completely dislodge) ingrained attachments, investments, and complicities in the systemic inequities and the forms of structural violence driving current planetary devastations. Rather than sidestepping all these bewildering and existential questions, my students taught me how rigorous confrontations with global ecological crisis—in part, *because* of its distinctly unsettling nature—could, with careful scaffolding, prove vital to their efforts to foster more expansive political horizons and radical imaginations better proportioned to the vast environmental transformations currently in motion.

NOTES

1. Alighieri, *Divine Comedy*, Inferno 1:1–3.
2. Ibid., Paradiso, 25:26.
3. Ibid., Inferno, 3:9.
4. Bendell, "Deep Adaptation."
5. Solnit, *Hope in the Dark*.
6. Penny, "On Hope."
7. Wallace-Wells, "Uninhabitable Earth."
8. Solnit, *Hope in the Dark*, 108.
9. Ibid., 20.
10. Ibid., 20.
11. Hannah Arendt, as quoted in Mann, "Doom," 92.
12. Malm and Hornborg, "Geology of Mankind?"
13. Mann, "Doom," 92.
14. Solnit, *Hope in the Dark*, xxv.
15. Klein, *This Changes Everything*, 20.
16. Jamail and Cecil, "As the Climate Collapses."
17. Bringhurst and Zwicky, *Learning to Die*.
18. Solnit, "Letter to a Young Climate Activist."
19. Gordon, "Something More Powerful than Skepticism," 264.
20. Solnit, *Hope in the Dark*, 23.
21. Bragg et al., "Hope."
22. Solnit, *Hope in the Dark*, 23.
23. Back, "Blind Pessimism."
24. Solnit, *Hope in the Dark*, 110.
25. Ibid,, xiv.
26. Klein, *This Changes Everything*, 28.
27. Wakefield, "Inhabiting the Anthropocene" 85; Klein, *This Changes Everything*, 21.
28. Gordon, "Something More Powerful than Skepticism," 258.
29. Ibid., 271.
30. Ibid., 272.
31. Klein, *This Changes Everything*.
32. Wakefield, "Inhabiting the Anthropocene."
33. Wallace-Wells, "Uninhabitable Earth."
34. Scranton, *Learning to Die in the Anthropocene*.
35. Franzen, "What If We Stopped Pretending?"
36. Scranton, *Learning to Die in the Anthropocene*; Wallace-Wells, "Uninhabitable Earth."
37. Atkinson, "Eco-Grief: Our Greatest Ally?"
38. Solnit, *Hope in the Dark*, xii.
39. Ray, *Field Guide to Climate Anxiety*.
40. Bendell, "Deep Adaptation."
41. Atkinson, "Eco-Grief: Our Greatest Ally?"

42. Solnit, *Hope in the Dark*, 4.
43. Ibid., 4.
44. Ibid., xxiii.
45. Ibid., xxv.
46. Ibid., xxv.
47. Crist, "Is It OK to Have a Child?"
48. Nixon, "The Great Acceleration and the Great Divergence."
49. Atkinson, "Eco-Grief: Our Greatest Ally?"
50. Heglar, "Climate Change"
51. Ibid.
52. Solnit, *Hope in the Dark*, xix.
53. Ibid.
54. Penny, "On Hope."
55. Duggan and Muñoz, "Hope and Hopelessness," 280.

REFERENCES

Alighieri, Dante. *The Divine Comedy*. Toronto: Aegitas, 2017.
Atkinson, Jennifer. "Eco-Grief: Our Greatest Ally?" *Facing It* (podcast), 2020. https://www.drjenniferatkinson.com/facing-it.
Back, Les. "Blind Pessimism and the Sociology of Hope." *Discover Society,* December 1, 2015. https://archive.discoversociety.org/2015/12/01/blind-pessimism-and-the-sociology-of-hope.
Bendell, Jem. "Deep Adaptation: A Map for Navigating Climate Tragedy." *IFLAS Occasional Paper* 2 (2018): 36.
Bragg, Melvyn, Beatrice Han-Pile, Robert Stern, and Judith Wolfe. "Hope." *In Our Time,* November 22, 2018. https://www.bbc.co.uk/programmes/m00017vl.
Bringhurst, Robert, and Jan Zwicky. *Learning to Die: Wisdom in the Age of Climate Crisis*. Regina, SK, Canada: University of Regina Press, 2018.
Crist, Meehan. "Is It OK to Have a Child?" *London Review of Books,* March 5, 2020. https://www.lrb.co.uk/the-paper/v42/n05/meehan-crist/is-it-ok-to-have-a-child.
Duggan, Lisa, and José Esteban Muñoz. "Hope and Hopelessness: A Dialogue." *Women and Performance* 19, no. 2 (2009): 275–83. https://doi.org/10.1080/07407700903064946.
Franzen, Jonathan. "What If We Stopped Pretending the Climate Apocalypse Can Be Stopped?." *The New Yorker,* September 8, 2019. https://www.newyorker.com/culture/cultural-comment/what-if-we-stopped-pretending.
Gordon, Avery F. "Something More Powerful than Skepticism." In *Savoring the Salt: The Legacy of Toni Cade Bambara*, edited by Linda Holmes and Cheryl Wall, 256–76. Philadelphia: Temple University Press, 2007.
Heglar, Mary Annaise. "Climate Change Isn't the First Existential Threat." *Medium,* February 18, 2019. https://zora.medium.com/sorry-yall-but-climate-change-ain-t-the-first-existential-threat-b3c999267aa0.
Jamail, Dahr, and Barbara Cecil. "As the Climate Collapses, We Ask: 'How Then Shall We Live?'" *Truthout,* February 4, 2019. https://truthout.org/articles/as-the-climate-collapses-we-ask-how-then-shall-we-live.
Klein, Naomi. *This Changes Everything: Capitalism vs. the Climate*. New York: Simon & Schuster, 2014.
Malm, Andreas, and Alf Hornborg. "The Geology of Mankind? A Critique of the Anthropocene Narrative." *Anthropocene Review* 1, no. 1 (2014): 62–69. https://doi.org/10.1177/2053019613516291.

Mann, Geoff. "Doom." In *Keywords in Radical Geography: Antipode at 50,* edited by Antipode Editorial Collective, 90–95. Hoboken, NJ: John Wiley & Sons, 2019.

Nixon, Rob. "The Great Acceleration and the Great Divergence: Vulnerability in the Anthropocene." *MLA Profession,* March (2014): 3–8.

Penny, Laurie. "On Hope (in a Time of Hopelessness)." *Wired,* November 27, 2019. https://www.wired.com/story/laurie-penny-on-hope.

Ray, Sarah Jaquette. *A Field Guide to Climate Anxiety: How to Keep Your Cool on a Warming Planet.* Oakland: University of California Press, 2020.

Scranton, Roy. *Learning to Die in the Anthropocene: Reflections on the End of a Civilization.* San Francisco: City Lights Books, 2015.

Solnit, Rebecca. *Hope in the Dark: Untold Histories, Wild Possibilities.* Chicago: Haymarket Books, 2016.

———. "Letter to a Young Climate Activist on the First Day of the New Decade." Literary Hub, January 1, 2020. https://lithub.com/letter-to-a-young-climate-activist-on-the-first-day-of-the-new-decade.

Wakefield, Stephanie. "Inhabiting the Anthropocene Back Loop." *Resilience* 6, no. 2 (2017): 1–18. https://doi.org/10.1080/21693293.2017.1411445.

Wallace-Wells, David. "The Uninhabitable Earth." *New York Magazine,* July 17, 2017. https://nymag.com/intelligencer/2017/07/climate-change-earth-too-hot-for-humans.html.

EIGHT

Unfucking the World

LEIF TARANTA

This essay was written by an undergraduate in 2018 for a college course called "Environmental Justice in the Anthropocene." When asked to answer the prompt "Are we fucked?" in relation to climate futures, this student argued that it is time for the "great unfucking." This essay also contains a reflection by the author on the experiences of being part of "unfucking" work.

I first wrote this essay in 2018 as a final assignment for Professor Dan Suarez's "Environmental Justice in the Anthropocene" class. We were asked to respond to the following prompt:

"Is Earth Fucked?" That was the title of a paper presented by a complex systems researcher in 2012 at a conference of the American Geophysical Union held in San Francisco. His answer? "More or less." For your final paper, I will ask you to develop your own preliminary response to a variation of this question: *Are We Fucked?*

In phrasing it this way, I wish to emphasize the freighted politics and contingent nature of this question rather than its strictly geophysical dimensions. In short, this essay is about the part of the question that deals with human agency: it asks you to analyze and assess our collective capacities to meet "The Anthropocene's" conjoined crises and, based on the available evidence, to come to some provisional conclusions about what you believe is emerging from this momentous historical crossroads. Along the way, this prompt also forces you to operationalize one of this course's central organizing questions as you develop your argument: *Who Is "We"?*

Your task here is not only to convince me but to convince yourself—one way or the other—and to wrestle with the full import of your answer together with its practical, ethical, and political consequences. What is at stake in your answer is also personal in ways you will need to address: it contains rather important logical implications for how you understand your place in the world, what can (and should) be done about it, and your decisions about how to live in it.[1]

Looking back on my response nearly four years later, I'm reminded of the optimism and determination I felt as a student reading the works of liberation scholars. I still carry that fire with me every day, as well as a deeper understanding of just how hard *un*fucking is. In the past years, I've been to jail multiple times, dropped out of college, and been injured by police and white supremacists. I've also witnessed firsthand the depths of power that community care can build—whether it's through crisis mutual aid around housing and food access, or direct actions that stop "business as usual" in its tracks. And I've had the honor of learning from Black, Indigenous, and unhoused community leaders in collective struggle, and witnessing moments where another world, another way of relating to each other, is alive and possible. If one thing hasn't changed, it's my belief that the world is definitely fucked, and my commitment to help unfuck it. I'm sure as hell not ready to give up.

UNFUCKING THE WORLD

It is time to unfuck our world.[2] Climate change threatens life on Earth as we know it, yet the question "Are we fucked?" draws attention away from the ways in which a united "we" has not previously existed. Marginalized populations have been environmentally oppressed for centuries, and climate change is simply a continuation of this history of exploitation and injustice. Led by youth and grassroots activists of color, the climate justice movement presents an opportunity to redress these wounds. Politicians, technocrats, and cynicism will not save us. Instead of wondering whether our world is fucked, we must fight as hard as we can to unfuck it, moving with hope and determination into an anthropocene that works for everyone.

When discussing whether the Earth was fucked, a complex systems researcher recently replied, "more or less."[3] This answer was informed by knowledge of the planet's current global warming trajectory, which is currently on track to far surpass 1.5–2°C warming. Business as usual will lead to widespread fire, warfare, agricultural collapse, disease, refugee crises, smog, ocean acidification, and ecological extinction. As Scranton writes, "We're fucked. The only questions are how soon and how badly."[4] However, the impacts of coming disasters will not be felt equally and "it is the poorest and least privileged who are being most egregiously affected."[5] This pattern of environmental injustice is based on the long-term process of othering that necessitates the question of who is meant by "we."

The question "Are we fucked?" imposes a collective "we" on a history of oppression, ignoring those people who have been marginalized for centuries. Contemporary climate change can be viewed as an extension of the colonial and capitalist exploitation that has been at work for centuries. The fossil fuel economy itself relies on the cheapening of the lives of people of color to make the sacrifice of people and places acceptable in liberal spheres. The stories of Flint, Katrina, Maria, and countless others illustrate the ways in which environmental crises are experienced differently along lines of racial and economic inequality, challenging the idea of a united humanity.[6] The anthropocene is characterized by a "great divergence" in wealth that shapes vulnerability to climate change, and an exclusive "we" is foundational to the continued crisis.[7] Mainstream environmentalist conceptions of belonging are rooted in white American ideals of frontier expansion, meaning that proposals for change often ignore the experiences of Indigenous peoples and urban residents.[8] Ultimately, the question "Are we fucked?" is one asked by billionaires with secret bunkers and those protected by militarized borders; by those hoping that climate change will not touch them.

Top-down, technocratic, and capitalist solutions to the climate crisis can serve as a neoliberal distraction from action and change. At fault for both climate change itself and for decades of political denial and inaction, neoliberal ideology is widespread even amongst climate advocates. Individualized and market based solutions presented by neoliberal environmentalism represent a disavowal of the true scope and severity of the climate crisis. "Cleantech" and "planetary improvement" can serve as distractions from more disruptive change,[9] while the billionaire fetish of geoengineering only perpetuates inequality and the "shock doctrine" behind climate change.[10] The powerful only advocate for change that maintains the status quo. It will not be cleantech or green capitalism or top-down policy that saves us, as these paths only seek to protect the most elite "we" from the "are we fucked" question. For most others, the answer is, "We always have been." The reality of climate change suggests that white supremacist, capitalist subjugation must end. If "we" applies to the people of this planet, then it is time for the Great Unfucking.

It is time to change the meaning of the word "anthropocene." The term has been used as a way to blame a collective human folly for the climate crisis. But what if the name for our geologic era was a creative word and not a blaming word, a statement of collective goals rather than collective culpability? What if "anthropo" did not refer to who was at fault for our current crisis, but to the people helped by the resulting world? If we have been living in the age of the "capitalocene," a civilization structured by and for capital, then I want to live in the anthropocene—a world built by humans for all humans and non-human relatives.[11] The anthropocene is terrifying and tragic, but it is also an opportunity to construct the collective "we" that has so long been missing. This push can only be led by frontline communities.

Community resistance built from Black and Indigenous knowledge serves as the guiding force for the unfucking process. For example, Ranganathan calls for an abolition ecology that acts as "an approach to studying unjust urban natures that is informed by black radical thought, postcolonial (and decolonial) theory, and indigenous theory,"[12] while decolonization theory and native-rights provide an important strategic framework to the fight against fossil fuel infrastructure. As Simpson states: "That's something my ancestors had figured out. . . .

They had their own economy; they knew how to live in the world without being capitalist. . . . Another world is possible."[13] Indeed, the great unfucking is already led by the people most marginalized by colonization, exploitation, and now, climate change. Grassroots movements and ancestral knowledge are also opening pathways to safety—through land remediation, community healing, and shifting relationships. The chances of success are slim, but the Great Unfucking is no time for inaction.

> To face the Anthropocene, we need the courage to face the reality of our world and the willingness to work against the odds and make it better.

Now is not the time to give up. Solnit writes, "Hope should shove you out the door, because it will take everything you have to steer the future away from endless war, from the annihilation of the earth's treasures and the grinding down of the poor and marginal. . . . hope calls for action."[14] To face the anthropocene, we need an active, determined hope. We need the courage to face the reality of our world and the willingness to work against the odds and make it better. There is time for grief, of course, and for letting go of false capitalist promises and exploitative lifeways. But there is also time for rebuilding, for mitigating what is to come, and for replacing it with something beautiful. And every year for as long as we can, we must be fighting. Hope is a verb, and we are going to unfuck the world.

This week, I was arrested as part of a protest action demanding a just climate future. Marching through the halls of power with hundreds of other young people, I felt ready to fight for it. I came into this semester terrified about the future. Yet even as we've received horrific climate news and watched our so-called democracy crumble, we've also witnessed inconceivable progress become possible. I see the zeitgeist changing. This is one of the most important decades in the history of our planet. Everything we do will become part of the geologic record. We face two futures. One provides a habitable planet, mitigates oppression, and moves us towards a better world. The other is unthinkable. There are no options here. We must do this, and we will; with direct action and daily acts of Indigenous resistance, in floods of people rising up against the capitalist walls of power. This past weekend I stood in a church full of a thousand young people ready to put their bodies on the line for this fight. We sang and stomped our feet, and the future shook.

NOTES

1. Suarez, course assignment to author.
2. Finley, "Unfuck the World Day."
3. Klein, *This Changes Everything*, 449–50.
4. Scranton, *Learning to Die in the Anthropocene*, 16, 19.
5. Hern and Johal, *Global Warming and the Sweetness of Life*, 59–60.
6. Pulido, "Flint, Environmental Racism, and Racial Capitalism," 1–16.
7. Nixon, "The Great Acceleration and the Great Divergence."

8. Cronon, "The Trouble with Wilderness," 13–16, 21.
9. Goldstein, *Planetary Improvement*, 157–58.
10. Klein, *This Changes Everything*, 449–50.
11. Patel and Moore, *History of the World*, 1–3.
12. Ranganathan, "Thinking with Flint," 29.
13. Klein, "Dancing the World into Being."
14. Solnit, *Hope in the Dark*, 4.

REFERENCES

Cronon, William. "The Trouble with Wilderness: Or, Getting Back to the Wrong Nature." *Environmental History* 1, no. 1 (1996): 7–28. https://doi.org/10.2307/3985059.

Davis, Heather, and Zoe Todd. "On the Importance of a Date, Or, Decolonizing the Anthropocene." *ACME: An International Journal for Critical Geographies* 16, no. 4 (2017): 761–80. https://acme-journal.org/index.php/acme/article/view/1539.

Finley, Ron. "Unfuck the World Day—Garden Renegade Ron Finley." Unfuck the World, last modified 2023. http://unfucktheworld.net/ron-finley-garden-renegade.

Goldstein, Jesse. *Planetary Improvement: Cleantech Entrepreneurship and the Contradictions of Green Capitalism*. Cambridge, MA: MIT Press, 2017.

Hern, Matt, and Am Johal. *Global Warming and the Sweetness of Life: A Tar Sands Tale*. See esp. chap. 2, "Edmonton." Cambridge, MA: MIT Press, 2018.

Klein, Naomi. "Dancing the World into Being: A Conversation with Idle No More's Leanne Simpson." *Yes Magazine*, March 6, 2013. https://www.yesmagazine.org/peace-justice/dancing-the-world-into-being-a-conversation-with-idle-no-more-leanne-simpson.

———. *This Changes Everything: Capitalism vs. the Climate*. New York: Simon & Schuster, 2014.

Nixon, Rob. "The Great Acceleration and the Great Divergence: Vulnerability in the Anthropocene." *Profession*, March 2014. https://profession.mla.org/the-great-acceleration-and-the-great-divergence-vulnerability-in-the-anthropocene.

Patel, Raj, and Jason W. Moore. *A History of the World in Seven Cheap Things: A Guide to Capitalism, Nature, and the Future of the Planet*. See esp. "Introduction." Oakland: University of California Press, 2018.

Pulido, Laura. "Flint, Environmental Racism, and Racial Capitalism." *Capitalism Nature Socialism* 27, no. 3 (2016): 1–16. https://doi.org/10.1080/10455752.2016.1213013.

Ranganathan, Malini. "Thinking with Flint: Racial Liberalism and the Roots of an American Water Tragedy." *Capitalism Nature Socialism* 27, no. 3 (2016): 17–33. https://doi.org/10.1080/10455752.2016.1206583.

Scranton, Roy. *Learning to Die in the Anthropocene: Reflections on the End of a Civilization*. San Francisco: City Lights Books, 2015.

Solnit, Rebecca. *Hope in the Dark: Untold Histories, Wild Possibilities*. Chicago: Haymarket Books, 2016.

Suarez, Daniel. "Prompt." 2018. In possession of the author.

PART TWO

JUSTICE AS AFFECTIVE PEDAGOGY

NINE

Preparing Students to Navigate a Harrowing Educational Landscape

Accessibility and Inclusion for the Climate Justice Classroom

ASHLEY E. REIS

This chapter highlights the importance of accessibility and belonging in the climate change classroom. It offers suggestions for how educators can support students through classroom policy and structure.

Today's college and university-level students learning about the complexities of climate change and climate justice struggle with more than just grim content. In my experience teaching, whether in the Environmental Studies, Environmental Humanities, or the College Composition classroom, once I provide students with a basic lesson in fossil fuels, greenhouse gasses, infrared light, and the greenhouse effect, they quickly comprehend the physical processes that so many who hoard great wealth and power choose to repudiate.

Granted, our students have a unique stake in understanding climate change. For the most part, the students we teach today represent the climate generation, or those born between 1990 and the early 2000s, and thus are the first to spend their entire lives facing the looming threat of climate disruption.[1] It undeniably serves this generation to recognize and respond to the bleak "forecast about the viability of life on this planet"[2] that threatens them disproportionately. Accordingly, given the proper tools for comprehension, students rarely challenge the pronounced and tangible physical reality of climate change, which they see with their own eyes as they experience it within their own daily lives.

Instead, students struggle to manage the emotions that manifest for them once they begin to understand the social repercussions of climate change. As Ray points out in *A Field Guide to Climate Anxiety: How to Keep Your Cool on a Warming Planet* (2020), the climate generation is not only disproportionately impacted by climate change; they are also particularly concerned with social justice issues and are the most ethnically diverse generation in US history.[3] Accordingly, even a basic primer on climate justice can elicit a profound affective response in students who are concerned for their own health and the well-being of their communities. What's more, once students learn the basics of climate injustice—that nonindustrial and/or island nations and Indigenous territories in the Global South contribute the least to climate change and yet are typically the first and most affected by the outcome of industrialized nations' unchecked consumption and emissions—an affective response, that can present as anything from guilt to sadness to anger, is both likely and warranted.

The emotions that arise for my students, ranging from grief to anxiety to sadness and beyond, may be particularly challenging to navigate given that many had not anticipated the affective significance of the content with which they'd be engaging. In the case of the dual credit students in my Spring 2021 Project-Based Learning course focusing on the "wicked problem" of climate change, a majority (60 percent) of respondents[4] to an end-of-semester informal poll noted that they had originally imagined that the wicked problem of climate change would make them feel either "barely emotional" or "neither emotional nor unemotional." And yet, while they hadn't anticipated the affective reverberations of their climate change and climate justice studies, 100 percent of those respondents indicated that our topic of inquiry made them either "somewhat emotional," "very emotional," or "overwhelmed with emotion." The emotions my students reported indeed varied, although notably, students across the board identified *frustration* as the primary emotion they experienced, indicating the extent to which students understand that it is unjust to expect they clean up the mess older generations have created. And still, one thing was certain: students disclosed that they had not anticipated the emotions climate change and climate justice studies would make them feel, nor had they anticipated the intensity of these emotions.

Given that our empathetic, community-oriented, diverse students of the climate generation are not expecting to be in a position to prepare themselves for the inherently taxing pursuit of climate change and climate justice studies, it is imperative that we prepare them. As a professor rooted in the humanities with limited resources, I've spent a great deal of time considering the measures I'm even qualified to enact. I certainly do not possess the kind of degree in eco-psychology or sociology that might position me to walk my students, step by step, through the process of attending to the sentiments that could derail their best academic and advocacy efforts, nor have I been trained in psychology to the extent that I'm qualified to help my students process the climate anxiety they feel when they grasp that they are coming of age at the end of the world as we know it. And yet *I am* equipped to establish an accessible and inclusive classroom, wherein students feel secure and valued to the extent that they are prepared to face and interrogate the emotions that will inevitably arise as we navigate the affective landscape of ecological degradation and social injustice in the age of climate disruption.

This work can and should be in our purview as educators ushering our students through a challenging and complex educational experience. Many of our students, after all, face more institutional and systemic barriers to success than others. These barriers are emotionally taxing on students; in and of themselves, they can and do impede students' ability to engage both with the course material and the emotions inherent in the work we do in the climate change and climate justice studies classroom. By acknowledging and working towards eliminating these impediments, educators stand to offer our disproportionately overburdened students culturally sensitive and accessibility-minded access points to this emotionally charged field of study so that they are better positioned to engage with the material and their emotions that arise in light of it without the added stressors of navigating commonplace institutional and systemic barriers to success.

Accordingly, in my experience teaching climate change and climate justice studies, establishing a classroom space that prioritizes a practice of inclusion is fundamental. In its definition of inclusion, the Association of American Colleges and Universities (AACU) emphasizes the importance of the *active, intentional,* and *ongoing* nature of inclusion as essential to educators hoping to ensure equity for all students, regardless of their backgrounds and life experiences.[5] In other words, inclusion doesn't just *happen*—it is the result of *deliberate* and *continuous* evaluation, scrutiny, analysis, and effort. Removing archaic barriers to success and dismantling bureaucratic classroom management strategies that uphold existing structures of privilege is integral to creating equitable and inclusive classroom spaces for students navigating the affective landscape of climate justice work. This work is challenging enough in and of itself, and educators can eliminate logistical and emotional impediments to student success that can hinder students before they are even able to attend to the work of engaging with our course materials. Additionally, crafting and asserting deliberately inclusive classroom policies and procedures provides educators with the opportunity to guide students through important conversations about the reasoning behind these policies and procedures on the first day of class. In this way, educators can set a vitally important tone for the classroom that communicates to students who have varied life experiences and disparate educational backgrounds that diversity is an asset because the existing dominant structures are failing to attend to this new historical moment and are failing to adequately educate the next generation of change-makers.

My contribution to this collection lays out various ways in which we educators can examine, deconstruct, and redeploy classroom policies in order to ensure we provide our diverse cohorts of students with equitable experiences in the climate justice classroom. I will highlight as sites of opportunity: (1) assessment policies that have traditionally centered whiteness and advantaged white, socioeconomically privileged students above all others; (2) management policies that exclude undocumented students and other students without socioeconomic privilege, and (3) assignment policies that disregard students' extra-academic struggles and challenges. By focusing on classroom practicalities, we can work to ensure students an equitable experience that fosters student trust, the willingness to engage in emotionally charged course content, and the sense of security students need to talk openly about their emotions with their peers and instructors. My hope is that these ideas will spark further conversation about how the seemingly prosaic arena of classroom logistics are key to our work as educators in this new climate-aware landscape.

ASSESSMENT POLICIES

An appropriate place to direct our efforts is towards dismantling standard syllabus language, which we can replace with innovative new policy that supports student growth rather than disadvantaging certain subsets of the classroom population. In particular, certain forms of assessment center whiteness and advantage socioeconomically privileged students. Factoring grammar, style, and usage into students' grades is one of the ways in which my grading policies once rewarded some students over others for arbitrary skills. Yet, at a time when institutions are serving "students who traverse cultural and linguistic borders more fluidly than ever,"[6] educators must actively dismantle "deficit model" policies that make students feel they are functioning academically at a deficit. Unfortunately, educators often perceive the varied linguistic abilities of students from diverse racial and linguistic backgrounds as limitations to learning rather than as "natural dimension of human social and cultural identities."[7] In order to celebrate rather than penalize these variations, I have revised my grading policies, for instance, to no longer factor grammar, style, and usage into my assessment of student writing. After all, not all students come to my classroom with the same experience with and proficiency using Standard American English and it is critical that I demonstrate to my students my respect for the linguistic diversity they bring to the classroom. Accordingly, I include the following language in my syllabi to outline my assessment policy as it relates to grammar, usage, and style:[8]

> Our class is rooted in the study of language, literature, argumentation, logic, and persuasion. I will not comment on grammar, usage, and style unless they obstruct the clarity of your arguments, logic, or ability to persuade your intended audience. Grammar and style "rules" are based on a command of one dialect used in the U.S.: Standard American English (SAE). However, students have had varied access to this variety of the English language depending on upbringing and experience. . . . [Because] not all students come to this class with the same level of access to an education rooted in SAE, it would not be fair for me to grade based on extent of access and thus the ability to deploy the conventions of SAE. Accordingly, I will not be holding you accountable for grammar rules to which you may not have been exposed due to your cultural, racial, ethnic, national, or socioeconomic backgrounds.
>
> While I'll be grading you based on your ability to develop logical and persuasive arguments, if you would like to seek out help in improving your grammar and style—it will be important for you to develop proficiency in this area if you plan to enter a workforce that is not yet appreciative of the ways in which nuanced modes of the English language allow a diverse array of individuals to communicate—you can schedule time to meet with me, or you can access free appointments at any time with the Writing Center, which has online tutoring available Monday–Friday.

Culturally relevant pedagogy entails honoring students' cultural expertise,[9] and this mode of assessment reflects my appreciation for cultural diversity and its importance to the climate justice movement.

An overview of this policy on the first day of class, furthermore, creates space for a critical conversation about linguistic diversity. Students should not change who they are when they enter the classroom. They should expend their emotional energy instead where it is most needed: navigating the affective landscape of our climate justice studies. One student noted accordingly in their post-semester feedback that in light of the aforementioned assessment policy they "felt safe about expressing [their] opinions and research knowing that [their] work would be graded more on content than 'proper' grammar." Our diverse cohorts of students have relevant and applicable experience to contribute to our class discussions, the teamwork in which they engage, and the projects they complete. For instance, amidst our semester dedicated to climate justice, Texas experienced an unprecedented winter freeze that rendered our power grid inaccessible to millions of residents across the state. Students suddenly had firsthand experience with extreme weather, the disproportionate nature of the storm's effects on low-income communities, and the injustice of relief efforts. Rather than struggling through the discouragement, aggravation, and disheartenment of having to master an entirely new-to-them written dialect, culturally relevant assessment policies positioned students to make bold statements in favor of impactful change using the kind of meaningful language, dialect, and turns of phrase that resonate within their families and communities, where they hoped to make a difference.

MANAGEMENT POLICIES

Our classroom management customs also often exclude undocumented students and other students without socioeconomic privilege. For instance, while serving as a teaching fellow, my course's attendance policies were to reflect those laid out by my university's Writing Program. This policy allowed students a certain number of unexcused absences from class, which communicated to students that there existed such a thing as an "excused" absence; in reality, only university-sponsored events such as athletic meets, club outings, etc., were "excused." As a result, when students missed class, upon their return, they would often present me with a note from a medical doctor. While the COVID-19 pandemic has resulted in a holistic re-envisioning of my attendance policy, even prior to the pandemic, it was important for me to draw attention to the healthcare inequities from which many of my students benefit, while others are denied access. Accordingly, I included the following language in my syllabus:

> Please note that I do not excuse absences for illness, even when students provide medical doctor's notes. Due to healthcare inequities in the U.S., not all students have access to doctor's visits and accordingly I cannot accommodate those students who do over those who do not. Accordingly, please save up and use wisely your allotted absences in the case that you become ill.

This policy created an important opportunity for discussion with the class and positioned me to ask students if they understood the difficulties of accessing healthcare in the US, that healthcare is most often tied to employment, and that students enrolled in DACA or those without documentation may face undue hardship when it comes to obtaining medical treatment. Hopefully this eased any anxieties they brought into the classroom and communicated my investment in equitable treatment of all students, allowing students to focus their mental energy on the emotionally challenging task at hand: climate change and climate justice studies.

I found myself questioning my own attendance policies as the COVID-19 pandemic raged on, however, and have taken to heart the generous musings of disability scholars and academics with disabilities who argue that attendance policies disadvantage students with chronic illnesses and/or disabilities. Accordingly, for the time being I have revised my attendance policy to read:

> Under "typical" circumstances, students who take this course benefit from regular attendance and active participation on a daily basis. However, as we contend with a global pandemic, we can only do our best. This means that so long as you are physically healthy and aren't contending with undue hardship (family illness, childcare, mental health challenges and/or concerns, etc.), I urge you to make class attendance a priority. Please, however, be in touch as soon as possible if you foresee extenuating circumstances that will make meeting this requirement difficult for you. I am always happy to work with students to make our class as accessible as possible—reaching out as soon as you're able will help position me to best accommodate you.

Because I have the flexibility in my role to set my own course policies in this way, I imagine I will continue my use of this policy into the future. After all, students *will always be* contending with undue hardship beyond the global pandemic: climate change is coming to bear on individual students differently and it is important to support them accordingly.

Additional management policies beyond attendance that offer sites of accessibility include offering accommodations—as best as possible—for students who do not require disability accommodations but may still benefit from reasonable course policy adjustments the two of you imagine together.

Better yet is to invert this deficit/asset framing of "accommodation" entirely, by treating students' hardships not as *threats* to their academic success, but as *qualifications* for it. This can be done in assignments, which I will discuss in the next section.

ASSIGNMENT POLICIES

The way we craft our assignments offers yet another opportunity to implement more inclusive classroom practices. As many of us learned during our first year of teaching during a global pandemic, our students contend with innumerable obstacles and hardships in their everyday lives outside of the classroom, which affect students' preparation for class, ability to concentrate during group discussions, and capacity to complete their work to the best of their ability. For this reason, I implemented a time bank model that affords students a degree of flexibility when it comes to due dates, which I communicate using the following language in my syllabus:

> While all essays are due by the start of class time on the due date, we are all human and I understand that things come up that may impede your ability to submit your work on time. Accordingly, I will be employing the "time bank" model this semester. This means that each student will have one two-day grace period for one assignment, *or* two one-day extensions for two different assignments. In order to provide the most helpful feedback possible, I cannot comment on or grade subsequent assignments from any students until all previous assignments have been completed.

I ask my students who plan to withdraw a day or two from their time bank to email me at least 24 hours before the assignment's due date to notify me of their plan to utilize the time bank option, and how many days they will be withdrawing. This requirement offers the opportunity for a discussion with the student regarding their health and well-being. I'm able to inquire if they need any extra help from me or access to other resources on campus, and to even discuss with them the option of a longer extension, should extenuating circumstances warrant one. One hundred percent of my student respondents reported after the semester's end that flexibility with due dates by way of the time bank model helped them to manage their emotions related to course content. The "relaxed" approach to assignments, one student noted, "allowed [students] to think things through without added stress and emotional overload." Another reported that the time bank model, "gave [them] an extra day of two to write how [they] really felt and made [them] able to express [their] disappointment and frustration with the issue. It allowed [them] the extra time to be straightforward and truthful." The time bank model can be tailored to the needs of any student population, and the one- and two-day extensions are mere examples of what is possible for educators seeking to create flexibility in policies, as well as policies that don't simply reward students with access to the most resources. which may help them to cope with hardship.

Another option beyond the time bank model is to simply utilize due dates as *guidelines,* a policy with which I am currently experimenting in one section of College Writing. Because late penalties tend to disproportionately impact already overburdened students, I have explained to this class that due dates are in place to help both students and me, the instructor, distribute our workload across the semester. These students have agreed to communicate with me as soon as possible if they foresee difficulty meeting any deadline so that we can discuss strategies for them to manage their workload and complete their work.

AN EDUCATOR'S RESPONSIBILITY

While we educators cannot necessarily prepare our students for the affective responses they will experience as they learn about climate change and climate injustice, we are by no means powerless to support them. And as one student mused in their final response to my informal poll, educators have a responsibility to do this. They wrote, "[T]eachers have one of the largest [impacts] on our emotions and how we manage climate justice work. Teachers, at least for me, are the people I look to when I feel lost and am unsure how to resolve a problem, the climate crisis issue is no different. When teachers invalidate or shy away from students' feelings it doesn't encourage growth and development, it shuts us down." Helping students first identify and then deal directly with their emotions is critical. But developing course structures that will support students holistically and assist them in deciphering the "hidden curriculum"[10] of higher education can cultivate the kind of trust between students, their peers, and us, their instructor, that is essential to climate change and climate justice education.

NOTES

1. Ray, *Field Guide to Climate Anxiety*, 4.
2. Ibid., 4.
3. Ibid., 4.
4. Respondents, who range in age from 17 to 18, are a part of a Project-Based Learning cohort of dual credit students from the suburbs of Dallas, Texas, enrolled in The University of North Texas at Frisco's North Texas NOW! program.
5. AACU, "Making Excellence Inclusive."
6. Artilo and Kozleski, "Beyond Convictions," 359.
7. Beneke and Cheatham, "Speaking Up for African American English," 128.
8. An abridged version of this language appears here.
9. Beneke and Cheatham, "Speaking Up for African American English," 128.
10. Margolis et al., "Peekaboo: Hiding and Outing the Curriculum," 8.

REFERENCES

American Association of Colleges and Universities (AACU). "Making Excellence Inclusive." Accessed July 29, 2021. https://www.aacu.org/making-excellence-inclusive.

Artilo, Alfredo J., and Kozleski, Elizabeth B. "Beyond Convictions: Interrogating Culture, History, and Power in Inclusive Education." *Language Arts* 84, no. 4 (2007): 357–64.

Beneke, Margaret, and Gregory A. Cheatham. "Speaking Up for African American English: Equity and Inclusion in Early Childhood Settings." *Early Childhood Education Journal* 43, no. 2 (2015): 127–34.

Margolis, Eric, Michael Soldatenko, Sandra Acker, and Marina Gair. "Peekaboo: Hiding and Outing the Curriculum." In *The Hidden Curriculum in Higher Education,* edited by Eric Margolis, 1–20. New York: Routledge Press, 2001.

Ray, Sarah Jaquette. *A Field Guide to Climate Anxiety: How to Keep Your Cool on a Warming Planet.* Oakland: University of California Press, 2020.

TEN

Photovoice for the Climate Justice Classroom

Inviting Students' Affective and Sociopolitical Engagement

CARLIE D. TROTT

This chapter provides a modifiable step-by-step guide to photovoice for the climate justice classroom—an approach that simultaneously attends to the "more-than-science" dimensions of the climate crisis and invites learners' affective engagement with the topic from their own position and point of view. By centering students' perspectives and inviting their critical reflection and action, photovoice is presented here as a decolonizing approach to climate justice education that challenges "top-down," science-centric educational approaches and better positions students to envision and enact climate justice–driven alternatives in their classrooms and communities.

Demands for climate justice (CJ) are rooted in the recognition that climate change burdens are falling disproportionately on the most marginalized and dispossessed groups around the world who have contributed the least to the problem.[1] Further, CJ advocates recognize that unless the specific needs and interests of these groups are at the center of climate change mitigation and adaptation efforts, as climate change worsens, so too will existing social, economic, and environmental injustices.[2] Research has documented that people actively involved in climate change mitigation and education are as, or perhaps more, motivated by justice concerns as biospheric concerns,[3] suggesting that, in the classroom, discussing CJ could be a pathway to learners' motivated engagement.

Despite this, climate change education (CCE) has disproportionately focused on the scientific and environmental dimensions of the climate crisis over its social and political dimensions.[4] Moreover, mainstream CCE approaches have largely overlooked the deep and wide-ranging emotions experienced by climate change educators and learners alike.[5]

Avoiding the sociopolitical and affective dimensions of the climate crisis—particularly in the science classroom where most CCE takes place[6]—is a choice firmly rooted in the Western philosophy of rationalism, which views emotion as a hindrance to truth, objectivity, and reason in the quest for knowledge.[7] From this perspective, teachers must elevate facts over values and reason over emotions because doing otherwise would "politicize" the topic, thereby removing it from the realm of "rational" dialogue and debate. However, critical pedagogies rooted in alternative educational philosophies hold firmly that education is not a value-free enterprise,[8] and learners' affective experiences are core to—not separate from—the learning process.[9] As such, we must account for these realities as we invite students to question the unjust and non-sustainable societal status quo.

In CCE, there is a need for alternative approaches that deal directly with the sociopolitical facets of the climate crisis (e.g., climate injustice) as well as those that embrace learners' full range of emotions as they reflect on and discuss climate change in the classroom. In this essay, I introduce photovoice methodology as a promising pathway towards simultaneously attending to the "more-than-science" dimensions of climate change, while inviting learners' affective engagement with the topic from their own position and point of view. By centering students' perspectives and inviting their critical reflection and action, photovoice is presented here as a decolonizing approach to CCE that challenges "top-down," science-centric educational approaches and better positions students to envision and enact CJ-driven alternatives in their classrooms and communities.

> By centering students' perspectives and inviting their critical reflection and action, photovoice is presented here as a decolonizing approach to climate change education that challenges "top-down," science-centric educational approaches and better positions students to envision and enact climate justice-driven alternatives in their classrooms and communities.

PHOTOVOICE: A SOCIAL JUSTICE STRATEGY

For more than two decades, photovoice—a participatory action research (PAR) method—has been a social justice strategy used by community groups to "identify, document, and represent their community's strengths and concerns from their own perspective through the use of a specific photographic technique."[10] Specifically, photovoice participants capture images of their lived realities using photography, discuss photographs collectively to make meaning of the situation, and finally take action—often towards policy change.[11] Photovoice methodology has theoretical roots in: (1) critical pedagogy and the notion of critical consciousness;[12] (2) feminist theory and the concept of voice;[13] and (3) documentary photography, "the genre that allows the most vulnerable people within society to convey their own vision of the world."[14] As a PAR method, photovoice has most often been used with marginalized groups who have been silenced in the policymaking realm[15] and has been implemented in a range of community contexts and applied to numerous issues from healthcare to homelessness.[16]

Despite its origins as a research approach, photovoice is increasingly used in educational settings.[17] For example, photovoice has been used in higher education as a way to decolonize the college classroom and encourage learners to bring their own perspectives and experiences into conversation with course material.[18] Importantly, this is occurring across disciplines—in science, social science, and humanities classrooms.[19] Typically, students are asked to take photographs related to course content, then discuss the meaning and significance of their photographs together with peers in the classroom or others across the school or university as a platform for action.[20] Before offering a modifiable step-by-step guide to photovoice in CCE contexts, in the sections below, I draw out key strengths of the method towards supporting learners' affective and justice-driven climate change engagement. The following sections are based on the three chief aims of photovoice, which are to: (1) invite participants to engage with a topic on their own terms; (2) strengthen participants' critical consciousness; and (3) encourage participants to envision and enact change.

Students' Photography-Based Narratives

The first major component of photovoice is documenting reality from the perspective of photovoice participants—in this case, students. Traditionally in photovoice, participants are given cameras to take photographs of various facets of their lives (e.g., community strengths and concerns). This is the "photo" portion of photovoice. Later, in group discussions, participants describe the content and meaning of their photographs while identifying pathways towards addressing the issues identified. Often, the "voice" portion of photovoice asks participants to give a title and caption to a specific number of photographs, then explain each photograph's meaning and significance. Photovoice discussions invite learners' affective engagement through student-centered storytelling and connection. In CCE contexts, photovoice can open the door to learning about students' climate-related emotions, such as anxiety, grief, or desire for change.[21] Research shows that acknowledging difficult emotions helps to reduce their intensity and duration.[22] Moreover, encouraging students' affective engagement can deepen the learning process,[23] while helping students to acknowledge and regulate their emotions through shared sense-making and connection.[24]

Critical Reflection and Dialogue

The second major component of photovoice is sharing in critical reflection and dialogue about the issues uncovered in participants' photography. Photovoice discussions go beyond student-centered storytelling—addressing questions of who, what, and when—to a critical examination of the issues identified in participants' photographs—addressing questions such as how and why.[25] In CCE contexts, asking such questions may serve to initiate reflexive conversations about how climate change is perceived by students to affect their lives or why students feel the way they do about the topic. Importantly, such questions may ignite conversations around how we arrived at the present moment, and why such urgent action is needed today, opening the door to discussing the sociopolitical and justice dimensions of the climate crisis. For example, a study exploring how ten- to twelve-year-olds were making personal connections to climate change—through qualitative analysis of audio-recorded photovoice discussions—found that photovoice was a pathway not only towards deeper understanding, but also to children's affective processing and plans for action.[26] This indivisible mix of children's learning, caring, and action through photovoice was made possible through a process of *collective meaning-making* during which children shared in critical dialogue about their own and others' photographs.

An essential aim of photovoice is to raise participants' critical consciousness by discussing the structural dimensions of issues identified in their photographs as a first step towards identifying avenues for systemic change. In this way, photovoice is intended to promote social justice awareness, amelioration, and transformation of conditions.[27] In CCE contexts, emphasizing CJ themes may serve to strengthen students' engagement with the issue,[28] while providing students with a more comprehensive, realistic, and equity-driven understanding of climate change solutions and what it will take to accomplish them.[29] Moreover, inviting young people's perspectives and actions, through photovoice, can itself be an exercise in CJ because young people's voices are often silenced in the climate change policy arena.

Student-Led Social Action

The third major component of photovoice is taking action, usually towards policy change.[30] Students' affective engagement and critical dialogue together serve as a bridge towards action. Through the photovoice process, "[participants'] emotional engagement and collective introspection become keys to the shift in cognitive-emotional interpretations that recognize responsibility for change."[31] In CCE contexts, this can open the door to student-led CJ action, based upon their own values and interests, attributions of responsibility, and visions for change. A growing research base has documented that young people can be effective environmental communicators and catalysts of social change in their communities, including on climate change.[32] Engaging cognitive, affective, and behavioral domains through photovoice-driven social action can be a transformative process for learners, allowing them to see and to move through the world in ways that reflect their values and goals.[33] The mere act of envisioning a desirable outcome—a process that "engages the affect and the imagination as much as the intellect" can be supportive of young learners' sense of hope[34] (for more on the role of the imagination, see also essays by Gray, Anson, and Stabinsky and Oesterby in this volume). Moreover, working towards desired outcomes can have psychological as well as tangible benefits in community settings.[35] CCE research suggests that taking action, including through photovoice, can be an important avenue towards ameliorating the negative psychological consequences of climate change awareness by supporting students' sense of agency and constructive hope.[36]

> Engaging cognitive, affective, and behavioral domains through photovoice-driven social action can be a transformative process for learners, allowing them to see and to move through the world in ways that reflect their values and goals.

PHOTOVOICE FOR THE CLIMATE JUSTICE CLASSROOM

A recent review article found that photovoice has been applied in a range of environmental education settings with children as well as adults, at times focusing explicitly on climate change awareness and action.[37] In contrast to traditional photovoice methodology—which leaves the subject matter of participants' photography very open-ended—when applied as a pedagogical tool, oftentimes a predetermined focus must be chosen to guide students' photo-taking.[38] For example, in an after-school program I designed and implemented with ten- to twelve-year-olds, children were asked to capture images representing how they "think and feel" about climate change following several hands-on activities demonstrating local ecosystem impacts and emphasizing the importance of human action to address the issue.[39] As photovoice has gained in popularity, not all such modifications to the original method have adhered to the emancipatory intentions of photovoice originators Wang and Burris.[40] Specifically, photovoice components involving students' photography and group dialogue are more widely practiced than encouraging students to engage in advocacy (e.g., with policy-makers) or other forms of student-led action.[41] In CCE contexts, educators must decide for themselves what makes the most sense for their specific situation.

Fortunately, a key strength of photovoice for the CJ classroom is its wide accessibility across age groups as well as its versatility to a range of classroom and community contexts. For this reason, there is no single recipe for doing photovoice in CCE settings. Still, there are several common steps to consider, each with their own flexibility of application. Figure 10.1 offers a step-by-step guide to implementing photovoice in the CJ classroom, based on my own community-engaged research[42] and the work of Wang,[43] Sutton-Brown,[44] and Goodhart and colleagues.[45] In many cases, educators will need to introduce the topic of climate change before commencing the photovoice process, ideally emphasizing CJ themes. Educators may also wish to provide basic photography instruction before asking students to take photographs. Finally, after students have engaged in critical reflection and dialogue, a key step is to identify common themes in students' photo-narratives that may shape the direction of student-led action. A traditional photovoice outcome would be a student-led community photography exhibition to display their photographs, raise awareness about climate injustice, and demand action from policymakers and decision-makers. Other examples of student-led, photovoice-driven action from my own work include a city council presentation, tree-planting campaign, and student-designed community garden and outdoor learning space.[46] In photovoice, the possibilities for action are bounded only by students' imaginations, available resources, and other contextual considerations.

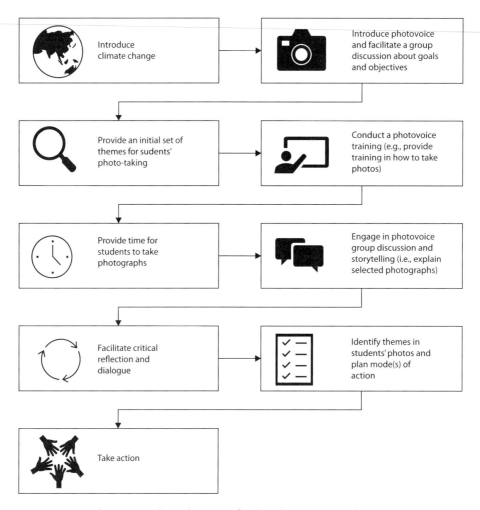

FIGURE 10.1 Step-by-Step Guide to Photovoice for the Climate Justice Classroom.

CONCLUSION

In its earliest applications, photovoice was a resource- and time-intensive method, often requiring facilitators to provide cameras to each participant, and later wait for physical photographs to be developed from film and hand-delivered back to participants for discussion. As a result, the method was out of reach for low-resource community groups and classrooms. Importantly for CJ action, today the method is much more widely accessible due to the ubiquity of digital photography, mobile devices, and online file-sharing platforms. That photovoice has withstood the test of time speaks to its strengths as a powerful tool for community dialogue and social change action. Diverging sharply from "top-down," science-centric CCE approaches, the potential "bottom-up" outcomes of photovoice are as wide-ranging and multifaceted as the perspectives and aspirations of its participants. As a pedagogical pathway to climate justice, photovoice can invite students to recognize and collectively process their emotions, critically reflect on the sociopolitical dimensions of present-day challenges, and imagine and bring into being better futures through their own actions.

> Photovoice can invite students to recognize and collectively process their emotions, critically reflect on the sociopolitical dimensions of present-day challenges, and imagine and bring into being better futures through their own actions.

NOTES

1. Burnham et al., "Extending a Geographic Lens," 239.
2. Perkins, "Climate Justice, Commons, and Degrowth," 183.
3. Howell and Allen, "Significant Life Experiences, Motivations and Values," 813.
4. Stevenson et al., "What is Climate Change Education?" 67.
5. Jones and Davison, "Disempowering Emotions," 190.
6. Monroe et al., "Identifying Effective Climate Change Education Strategies," 791.
7. Alsop and Watts, "Science Education and Affect," 1043.
8. Freire, *Education for Critical Consciousness*.
9. Pekrun and Linnenbrink-Garcia, *Introduction to Emotions in Education*.
10. Sutton-Brown, "Photovoice: A Methodological Guide," 169.
11. Wang and Burris, "Empowerment through Photo Novella," 171; Wang and Burris, "Photovoice: Concept, Methodology, and Use," 369.
12. Freire, *Education for Critical Consciousness*.
13. hooks, *Ain't I a Woman*.
14. Wang and Burris, "Empowerment through Photo Novella," 177.
15. Mitchell et al., "Utilizing Photovoice," 51.
16. Sutton-Brown, "Photovoice: A Methodological Guide," 169.
17. Schell et al., "Photovoice as a Teaching Tool," 340.
18. Bissell and Korteweg, "Digital Narratives," 1; Cook, "Grappling with Wicked Problems," 581.
19. Chio and Fandt, "Photovoice in the Diversity Classroom," 484–86; Lichty, "Photovoice as a Pedagogical Tool," 89.
20. Cook and Quigley, "Connecting to our Community," 339–43.
21. Trott, "Climate Change Education for Transformation," 1023.
22. Torre and Lieberman, "Putting Feelings into Words," 116–18.
23. Pekrun and Linnenbrink-Garcia, *Introduction to Emotions in Education*.
24. Macfarlane et al., "Implications of Participatory Methods," 33–35.
25. Ibid.
26. Lam and Trott, "Children's Climate Change Meaning-Making through Photovoice," 1–25.
27. Sanon et al., "Exploration of Social Justice Intent," 212.
28. Howell and Allen, "Significant Life Experiences, Motivations and Values," 813.
29. Schindel Dimick, "Supporting Youth to Develop Environmental Citizenship," 390.
30. Wang and Burris, "Photovoice: Concept, Methodology, and Use," 369.
31. Carlson et al., "Photovoice as a Social Process," 836.

32. Tanner, "Shifting the Narrative," 339; Trott, "Reshaping Our World," 42.

33. Sipos et al., "Achieving Transformative Sustainability Learning," 68; Trott, "Climate Change Education for Transformation," 1023.

34. Hicks, *Educating for Hope in Troubled Times*, 139.

35. Wang, "Youth Participation in Photovoice," 147.

36. Ojala, "Young People and Global Climate Change," 1–19; Trott, "Reshaping Our World," 42; Trott, "Climate Change Education for Transformation," 1023.

37. Derr and Simons, "Review of Photovoice Applications," 359–80.

38. Bissell and Korteweg, "Digital Narratives," 1–25.

39. Trott, "Reshaping Our World," 42–62; Trott, "Children's Constructive Climate Change Engagement," 532.

40. Wang and Burris, "Photovoice: Concept, Methodology, and Use," 369–87.

41. Derr and Simons, "Review of Photovoice Applications," 359–80.

42. Trott, "Reshaping Our World," 42–62.

43. Wang, "Photovoice," 185.

44. Sutton-Brown, "Photovoice: A Methodological Guide," 169.

45. Goodhart et al., "View through a Different Lens," 53–56.

46. Trott, "Reshaping Our World," 42–62.

REFERENCES

Alsop, Steve, and Mike Watts. "Science Education and Affect." *International Journal of Science Education* 25, no. 9 (2003): 1043–47. https://doi.org/10.1080/0950069032000052180.

Bissell, A., and L. Korteweg. "Digital Narratives as a Means of Shifting Settler-Teacher Horizons toward Reconciliation." *Canadian Journal of Education* 39, no. 3 (2016): 1–25.

Burnham, Morey, Claudia Radel, Zhao Ma, and Ann Laudati. "Extending a Geographic Lens Towards Climate Justice, Part 1: Climate Change Characterization and Impacts." *Geography Compass* 7, no. 3 (2013): 239–48. https://doi.org/10.1111/gec3.12034.

Carlson, Elizabeth D., Joan Engebretson, and Robert M. Chamberlain. "Photovoice as a Social Process of Critical Consciousness." *Qualitative Health Research* 16, no. 6 (2006): 836–52. https://doi.org/10.1177/1049732306287525.

Chio, Vanessa C. M., and Patricia M. Fandt. "Photovoice in the Diversity Classroom: Engagement, Voice, and the 'Eye/I' of the Camera." *Journal of Management Education* 31, no. 4 (2007): 484–504. https://doi.org/10.1177/1052562906288124.

Cook, Kristin, and Cassie F. Quigley. "Connecting to our Community: Utilizing Photovoice as a Pedagogical Tool to Connect College Students to Science." *International Journal of Environmental and Science Education* 8, no. 2 (2013): 339–57. https://doi.org/10.12973/ijese.2013.205a.

Cook, Kristin. "Grappling with Wicked Problems: Exploring Photovoice as a Decolonizing Methodology in Science Education." *Cultural Studies of Science Education* 10, no. 3 (2015): 581–92. https://doi.org/10.1007/s11422-014-9613-0.

Derr, Victoria, and Jordin Simons. "A Review of Photovoice Applications in Environment, Sustainability, and Conservation Contexts: Is the Method Maintaining its Emancipatory Intents?" *Environmental Education Research* 26, no. 3 (2020): 359–80. https://doi.org/10.1080/13504622.2019.1693511.

Freire, Paulo. *Education for Critical Consciousness*. New York: Continuum, 2007.

Goodhart, Fern Walter, Joanne Hsu, Ji H. Baek, Adrienne L. Coleman, Francesca M. Maresca, and Marilyn B. Miller. "A View through a Different Lens: Photovoice as a Tool for Student Advocacy." *Journal of American College Health* 55, no. 1 (2006): 53–56. https://doi.org/10.3200/JACH.55.1.53-56.

Hicks, David. *Educating for Hope in Troubled Times: Climate Change and the Transition to a Post-Carbon Future*. London: Institute of Education Press, 2014.

hooks, bell. *Ain't I a Woman: Black Women and Feminism*. Boston: South End Press, 1981.

Howell, Rachel A., and Simon Allen. "Significant Life Experiences, Motivations and Values of Climate Change Educators." *Environmental Education Research* 25, no. 6 (2019): 813–31. https://doi.org/10.1080/13504622.2016.1158242.

Jones, Charlotte A., and Aidan Davison. "Disempowering Emotions: The Role of Educational Experiences in Social Responses to Climate Change." *Geoforum* 118 (2021): 190–200. https://doi.org/10.1016/j.geoforum.2020.11.006.

Lam, Stephanie, and Carlie D. Trott. "Children's Climate Change Meaning-Making through Photovoice: Empowering Children to Learn, Care, and Act through Participatory Process." *Educação, Sociedade & Culturas* 62 (2022): 1–25. https://doi.org/10.24840/esc.vi62.478.

Lichty, Lauren F. "Photovoice as a Pedagogical Tool in the Community Psychology Classroom." *Journal of Prevention and Intervention in the Community* 41, no. 2 (2013): 89–96. https://doi.org/10.1080/10852352.2013.757984.

MacFarlane, Elizabeth K., Renu Shakya, Helen L. Berry, and Brandon A. Kohrt. "Implications of Participatory Methods to Address Mental Health Needs Associated with Climate Change: 'Photovoice' in Nepal." *BJPsych International* 12, no. 2 (2015): 33–35. https://doi.org/10.1192/S2056474000000246.

Mitchell, Felicia M., Shanondora Billiot, and Stephanie Lechuga-Peña. "Utilizing Photovoice to Support Indigenous Accounts of Environmental Change and Injustice." *Genealogy* 4, no. 2 (2020): 51. https://doi.org/10.3390/genealogy4020051.

Monroe, Martha C., Richard R. Plate, Annie Oxarart, Alison Bowers, and Willandia A. Chaves. "Identifying Effective Climate Change Education Strategies: A Systematic Review of the Research." *Environmental Education Research* 25, no. 6 (2019): 791–812. https://doi.org/10.1080/13504622.2017.1360842.

Ojala, Maria. "Young People and Global Climate Change: Emotions, Coping, and Engagement in Everyday Life." In *Geographies of Global Issues: Change and Threat*, edited by Nicola Ansell, Natascha Klocker, and Tracey Skelton, 1–19. Singapore: Springer Science + Business Media, 2016. https://doi.org/10.1007/978-981-4585-95-8_3-1.

Pekrun, Reinhard, and Lisa Linnenbrink-Garcia, eds. *Introduction to Emotions in Education*. New York: Routledge, 2014.

Perkins, Patricia E. Ellie. "Climate Justice, Commons, and Degrowth." *Ecological Economics* 160 (2019): 183–90. https://doi.org/10.1016/j.ecolecon.2019.02.005.

Sanon, Marie-Anne, Robin A. Evans-Agnew, and Doris M. Boutain. "An Exploration of Social Justice Intent in Photovoice Research Studies from 2008 to 2013." *Nursing Inquiry* 21, no. 3 (2014): 212–26. https://doi.org/10.1111/nin.12064.

Schell, Kara, Alana Ferguson, Rita Hamoline, Jennifer Shea, and Roanne Thomas-Maclean. "Photovoice as a Teaching Tool: Learning by Doing with Visual Methods." *International Journal of Teaching and Learning in Higher Education* 21, no. 3 (2009): 340–52.

Schindel Dimick, Alexandra. "Supporting Youth to Develop Environmental Citizenship Within/Against a Neoliberal Context." *Environmental Education Research* 21, no. 3 (2015): 390–402. https://doi.org/10.1080/13504622.2014.994164.

Sipos, Yona, Bryce Battisti, and Kurt Grimm. "Achieving Transformative Sustainability Learning: Engaging Head, Hands and Heart." *International Journal of Sustainability in Higher Education* 9, no. 1 (2008): 68–86. https://doi.org/10.1108/14676370810842193.

Stevenson, Robert B., Jennifer Nicholls, and Hilary Whitehouse. "What Is Climate Change Education?" *Curriculum Perspectives* 37, no. 1 (2017): 67–71. https://doi.org/10.1007/s41297-017-0015-9.

Sutton-Brown, Camille A. "Photovoice: A Methodological Guide." *Photography and Culture* 7, no. 2 (2014): 169–85. https://doi.org/10.2752/175145214X13999922103165.

Tanner, Thomas. "Shifting the Narrative: Child-led Responses to Climate Change and Disasters in El Salvador and the Philippines." *Children & Society* 24, no. 4 (2010): 339–51. https://doi.org/10.1111/j.1099-0860.2010.00316.x.

Torre, Jared B., and Matthew D. Lieberman. "Putting Feelings into Words: Affect Labeling as Implicit Emotion Regulation." *Emotion Review* 10, no. 2 (2018): 116–24. https://doi.org/10.1177/1754073917742706.

Trott, Carlie D. "Children's Constructive Climate Change Engagement: Empowering Awareness, Agency, and Action." *Environmental Education Research* 26, no. 4 (2020): 532–54. https://doi.org/10.1080/13504622.2019.1675594.

———. "Climate Change Education for Transformation: Exploring the Affective and Attitudinal Dimensions of Children's Learning and Action." *Environmental Education Research* 28, no. 7 (2021): 1023–42. https://doi.org/10.1080/13504622.2021.2007223.

———. "Reshaping our World: Collaborating with Children for Community-Based Climate Change Action." *Action Research* 17, no. 1 (2019): 42–62. https://doi.org/10.1177/1476750319829209.

Wang, Caroline C. "Photovoice: A Participatory Action Research Strategy Applied to Women's Health." *Journal of Women's Health* 8, no. 2 (1999): 185–92. https://doi.org/10.1089/jwh.1999.8.185.

———. "Youth Participation in Photovoice as a Strategy for Community Change." *Journal of Community Practice* 14, no. 1–2 (2006): 147–61.

Wang, Caroline, and Mary Ann Burris. "Empowerment through Photo Novella: Portraits of Participation." *Health Education Quarterly* 21, no. 2 (1994): 171–86.

———. "Photovoice: Concept, Methodology, and Use for Participatory Needs Assessment." *Health Education & Behavior* 24, no. 3 (1997): 369–87.

ELEVEN

Leveraging Affect for Climate Justice

MICHELLE GARVEY

In this course activity, students learn what constitutes climate justice (CJ) methods and leverage points. Its goals are to distinguish the movement of CJ, suggest creative ways to experiment with CJ actions, harness action as a salve for despair, and make measurable contributions to the CJ movement.

As an educator who seeks to advance climate justice movements through project-based, experiential, community-engaged teaching, I offer this activity to strengthen the related goals of climate justice education and affective learning. In my experience (which is supported by scholarship on best practices), action is a salve for climate despair, and education is "stickier"—more memorable and meaningful—when approached experientially.

The climate justice (CJ) movement appreciates our global reality of climate change-induced suffering and disparity from a decolonial lens. That is, uneven control of resources perpetuated by capitalist economic systems and hegemonic patterns of thought—e.g., white supremacy, patriarchy, egoism—are root causes of climate chaos and injustice initiated by European colonialism (from the fifteenth century to the present). CJ, then, necessarily uplifts decolonial ideologies, both in *how* we labor to achieve CJ as well as *which* CJ goals we seek to advance.

In utilizing CJ's distinct methods to ascertain insights into its distinct leverage points, CJ's distinct content is deliverable accessibly and impactfully. At a personal level, you are invited to bring your whole self to your learning. At a systemic level, you are shown how to "plug in" to achieve systems change in ways that resonate with your unique skills and insights.

CLIMATE JUSTICE METHODS

A method is a process for attaining knowledge, and most academic disciplines rely on specific procedures for gathering data to draw conclusions that embody a field's content. Because CJ scholarship occurs across many disciplines, and because the disciplinary nature of academia itself runs contrary to decolonial knowledge practices (pluralistic and intersectional as they are), this activity focuses on processes for attaining knowledge that cut across traditional academic fields and embrace the affective modes of learning that are marginalized in the colonial University.

Whereas sanctioned scholarly methods might include peer-reviewed surveys, ethnographies, or experiments—none of which are inherently colonial—the activity outlined in this essay familiarizes students with often marginalized ways of knowing critical to cultures of people historically oppressed by colonial power. These ways of knowing show up in Traditional Ecological Knowledges and fields that recuperate suppressed epistemologies, such as ethnic, feminist, and queer studies. To be sure, these methods are only decolonial to the extent that they are relevant to stakeholders on the frontlines of climate injustice; inclusive of community ways of knowing, including cultural and spiritual considerations; participatory at every stage of development; transparent about researcher bias; accountable to the community within which knowledge is acquired; and inclusive of more-than-human needs.

Further, what these interdisciplinary methods also embrace are affective ways of knowing that privilege first-person, place-based experience, which continue to inform us that (a) environmental harms and benefits are inequitably distributed and (b) a multitude of approaches to being in the world exist. Knowledge-gatherers are invited to employ whole-body sensing, to be as emotional as they are rational, and thus enter knowledge pursuits as full humans always situated in context, with valuable yet partial experiences to contribute within a diverse tapestry of other knowers. These features form the basis of CJ's methodology.

CLIMATE JUSTICE LEVERAGE POINTS

As defined by Donella Meadows, a leverage point is a place within a complex system where a shift in one thing produces changes in everything.[1] There are different places to intervene in a system: a high leverage point is more radical and produces a big change, but is usually difficult to implement; a low leverage point is more reformist and produces a smaller change, but is usually easy to implement.

Today's CJ movements employ a plethora of effective and innovative leverage points to intervene in unjust systems. Gaining in popularity and legitimacy, many succeed in transforming communities through power redistribution and ecological resilience. All tend to pressing contextual needs of the moment, and thus emerge in response to changing conditions.

At minimum, any CJ leverage point redistributes power inequities ensuring greater parity in decision-making and/or resource attainment, while at its most efficient, a leverage point dismantles an injustice in the first place. For example, some NIMBY ("not in my back yard") approaches draw attention to the disproportionate siting of carbon offset schemes in historically exploited nations in the Global South. These have been shown to disrupt vital local economies and ecologies, while preserving the ability for pollutive industries to emit in the first place.[2] A fairer distribution of carbon-emitting industries across wealthy and poor nations would embody a lower, reformist leverage point. On the other hand, a higher, more radical leverage point would tackle pollutive industry from the start, and might involve a moratorium on carbon-emitting technologies. This embodies a NOPE ("not on planet Earth") approach to CJ.[3]

To be sure, a CJ leverage point is not exempt from critique. As critical thinkers often responding to urgent injustices, we must continually ask such questions as

- Who does and doesn't benefit from this leverage point?
- When should, and shouldn't, it be employed?
- Where is it most appropriate and effective?
- How can it be achieved while expanding justice, equity, and inclusion of frontline communities?

Answers may reveal that an effective leverage point in one context will prove ineffective in another. One leverage point may be needed to stimulate another. A new leverage point may have to be created in order to adapt to a novel condition, or a time-tested leverage point may be most apt because resources to implement it already exist.

EXERCISE OUTLINE: INSTRUCTIONS FOR STUDENTS

This exercise braids CJ methods and leverage points together. From the "Method" options (described in Table 11.1), select or suggest a decolonial approach to knowledge acquisition in order to implement a CJ "Leverage Point" option (described in Table 11.2). Cursory explanations and examples of each are offered in the tables, but avoid allowing these to circumscribe your interpretation. A selected list of resources to initiate your explorations is provided.

After reflecting upon the questions provided below, *implement* the leverage point using your chosen method with classmates. Implementation could consist of any iteration of the leverage point or contribution to actualizing it. It could be large-scale or small; a minor or major project; and involve any number of students. Your educator(s) can support you by connecting you with local climate justice initiatives and leaders.

REFLEXIVITY QUESTIONNAIRE

To build capacity for critical evaluation, consider your selections for CJ methods and leverage points from the following perspectives before and after embarking on your activity.

TABLE 11.1 Climate Justice Methods

	Method
Feel	Lean into your emotional response to climate-related loss, change, or opportunity. Reflect upon a first-person narrative from the frontlines. Isolate a common emotion motivating seemingly disparate approaches to climate policy in order to find common ground.
Taste	Embark upon a foraging tour to taste the abundance of your local landscape. Prepare and share a meal comprised of ingredients from a local garden/farm. Patronize a BIPOC-owned market or restaurant. Cook a recipe from a frontline author using local ingredients.
Dance	Bear witness to a dance performance expressive of cultural ties to land or water. Choreograph your own interpretation of a climate story. Choreograph symbolic movement and perform at a direct action.
Paddle	Experience the history, geology, and hydrology of a place via boat, responding to the current, wind, and bends of the water via self-propulsion. Consider how a well-known perspective changes when on the water. Paddle with a mission to educate, raise funds, collect data, or blockade.
Storytell	Study the myths and stories of cultures indigenous to your area, or listen to a Native elder tell those stories. Interpret the murals your community has co-created or graffitied. Co-create stories of climate loss, resilience, or justice and share them.
Play	Learn about ecology and phenology while enjoying stress relief through activities typically designed for children. Resurrect your own favorite activities played as a child. Create a new game designed to open channels of climate-related learning. Reflect upon the emergence, cooperation, and surprises play often brings.
Build	Assist a local initiative building solar arrays or installing wind-powered electricity by and for communities made vulnerable to climate change. Join a Water Protector camp to build infrastructure along oil pipeline routes. Break ground for an urban farm.
Craft	Quilt a message of resistance or resilience and gift it for educational or practical use. Sew clothes made of compostable and/or reused materials and stage a climate-ready fashion show. Create supplies (cookware, soaps, etc.) from sustainable materials for distribution.
Map	Rewrite geological markers using the appropriate Indigenous language. Illustrate disparity trends or vital community resources. Create interpretive signs depicting notable ecological, cultural, or restored features. Create a fantasy map of a climate-just future.
Plant	Purchase or propagate—then save—heirloom, native seeds. Grow seasonal, organic foods to share in a food apartheid zone. Restore native plants and/or remove displaced plants. Steward a local community garden. Start a garden or food forest at your school.
Ritual	Develop a repetitive (daily, weekly, seasonal) practice of gratitude or mindfulness via journaling, meditation, prayer, body movement, or gardening. Notice how the focus of your practice shifts in response to personal and political influences.
Listen	At a particular location (above ground or underwater), preferably of personal or cultural significance, focus your attention on its sounds, e.g., industry, pollution, traffic, laughter, music, bird songs. Note how the sounds change at different times and consider the broad impact of sound. *What else would you add?*

TABLE 11.2 Climate Justice Leverage Points

	Leverage Point
Land Back	Part of the broad struggle for decolonization, Land Back means honoring treaties and returning colonized land to Indigenous hands. It may take a diversity of forms, e.g.. community land trusts, voluntarily paying taxes to tribes, free education tuition for Native students, and subsidized housing for Native residents.
Green New Deal	The GND is far-reaching municipal, state, or federal policy to redistribute power and resources from those historically advantaged to those disadvantaged (and thus most threatened by climate change). Intended to transition a community—justly—to a climate-ready future, especially though green jobs and infrastructure.
Divestment	Divestment is the removal of an institution's investment capital from stocks, bonds, or funds in fossil fuel companies or other industries/nations creating/exacerbating climate injustice.
Direct Action	Direct action is the employment of public forms of protest (rather than negotiation) to obstruct an objectionable practice and/or solve an injustice that a normative outlet cannot.
Police Abolition	One of the many institutions of white supremacy, policing is the "fast violence" disproportionately harming BIPOC communities already suffering the "slow violence" of environmental injustice. Police abolition confronts both, and reveals financial ties between fossil fuel companies and police groups.
Reparations	Responsive to victim-defined needs, reparations are acts of repair or compensation for wrongs committed by a society or individual in accompaniment with a reckoning for those wrongs.
Rights of Nature	Rights of Nature are a collection of legal strategies, employed contextually across the globe, that are designed to recognize harm to the more-than-human world, critique Anthropocentric norms, and protect ecological commons as well as particular species.
Animal Liberation	In recognition of radical intersectionality—that all oppression is linked—animal liberation opens channels for cross-movement collaboration and responds to the particular, overwhelming suffering nonhuman animals are made to endure, especially due to CAFOs and habitat destruction, using a variety of radical to reformist strategies.
Agroecology	Agroecology is farming based in ecology. Practices enhance soil fertility, recycle nutrients, optimize energy and water use, and increase beneficial organism-ecosystem interactions. All of this works to stabilize climate by capturing carbon, eliminating synthetic inputs, and increasing biodiversity.
Resilience Hubs	Resilience hubs are sites of local community resourcing in times of normalcy, disruption, and recovery. They're designed to respond to the contextualized needs of specific communities, building the capacity for communities to withstand, adapt to, and direct climate-related changes.
Ecological Restoration	Restoration is the intentional, sustained attempt to compensate for damaging influences on an ecosystem and manage it for the self-sustenance of either (a) species with a long coevolutionary history on a site or (b) species adapted to current and/or anticipated climate conditions.
Mutual Aid	Borne of an acknowledgement of systemic failure to meet peoples' needs, mutual aid is about collectively meeting needs for resources, mobilization, communication, and/or education. *What else would you add?*

- What knowledge can be attained from bringing our whole selves—rational and emotional—to our education? What's do we miss when emotions are suppressed? Is emotional connection to our subjects of study beneficial or detrimental?
- How do we productively tap into emotions to fuel—and sustain—the CJ movement? How can emotions be used to bridge political and educational divides?
- Why did you select your method? Could it be adapted to your field's traditional methodology? Are there gaps in experiencing that your field's normative methods produce?
- Does greater potential exist for all of us as knowers to feel seen and valued in using these decolonial methods? Why or why not?
- Which methods offered in this activity might be usefully—or creatively—paired?
- Why did you select your leverage point? What social change can it bring about? How efficacious is it in which contexts?
- Can enacting your chosen leverage point form the basis of your scholarship, so that you are contributing to the CJ movement even as you study it? If so, how?
- With which other leverage point initiatives could you collaborate?
- What did your method reveal about your leverage point that traditional scholarship could not convey? Which other combinations need to be employed in the global CJ movement?

NOTES

1. Donella Meadows, "Leverage Points: Places to Intervene in a System" (Academy for Systems Change, 1996).

2. See examples in Larry Lohmann, *Carbon Trading* (Uppsala: Dag Hammarskjöld Foundation, 2006).

3. David Naguib Pellow, "The Global Village Dump," in *Resisting Global Toxics* (Cambridge: MIT Press, 2007), 126.

SELECTED RESOURCES

Climate Justice

Bhavnani, Kum-Kum, John Foran, Priya A. Kurian, and Debashish Munshi, eds. *Climate Futures: Reimagining Global Climate Justice*. London: Zed Books, 2019.

Johnson, Ayana Elizabeth, and Katharine Wilkinson, eds. *All We Can Save: Truth, Courage, and Solutions for the Climate Crisis*. New York: One World, 2020.

Decolonial Methodology

Kimmerer, Robin Wall. *Braiding Sweetgrass: Indigenous Wisdom, Scientific Knowledge, and the Teachings of Plants*. Minneapolis: Milkweed, 2013.

Smith, Linda Tuhiwai. *Decolonizing Methodologies: Research and Indigenous Peoples*. London: Zed Books, 1999.

Wilson, Shawn. *Research Is Ceremony: Indigenous Research Methods*. Black Point, NS, Canada: Fernwood Publishing, 2008.

Feel

Bacon, J. M., and Kari Marie Noorgaard. "Emotions of Environmental Justice." In *Lessons in Environmental Justice: From Civil Rights to Black Lives Matter and Idle No More*, edited by Michael Mascarenhas, 110–26. Thousand Oaks, CA: Sage, 2021.

Taste

Nikole, Alexis. blackforager. Instagram. https://www.instagram.com/blackforager/?hl=en.
Sherman, Sean, with Beth Dooley. *The Sioux Chef's Indigenous Kitchen*. Minneapolis: University of Minnesota Press, 2017.

Dance

Ananya Dance Theatre. "Pipaashaa: Extreme Thirst." Minneapolis, 2007. https://www.ananyadancetheatre.org/dance/pipaashaa.
Don't You Feel It Too? Accessed January 2022. https://www.dyfit.org.

Paddle

Underhill, Joe. "What We Learned from the River." *Open Rivers*, Spring 2017. https://editions.lib.umn.edu/openrivers/article/what-we-learned-from-the-river.

Storytell

Dembicki, Matt. *Trickster: Native American Tales: A Graphic Collection*. Golden, CO: Fulcrum Books, 2010.
Power of Vision Mural Project/Hope Community. "Defend, Grow, Nurture Phillips." 2020. https://www.olivialevinsholdenart.com/murals.html.

Play

Powers, Julie, and Sheila Williams Ridge. *Nature-Based Learning for Young Children: Anytime, Anywhere, on Any Budget*. St. Paul: Redleaf Press, 2019.

Build

Blake, Bob. "Red Lake Solar Project: Clean Energy Success Stories." YouTube, October 21, 2020. https://www.youtube.com/watch?v=MVqEYCWLpYs&t=154s.

Craft

Zaleha, Sarita. *Global Warming Blanket*. 2013. https://www.saritazaleha.com/global-warming-blanket.

Map

CEED Environmental Justice Atlas. 2021. https://umn.maps.arcgis.com/apps/OnePane/basicviewer/index.html?appid=a826e71660804b97afd942c1d5001c22.
Mapping Prejudice. University of Minnesota Libraries. Accessed January 2022. https://mappingprejudice.umn.edu.

Plant

Maathai, Wangari. *The Green Belt Movement: Sharing the Approach and the Experience*. Brooklyn, NY: Lantern, 2003.

Ritual

Jordan, William. "Sacrifice and Celebration: Restoration as Performing Art." In *The Sunflower Forest: Ecological Restoration and the New Communion with Nature*, 160–94. Berkeley: University of California Press, 2012.

Listen

Mendes, Margarita. "Sounding the Mississippi." Anthropocene Curriculum, 2020. https://www.anthropocene-curriculum.org/project/mississippi/anthropocene-river-journey/sounding-the-mississippi.

Climate Justice Leverage Points

Meadows, Donella. "Leverage Points: Places to Intervene in a System." The Donnella Meadows Project: Academy for Systems Change, 1996. https://donellameadows.org/archives/leverage-points-places-to-intervene-in-a-system.

Land Back

Briarpatch. "The Land Back Issue." September/October 2020. https://briarpatchmagazine.com/issues/view/september-october-2020.

Penniman, Leah. "Black Gold." In *All We Can Save: Truth, Courage, and Solutions to the Climate Crisis*, edited by Ayana Elizabeth Johnson and Katharine K. Wilkinson, 301–10. New York: One World, 2020.

Green New Deal

Klein, Naomi. *On Fire: The Burning Case for a Green New Deal*. London: Penguin Books, 2019.

———. "A Message from the Future with Alexandria Ocasio-Cortez." *The Intercept*, April 17, 2019. https://theintercept.com/2019/04/17/green-new-deal-short-film-alexandria-ocasio-cortez.

Divestment

O'Brien, Susie. "Fossil Fuel Divestment Is the Road to Climate Justice." *The Conversation*, May 24, 2021. https://theconversation.com/fossil-fuel-divestment-is-the-road-to-climate-justice-159095.

Direct Action

Pellow, David Naguib. "Direct Action." In *Total Liberation: The Power and Promise of Animal Rights and the Radical Earth Movement*, 127–61. Minneapolis: University of Minnesota Press, 2014.

Police Abolition

alex_draws_good. "What Does the Fossil Fuel Industry Have to Do with Police Brutality?" Instagram, comic, September 16, 2020.

Reparations

Sheller, Mimi. "The Case for Climate Reparations." *Bulletin of the Atomic Scientists*, November 6, 2020. https://thebulletin.org/2020/11/the-case-for-climate-reparations.

Rights of Nature

Pribanic, Joshua B., and Melissa A. Troutman. *Invisible Hand*. Grant Township, PA: Public Herald Studios, 2020. https://www.invisiblehandfilm.com.

Animal Liberation

Pellow, David Naguib. *Total Liberation: The Power and Promise of Animal Rights and the Radical Earth Movement*. Minneapolis: University of Minnesota Press, 2014.

Agroecology

Montenegro, Maywa. "Agroecology Can Help Fix Our Broken Food System. Here's How." *Ensia,* June 17, 2015. https://ensia.com/voices/agroecology-can-help-fix-our-broken-food-system-heres-how.

Resilience Hubs

Urban Sustainability Directors Network. "Guide to Developing Resilience Hubs." 2019. http://resilience-hub.org/wp-content/uploads/2019/10/USDN_ResilienceHubsGuidance-1.pdf.

Ecological Restoration

Garvey, Michelle. "Novel Ecosystems, Familiar Injustices: The Promise of Justice-Oriented Restoration." *Darkmatter,* April 2016. http://www.darkmatter101.org/site/2016/04/02/novel-ecosystems-familiar-injustices-the-promise-of-justice-oriented-ecological-restoration.

Mutual Aid

Spade, Dean. *Mutual Aid: Building Solidarity during This Crisis (and the Next)*. London: Verso, 2020.

TWELVE

Infrastructure Affects

Registering Impressions of Mega-Dams

RICHARD WATTS

Some agents of environmental harm are less visible than others. This chapter aims to provide guidance on developing course modules and assignments regarding mega-dams, which remain remote and strangely obscured in our imaginations, the affective responses they can elicit, and the justice-oriented relation to dams and other forms of hidden infrastructure that visual and text-based media can foster.

This assignment encourages reflection on mega-dams, their complicated relation to a range of social-ecological crises, and the ambivalent affective responses these imposing, anonymous structures provoke ... or might provoke, if they were ever given consideration.[1] Dams are everywhere, yet barely visible in the popular imagination. Globally, it is estimated that there are over 36,000 large dams in major river basins; the United States is home to approximately 8,000 of those large dams.[2] If mega-dam construction in the United States has slowed in recent years, it has accelerated in the rest of the world: in 2019, more than 3,700 medium to large hydropower dams were under construction or planned, the vast majority of them in South America, East and South Asia, and Africa.[3] Yet how many know the names of dams in their regions, the waters they gather and impound, and the human and other-than-human communities they currently serve or previously displaced? In higher-education contexts, outside such fields as Civil Engineering and Fisheries, how often are the dams that were and continue to be built on the seemingly uncontroversial promises of flood control, managed irrigation, improved navigation, recreation, and hydropower given serious consideration in relation to the rapidly changing environment and upstream and downstream justice? And when dams—as well as other forms of "invisible" infrastructure (industrial canals, flight paths, pesticide overspray zones) with environmental justice implications—are considered in this way, is the vocabulary for describing how they make us feel available? Just as in Animal Studies where scholars have identified the focus on "charismatic megafauna" (polar bears, orcas) and "flagship species" (spotted owls, chinook salmon) as a threat to the elision of less visible creatures,[4] it is imperative in environmental justice pedagogy to name hidden risks and ponder how to mobilize affect around them. This is a very different exercise indeed from reflecting on and "managing" the excess of affect that so many of the environmental crises of our times provoke and that many of the contributions to this volume address.

> It is imperative in environmental justice pedagogy to name hidden risks and ponder how to mobilize affect around them.

THE COURSE

For the past ten years, I have taught "The Water Crisis in Literature and Cinema," an environmental humanities course that engages students from a wide range of disciplines in understanding dramatic changes in the relation to water as expressed through literature and cinema, as well as sculpture, painting, song, landscape design, and other cultural forms. Dams are, of course, major disruptors of waterways, but also of foodways, ways of knowing, and ways of being in a place, and our engagement with them typically occupies one week of the ten-week, quarter-system course (two sessions, whether online or in person, synchronous or asynchronous, of approximately an hour each, not including student time for preparation), with the other nine weeks devoted to representations of other sites and manifestations of the water crisis (pollution, privatization, commodification, etc.). This module, which pairs two films—one that explores in an elegiac register the cultural losses occasioned by mega-dams and another that channels activist fervor in the name of dam removal—can be used in a variety of instructional settings: I have guest lectured on several occasions using this module in a class in the University of Washington College of the Environment, where reflections on affect are less immediately a part of the curriculum, and each time my introduction to the environmental humanities has elicited a strong response. In brief, enabling students across disciplines to reflect on storytelling about and affective responses to obscured elements of the built environment that have significant social-ecological impacts gives them a path to intervening in debates that can otherwise feel remote.[5]

> Dams are, of course, major disruptors of waterways, but also of foodways, ways of knowing, and ways of being in a place.

THEORETICAL SCAFFOLDING

Dams, which can be seen in one light as neutral, utilitarian masses of earth, masonry, concrete, and steel, nonetheless have what Jennifer Ladino calls *affective agency*, the capacity of seemingly inert matter "to generate felt impressions on other bodies even while remaining recalcitrant."[6] Historically, that felt impression at the sight of the large dams of the midcentury, those "modern marvels," was primarily one of awe and pride.[7] More relevant still to this module and to the conflicted experience of mega-dams today, which, for instance, provide "clean" power but have been shown to release significant quantities of the greenhouse gas methane,[8] is Ladino's term *affective dissonance,* "the unsettled state in which we experience more than one feeling at the same time, often with a sense of conflictedness or irony."[9] Ladino coins the term to describe the unintended effects of National Park Service memorials—sites where memory is solicited and affective response encouraged—but applies equally well to dams, especially large ones, which are tangled knots of engineers' problem-solving intentions and un/foreseen social-ecological consequences. Students whose family histories have not been directly impacted by dams might feel sadness at the lives displaced and ecological harm inflicted by this largely hidden infrastructure, astonishment, and dread at reservoir-triggered seismicity,[10] but also a sense of wonder and perhaps even hope at the "carbon-free" energy they generate. For students with lived experience or cultural memory of the harm dams produce, the range of affects may be quite different; here, particular attention needs to be paid to the positionality of both instructors and students, since, in the case of Indigenous students, dams may be a source of severe distress (as a result of dams having been built in the ostensibly "unpeopled" areas, as Franklin Delano Roosevelt famously put it,[11] that many Indigenous communities call home). And then there are the anxiogenic geopolitics of a project such as the Grand Ethiopian Renaissance Dam, which stirs up ethno-nationalistic fervor, especially in neighboring countries.[12] Regardless, because students are rarely put in a position to contemplate dams—to think about their emplacement on the land and in history and their status as producers simultaneously of "clean" energy and social-ecological harm—cultural texts that narrativize dams provide the opportunity to contemplate and feel their presence.

> Dams, especially large ones, are tangled knots of engineers' problem-solving intentions and un/foreseen social-ecological consequences.

THE MODULE

This module therefore aims to engage students on the affects that adhere to mega-dams and their social-ecological consequences, which can be as dramatic as those identified with a warming world and other manifestations of environmental change: mass forced migration, species die-offs, habitat fragmentation, degradation of Indigenous cultural webs (e.g., the elimination of traditional fishing sites), among other forms of harm. By closely analyzing two quite different films—Jia Zhangke's 2006 feature film *Still Life* centered on a town in east-central China about to be erased by the rising waters behind the Three Gorges Dam and the 2014 documentary film *DamNation* by Ben Knight and Travis Rummel on dam-removal activism in the western United States—the module encourages working through the range of affects that these mega-structures generate. This pairing also allows for the consideration of cinematic genre (documentary-realist fiction vs. documentary) and its effects on the mobilization of affect.

In some respects, *Still Life* is a difficult film. The pacing is slow, there is little in the way of plot, and—in spite of the film's predominantly documentary-realist aesthetic—surreal moments punctuate an already complex narrative structure (a UFO briefly appears in the sky; a building inexplicably takes off like a rocket). This presents an opportunity to discuss looking/viewing/reading and therefore *feeling* otherwise: What work do the narrative slowness and visual "stillness" of the film accomplish? How can we think with the film about different approaches to seeing and "feeling" dams? In other ways, though, the story is straightforward: in disconnected, but adjacent plotlines, Han Sanming, a miner, and Shen Hong, a nurse, look for their estranged spouses in Fengjie, a town that is slowly being submerged (Sanming exclaims "But it's under water!" when taken to his wife's last known address behind the dam). In a China wracked by breathtakingly rapid technological, economic, and social transformation, everyone feels displaced, and viewers of the film are invited to associate the forced removal of *at least* 1.3 million people provoked by the rising waters behind the Three Gorges Dam with other forms of dislocation. Questions of justice therefore loom over the film, even if they remain unresolved.[13]

DamNation, for its part, foregrounds questions of justice as it advocates for dam removal. The documentary is organized around the narrator Ben Knight's journey from skate photographer to environmental activist. In the interviews, archival footage, and stunning cinematography it deploys, *DamNation* highlights the dangers that dams pose to rivers, other-than-human animals, and especially the Indigenous communities whose stories are the documentary's most emotionally impactful. The film even (and quite effectively) mobilizes the affects of wild salmon, curious animals who are capable of "conniption fit(s)" (00:43:52) and who are imagined to utter "f*#% my life" (00:50:47) as they compete with hatchery fish. In so doing, it extends the discussion of justice to other-than-human animals and dramatizes how water infrastructure braids together the fates of many planetary partners. For all of its commitment to redressing harm done to Indigenous communities, though, *DamNation* seems equally driven by the potentially contradictory aim of "liberating" rivers for kayakers, fishers, and other recreationists, in line with the film's production arrangement with the outdoor clothing and equipment brand Patagonia. A question, then, for class discussion: does the film's ambivalence (or, less charitably, duplicity) regarding those for whom it is advocating trouble affective responses to it?

THE ASSIGNMENT

Having already worked with the vocabulary of affect in other modules, students are asked to engage in the following activities:

Pre-class: For each film and before each of the two class sessions, students create a two-column list registering on one side their emotional responses to particular scenes and in the other the range of affects expressed on screen (for a total of ten, which do not have to be distributed evenly between the lists; this number has proven to be sufficient to stimulate effective discussion).

In-class: After discussion of some of the questions enumerated in the film synopses above, students are asked to gather in groups to contemplate the alignment or disconnect between their two columns. Because the disjuncture between their own affective responses and those of the characters in *Still Life* is generally pronounced, I have found it more effective to teach that film first, which subsequently leads to questions about the reasons for students' self-perceived affective alignment with *DamNation*. At this juncture, we discuss the positionality of the filmmakers and, in the case of *DamNation*, the narrator, and of our own racial, social, and linguistic positionalities as they relate to the two films. Students are then asked to consider what Sianne Ngai calls "ugly feelings," especially what they often perceive as the "affective opacity" of characters in *Still Life* and the "irritation" it can provoke in them,[14] as well as the idea that we cannot dwell only in the "cathartic" affects like anger,[15] even when addressing questions of justice.

Post-class: We end the module with a short (300–500 word) writing activity. Students choose one of three prompts:

1. Write a review of the two films that focuses on how each orients affective responses to the social-ecological consequences of mega-dams. How do they mobilize affect differently? How might one compare the *impact* of the films (the feelings they inspire) and their likely *effectiveness* as change agents? Can we even presume that the films have similar aims with regard to the social meaning and implications of dams?
2. Write a brief "pitch" for a film, novel, or other creative work that dramatizes dams and the communities impacted by them. How might the dam function as a character? How does your proposed work manage to generate affective responses to the complex cultural and political issues surrounding dams, and to what putative end?
3. Select an existing dam from a place you know (here or elsewhere) and write the text and select the images for an infographic that would be put in place after dam removal (see Glines Canyon example, Figure 12.1). What rationale for

FIGURE 12.1 Infographic at Glines Canyon Dam Removal Site, Elwha River, Washington.

removal do you provide? Which affects do you aim to mobilize in the visitor to the dam removal site (in consciousness of the diversity of visitors)?

For many students, one of these three short writing activities becomes the basis of a longer, end-of-term essay or creative project.

The nature and contour of these assignments has varied over the years, but their central aim has not. What matters is that students be asked to hold in tension a number of perspectives on dams, none of them unassailable, and to register a range of complex, contradictory responses. Likewise for the choice of texts: many other texts that engage dams on an affective level could be part of the module, from the 2016 documentary film *Belo Monte: After the Flood* that centers eco-social harm to Indigenous populations caused by a mega-dam project in the Brazilian Amazon to Brenda Hillman's poem "Hydrology of California," which uses punctuation (in particular, slashes) to indicate the disruption of river flows by dams and, by extension, the mixed feelings that rivers (and river poetry) regrettably inspire today. However loosely or closely a module is modeled on this one, the point is to allow students to confront the ostensible silence of dams and other forms of hidden infrastructure with narratives—documentary or creative, testimonial or speculative—that oblige these mute structures to speak in the language of affect of their past, present, and future. This enables students to have a meaningful response to this largely hidden infrastructure and, perhaps, mobilize those affects in an ongoing, justice-oriented engagement with dams.

NOTES

1. I would like to thank María Elena García, Jennifer Ladino, and Jennifer Atkinson for their comments on earlier drafts and all of those who have hiked with me up to the Glines Canyon Dam removal site, a great place for thinking and feeling differently about dams.
2. Zhang et al., *Power of the River*, 1.
3. Zarfl et al., "Future Large Hydropower Dams," 1.
4. Douglas and Winkel, "Flipside of the Flagship," 979.
5. On storytelling's exclusion from academic discourse, see Dian Million, "Felt Theory." "(W)e *feel* our histories as well as think them" (54), writes Million regarding Indigenous communities' engagement with the past (and present). The course's emphasis on storytelling is therefore intended both as a means of accessing affect and, following Million, as a corrective to the overinvestment in dispassionate "argument" in student writing.
6. Ladino, *Memorials Matter*, 16.
7. Those affects persist: *Popular Mechanics* recently published a list of the "The 8 Most Awe-Inspiring Dams in America." It notes breathlessly that "the Grand Coulee Dam contains enough concrete to build [. . .] a sidewalk around the Equator—twice" (Newcomb, n.p.).
8. Methane is released when the biomass in reservoirs—namely, trees submerged behind a dam—begin to decompose. See Deemer et al., "Greenhouse Gas Emissions from Reservoir Water Surfaces," 949.
9. Ladino, *Memorials Matter*, 22.
10. Gupta, "Studies of Artificial Water Reservoir Triggered Earthquakes," 1556.
11. Roosevelt, "Address at the Dedication of Boulder Dam (1935)," n.p.
12. In October 2020, then-President Trump stoked those anxieties by claiming that Egypt would "blow up that dam" (Neumann, "Hydropower," n.p.). Most dangers related to dams have a slow, deliberate quality and generate a diffuse disquiet that can be difficult to name; this threat issued on behalf of Egypt is clearly intended to provoke intense or so-called "cathartic" emotions, namely fear, anger, and anxiety.
13. *Still Life* toggles between open critique of the dam project and veiled, opaque suggestions of the harm it causes. Perhaps this is what allowed it be recognized as "official cinema" by the Chinese central government, unlike some of Jia Zhangke's earlier films (e.g., his 1997 *Pickpocket*), which were banned from distribution in China.
14. Ngai, *Ugly Feelings*, 175.
15. Ibid., 6.

REFERENCES

Deemer, Bridget R., et al. "Greenhouse Gas Emissions from Reservoir Water Surfaces: A New Global Synthesis." *BioScience* 66, no. 11 (November 2016): 949-64.
Douglas, Leo R., and Gary Winkel. "The Flipside of the Flagship." *Biodiversity and Conservation* 23, no. 4 (2014): 979–97.
Gupta, Harsh K. "Studies of Artificial Water Reservoir Triggered Earthquakes at Koyna, India: A Summary." *Journal of the Geological Society of India* 97, no. 12 (2021): 1556–64.
Hillman, Brenda. "Hydrology of California." In *Practical Water*. Middletown: Wesleyan University Press, 2009.
Knight, Ben, and Travis Rummel, dirs. *DamNation*. Oley, PA: Bullfrog Films, 2014. DVD video.

Ladino, Jennifer. *Memorials Matter: Emotion, Environment, and Public Memory at American Historical Sites*. Reno: University of Nevada Press, 2019.

Million, Dian. "Felt Theory: An Indigenous Feminist Approach to Affect and History." *Wicazo Sa Review* 24, no. 2 (2009): 53–76.

Neumann, Ann. "Hydropower: A Dam on the Nile Roils Democratic Relations in the Horn of Africa." *The Baffler* 67 (March 2023). https://thebaffler.com/salvos/hydropower-neumann.

Newcomb, Tim. "The 8 Most Awe-Inspiring Dams in America." *Popular Mechanics*, October 9, 2020. https://www.popularmechanics.com/technology/infrastructure/g2837/7-most-serious-dams-us.

Ngai, Sianne. *Ugly Feelings*. Cambridge, MA: Harvard University Press, 2005.

Roosevelt, Franklin D. "Address at the Dedication of Boulder Dam (1935)." The American Presidency Project, edited by Gerhard Peters and John T. Woolley. https://www.presidency.ucsb.edu/node/209220.

Zarfl, Christian, Jürgen Berlekamp, Fengzhi He, Sonja Jähnig, William Darwall, and Klement Tockner. "Future Large Hydropower Dams Impact Global Freshwater Megafauna." *Scientific Reports* 9, no. 18531 (2019). https://doi.org/10.1038/s41598-019-54980-8.

Zhang, Alice Tianbo, Johannes Urpelainen, and Wolfram Schlenker. *Power of the River: Introducing the Global Dam Tracker (GDAT)*. New York: Columbia/SIPA Center on Global Energy Policy, 2018. https://www.energypolicy.columbia.edu/research/report/power-river-introducing-global-dam-tracker-gdat.

Zhangke, Jia, dir. *Still Life*. 2006; New York: New Yorker Video, 2008. DVD video.

THIRTEEN

From Principles to Praxis

Exploring the Roots and Ramifications of the Environmental Justice Movement

SHANE D. HALL

This chapter describes a scalable class project that fosters individual and collective understandings of a major taproot of the climate justice movement: the "Principles of Environmental Justice" document issued at the culmination of the 1991 First National People of Color Environmental Leadership Summit.

The phrase, "It takes roots to weather the storm," is emblematic of the existential wisdom of the climate justice movement. While the climate generation is putting down new roots of social resilience for just transitions, it also draws strength from the deep roots of other social movements, especially the environmental justice movement. In this chapter I describe a scalable class project that fosters individual and collective understandings of a major taproot of the climate justice movement: the "Principles of Environmental Justice" document issued at the culmination of the 1991 First National People of Color Environmental Leadership Summit. In the project's essays and discussions, students unpack the dynamic legacy of these environmental justice (EJ) principles in current activism and scholarship.

As Julie Sze notes, "these principles collectively ask the questions: What and where is justice for people of color and colonized peoples? What are the sources of environmental racism and injustice, and what can be done to promote environmental justice?"[1] To answer Sze's questions, each student takes responsibility for one principle affirmed and adopted by the First National People of Color Environmental Leadership Summit. If the class is larger than 17 students, classes may work in teams or incorporate additional key movement documents such as the "Jemez Principles" or the "Bali Principles of Climate Justice." Each student writes three interrelated essays on their chosen principle: one unpacking the ethical framework of the principle, another explicating an artistic representation of the principle, and the last tracing the principle in grassroots praxis. Each essay thus corresponds to a different aspect of the "head, heart, hands" model of social transformation. Students also meet in class "summits" to discuss the connections between the principles and EJ movements they are studying. The essays and summits help students better understand the influence of EJ activist traditions and the 1991 environmental justice principles on today's climate justice studies and struggles.

The ability to recognize the epistemological validity of activist experience is in and of itself an existential tool for "weathering the storms" of climate injustice. To recognize the "wisdom of communities on the frontlines" of the climate crisis requires we learn to think past the artificial academic distinctions between art, activism, and critical theorizing.[2] This assignment design scaffolds different kinds of thinking through environmental justice—as ethics, creative expressions, and grassroots praxis. The basic approach can be applied to virtually any archive of social movement work, as the actions of all social justice movements are rooted in new visions, values, and actions. The project requires students to take grassroots movements as serious repositories of knowledge.

This treatment of the First National People of Color Environmental Leadership Summit is not, however, hagiographic. By reading supplementary materials such as the summit's proceedings, especially the floor discussions of the draft principles, students see the various delegates' diverging interpretations, idiosyncratic foci and aporia. In short, the practitioners of environmental and climate justice are humanized. Humanizing activists and activism can be a prerequisite to students finding kinship and pathways into social movements. The climate justice movement isn't something wholly new under the sun, and extraordinary social change often comes from ordinary people working together. Exploring the roots of these truths in turn helps to teach students their own rootedness in their own cares and commitments.

ASSIGNMENT DESCRIPTION (ADDRESSED TO STUDENTS)

Overview and Goals of the Project

Over 650 grassroots activists gathered for the First National People of Color Environmental Leadership Summit from October 24–27 in 1991. The summit marked a watershed moment in the environmental justice movement. As chronicler of the summit, Charles Lee, describes it:

> ... [the] Summit was the defining event of the emerging movement for environmental justice. As the Summit's Call for Action stated, "Unlike traditional mainstream environmental and social justice organizations, this multiracial, multicultural movement of peoples of color is evolving from the bottom up and not from the top down. it seeks a global vision based on grassroots realities.[3]

That "global vision" Lee cites is embodied in the "Principles of Environmental Justice," a manifesto-like document modeled on the U.S. constitution. A subcommittee of delegates worked for months on the principles prior to the summit and then hashed out the final document through democratic debate during the summit. As a class we are going to return to this influential document to dig deeply into the philosophical, practical, and artistic ramifications of environmental justice. In doing so, we'll see how principles inform the practice and study of this "movement of movements" and interdisciplinary field of study.

The Task

Each student will take responsibility for one of the principles of environmental justice. You will write three essays: (1) a vision paper on the principle, (2) a close reading of an artistic depiction of the principle, and (3) an examination of an example of this principle "in the real world." Below you will find a description of each essay.

EJ Vision Paper: By "vision paper" I mean a concise essay that explains the philosophical and ethical basis of your particular principle. In the words of American Studies scholar George Lipsitz, "social movements shake up social life; they reconfigure the horizons of individuals and groups by challenging old forms of knowledge and advancing new ones."[4] In this essay your task is to explore how your particular principle "shakes up" or "reconfigures" old assumptions or advances new knowledge or understandings.

For your first EJ principles essay, you will write an essay on your assigned principle that explains the following: (1) what exactly—in the context of environmental justice studies in social movements—does your assigned principle mean? (2) How does your principle relate to the other principles of environmental justice? (3) Why is your principle significant to environmental justice studies in social movements?

This vision paper is your opportunity to dig deeply into the philosophical and ethical basis of your assigned principle. How do the specific principles and ideas fit into your understanding of what environmental justice is and why it's important?

EJ Art Essay: Art is a critical way that social movements form and sustain themselves. According to social justice artivist Favianna Rodriguez, "Art is a human right. Art is a practice that teaches us how to be critical thinkers, art teaches us how to have a voice, and teaches us to imagine the kind of world we wish to live in."[5] Therefore art *is* a form of an act, insofar as it can both imagine new ideas and act in the world. The ideas that form the principles of environmental justice are often communicated through art—paintings, murals, music, bumper stickers, posters, tagging, poems, movies, etc. More than any other form, it engages the emotional and affective dimensions of environmental justice.

For this essay you will find a piece of art—any kind of art—that communicates your principle. You will submit a copy of this art, correctly cited using MLA format, along with a short essay that performs a "close reading" of this piece of art. A close reading is a form of argumentative essay that analyzes a work of art. Everyone in this class will essentially be making the same argument: "X piece of art by Y artist(s) communicates Z principle of environmental justice. It does so by . . . "

The key to your success in this essay will be your ability to be clear and specific in your explication of the art. How does the form and context of the art communicate your EJ principle?

EJ Praxis Paper: The Oxford English Dictionary defines "praxis" as "that through which theory or philosophy is transformed into practical social activity; the synthesis of theory and practice seen as a basis for or condition of political and economic change. Also: an instance of this; the application of a theory or philosophy to a practical political, social, etc., activity or programme."[6]

In this EJ principles essay you will write a short essay that unpacks an example of your assigned EJ principle *in grassroots praxis.* In other words, your essay will identify an event, policy, or organization whose methods or goals of organizing provide evidence of your EJ principle "in action." Your task in the essay is to argue how this particular event or group mobilizes around your principle. Your essay must conclude by commenting on how this example of praxis adds to your own understanding of your EJ principle.

Perhaps the biggest key to your success in this essay will lie in your ability to research in order to get specific and provide evidence for your analysis. If, for example, you find an EJ organization that claims on the front of its webpage to "affirm the sacredness of mother Earth" (the first principle of EJ), it is not enough in this praxis essay to just say "I found this organization quotes the Principles of Environmental Justice on their website, therefore they must be implementing them accordingly." Such a claim fails to explain how the theory (the principle) is put into practice, or if the actions of that organization truly enact the principle rather than paying lip service to it. Note: it is rare indeed when on-the-ground struggles perfectly embody a movement's aspirational principle. Successful social movements don't just forge new ideas—they act to make concrete change in people's lives.

DESCRIPTIONS OF IN-CLASS "SUMMIT" DISCUSSIONS

The class periodically meets to discuss progress on the project and relate the EJ principles to other class learning. Here is a brief description of each "Summit day."

> **Early in Semester:** Instructor introduces project's goals, armature, and logistics. Students read aloud the "Principles," then work to identify common themes of the document.[7]
>
> **Post Vision Paper:** Each student reads aloud their principle and gives a short presentation of their analysis. This is followed by discussion.
>
> **Post Art Essay:** Half of the students share the art text they examined with class. Then small groups discuss how discreet art speaks to their own principles.
>
> **Post Praxis Essay:** Half of the students share their praxis case stories with class. Follow up discussions on the saliency of EJ principles to current case stories.

Near end of class: Discussion of principles in the climate justice movement. Instructor introduces final reflection essay. Initial discussions to spur reflection essay process.

RESOURCES

- Proceedings of the First National People of Color Environmental Leadership Summit, available online from the United Church of Christ: http://rescarta.ucc.org/jsp/RcWebImageViewer.jsp?doc_id=32092eb9-294e-4f6e-a880-17b8bbe02d88/OhClUCC0/00000001/00000070&pg_seq=1&search_doc=
- The EJ Atlas maps environmental justice conflicts globally. EJAtlas.org
- The Justseeds Artists' Cooperative is a great place to look for socially engaged art. Justseeds.org

NOTES

1. Sze, *Environmental Justice*, 30.
2. It Takes Roots, "Who We Are."
3. Lee, "Introduction," v.
4. Lipsitz, *American Studies*, xvi.
5. Rodriguez, "Be an Artist."
6. "praxis," *OED Online*.
7. In the proceedings for the summit (available online via the United Church of Christ), Charles Lee characterizes the principles as reflecting his own understanding of the main themes of the summit: (1) people of color speaking for themselves, (2) a redefinition of "environment" and "environmentalism," and (3) the connection of social and racial justice to environmental issues. I use Lee's framework to help jump-start student analysis.

REFERENCES

Climate Justice Alliance. "Just Transition." February 19, 2021. https://climatejusticealliance.org/just-transition.

It Takes Roots. "Who We Are." 2023. https://ittakesroots.org/about.

Lee, Charles. "Introduction." In *Proceedings of the First National People of Color Environmental Leadership Summit, October 24–27, 1991*. Washington, DC: United Church of Christ Commission for Racial Justice, 1992.

Lipsitz, George. *American Studies in a Moment of Danger*. Minneapolis: University of Minnesota Press, 2001.

"praxis, n." *OED Online*. New York: Oxford University Press, December 2021. www.oed.com/view/Entry/149425.

Rodriguez, Favianna. "Be an Artist: Work of Art by Favianna Rodriguez." 2021. www.favianna.com/artworks/be-an-artist.

Sze, Julie. *Environmental Justice in a Moment of Danger*. Oakland: University of California Press, 2020.

PART THREE

EMBODIED PEDAGOGIES

FOURTEEN

Working with Ecological Emotions

Mind Map and Spectrum Line

PANU PIHKALA

Before this course about eco-anxiety, I felt that I don't have any right to be anxious about ecological issues, because I felt that I didn't do enough pro-environmental behavior. To exaggerate only a bit, I thought that if one owns a car, one is not allowed to feel eco-anxiety. What has changed in me during this university course about eco-anxiety is that I've realized that I can allow myself permission to feel these difficult emotions.

STUDENT IN THE UNIVERSITY OF HELSINKI LECTURE COURSE "ECO-ANXIETY: INTERDISCIPLINARY PERSPECTIVES" (IN FINNISH), Spring 2021

This chapter introduces two practical methods developed in Finland for working with eco-emotions. The methods can be used in a variety of settings to support students and educators grappling with difficult environmental topics.

The student whose course feedback is cited above is actually quite active in pro-environmental behavior. The situation she describes is familiar for me. Many of the students in my eco-anxiety courses and workshops are diligent: they have made numerous lifestyle changes for environmental reasons, they support pro-environmental politics and environmental justice, and they often participate in some form of activism in their free time. But still, they constantly feel that this is not enough. Most of them are women, and gendered emotion norms make them even more vulnerable to unjust amounts of ecological guilt and shame. In other words, these students suffer from something we might understand as one kind of climate injustice—a culture that doesn't adequately support their emotions.

This is my main point: there are justice issues in climate education even in countries like Finland, where the direct impacts of the climate crisis are still mild when compared globally; and emotion-wise, education can help. In this article, I introduce Finnish initiatives and activities for discussing ecological emotions in schools and universities. I have been developing these materials over the past six years, and have facilitated them in various settings. I briefly set these activities into a wider framework of promises and problems with encountering existential climate issues in the classroom.

Two practical activities I focus on are the Mind Map of Ecological Emotions and the Spectrum Line of Ecological Emotions; both are freely available online at the Existential Toolkit for Climate Justice Educators website. Among many audiences, these have encouraged teachers and students to encounter their ecological emotions in general, making it possible to name particular emotions and discuss their practical and ethical dimensions. This has been especially important in relation to emotions that are seldom talked about due to a lack of safe spaces and encouragement.[1]

THE FINNISH CONTEXT AND PSYCHOSOCIAL DYNAMICS OF CLIMATE EMOTIONS

Justice issues are intertwined in these educational endeavors in several ways. On one hand, Finnish people are globally in a privileged position: democracy and the school system function well, which makes discussions of ecological emotions easier. Finland, a small nation with a population of roughly 5.5 million, has a rather large surface area. There is plenty of fresh water and lots of forests. While heatwaves and other extreme weather events are getting stronger and more common, the conditions are still much easier than even in Central Europe. Finland is situated at the northernmost part of Europe, almost on the latitude of Alaska, but the Atlantic Gulf Stream has so far provided enough warmth to make agriculture possible. One of the biggest physical threats of climate change for Finns is the slowing down of the Gulf Stream.

On the other hand, many of the Finnish people who feel difficult eco-emotions have encountered unethical treatment by others. This is often linked with existing sociodemographic factors and vulnerabilities. For example, young people and women have more often spoken about their eco-anxiety, and in response, many have been silenced or ostracized.[2] The movement of introducing activities about eco-emotions in schools results in normalization and gives voice to many people who felt voiceless before. Even in Finland, it has not been easy to make room for eco-emotions in schools and universities, but there has been improvement lately.[3]

In the very northern parts of Finland and neighboring countries, there still lives the Sámi people, the only thriving Indigenous people in Europe. Traditionally nomadic reindeer herders, the Sámi have suffered much from the imposed borders and discrimination of nation states, including Finland. The contemporary Finnish government (as of 2022), for all its ethical fame, has still not recognized the full rights to land ownership by the Sámi. They are not openly or aggressively discriminated against, but they do constantly suffer from structural injustice, and the effects of global warming are severe for them.[4] I work in Helsinki, which is on the south coast of Finland, and although many Sámi live in the metropolitan area, they very rarely participate in my university courses. However, I've had many fruitful conversations with Sámi in public events, and they resonate with the efforts to build a larger vocabulary for climate grief, anger, and other emotions, which are quite familiar indeed for them, in their own ways.

One of the strong cultural forms of climate anxiety and grief in Finland is related to how the winters are changing. Finns have traditionally loved snow and snowy winters are part of the cultural spirit. Now especially in southern Finland, climate change is making winters very unpredictable. Even in Lapland, in the north, the conditions are changing so that problems arise for the Sámi people. Naming this winter grief and snow sadness is the first step in building community support and peer support for all those people who feel it.[5]

USING A MIND MAP TO EXPLORE CLIMATE FEELINGS

Finnish people, like most people in industrialized countries, are generally bad at recognizing emotions. There are cultural reasons for this—emotional intelligence has been downplayed by petro-patriarchal structures of power for so long. There is wide alexithymia in a cultural sense: inability to name or even recognize emotions.[6] The international public discourse about ecological emotions is a new phenomenon; in Finland, that started to grow in Autumn 2017, and my own monograph about eco-anxiety also played a role in that shift.[7]

> Emotional intelligence has been downplayed by petro-patriarchal structures of power for so long that there is wide alexithymia in a cultural sense: inability to name or even recognize emotions.

Because of this, even relatively simple methods of emotional work, such as the Mind Map of Climate Emotions, can provide surprisingly powerful results, if the facilitator provides enough safety. Almost all people that I have used this activity with say that this was the first time that they were asked to name their climate emotions and to spend time with them. Public discussion about climate emotions makes them a socially accepted subject on a general level, although people naturally have different emotions and make different normative judgments about them. Naming difficult climate emotions can be especially important for those who feel rather isolated and lonely with their dark ecological emotions.[8] I provide a much wider discussion about eco-emotions and education in my academic articles,[9] but the following two figures and the table show many important emotions and feelings and a proposed process model of eco-anxiety and ecological grief.[10]

A THREE-STEP MODEL FOR EDUCATORS

I join the many scholars and thinkers who argue that educators have an ethical responsibility to help their students or other audiences constructively encounter difficult climate emotions.[11] I have constructed a three-step model to describe the tasks and possibilities at hand:

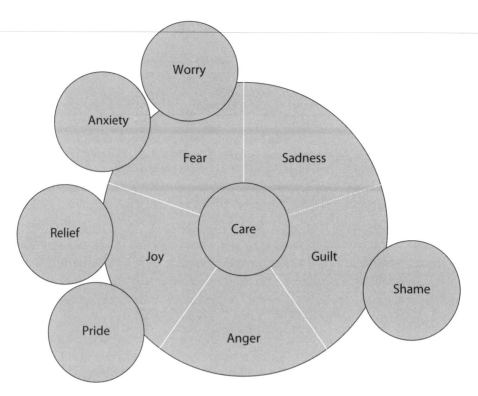

FIGURE 14.1 Examples of Key Emotions and Feelings in Relation to Ecological Issues.

1. Validating out loud the existence of difficult climate emotions
2. Providing opportunities to discuss climate emotions
3. Facilitating embodied activities to encounter climate emotions[12]

I regard Step 1 as something that all educators should do, regardless of their own position and skills. If and when the educator admits that the climate crisis is so severe that it sparks many kinds of emotions in people of all ages, this public recognition makes the subject valid and helps students find ways to approach it. It can also help in alleviating intergenerational tensions, because the educator, an adult, shows recognition of something that many young people feel. Since education always includes use of power, the educator in a position of power has a special responsibility to facilitate this validation and recognition.

> If and when the educator admits that the climate crisis is so severe that it sparks many kinds of emotions in people of all ages, this public recognition makes the subject valid and helps students to find ways to approach it.

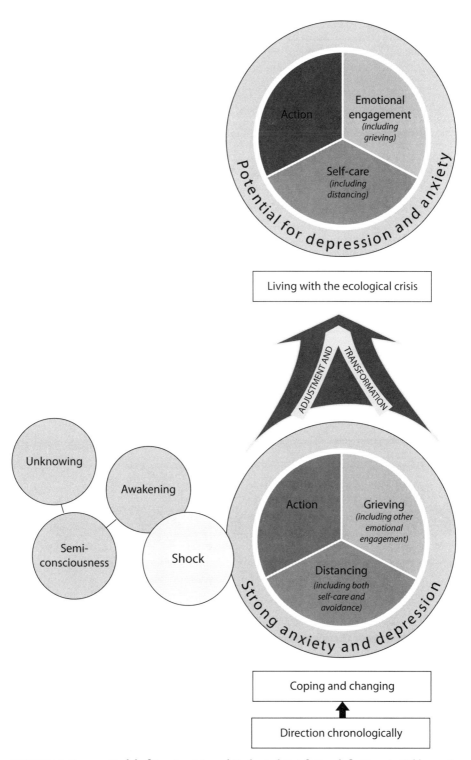

FIGURE 14.2 Process Model of Eco-Anxiety and Ecological Grief. (See definitions in Table 14.1.)

TABLE 14.1 Definition of Terms and Emotions Included in the Process Model (see Figure 14.2)

Semiconsciousness	A complex phase of both knowing and not knowing about the severity of the ecological crisis	**Action**	Doing something constructive in relation to the ecological crisis. Example: joining climate demonstrations
Awakening	A realization of the severity of the crisis. May happen either suddenly or as a result of a longer process	**Grieving**	Encountering explicitly the various loss-related emotions engendered by the ecological crisis
Shock	People become shaken and sometimes shocked or even traumatized	**Distancing**	Various means of taking distance from the ecological crisis, including both self-care and avoidance
Coping and Changing	For healthy adjustment and necessary transformation, all three dimensions are needed	**Strong Anxiety and/or Depression**	Various possible manifestations of strong and difficult mental states which are significantly impacted by the ecological crisis

Adjustment and Transformation refers to a long-term process where all three dimensions—action, grieving, distancing—are engaged with and people find ways to both adapt and change.

Living with the Ecological Crisis: When enough Adjustment and Transformation has happened, people enter this phase, where there is more awareness and control about the three dimensions.

The titles and subtitles of two dimensions have switched places, reflecting changes due to Adjustment and Transformation. While Grieving still continues, major aspects of it have been engaged with, and there is more room for encountering other emotions. Distancing has become more conscious and more balanced, and thus Self-care is the ruling form, although there may still be moments of stronger and more unconscious disavowal.

Potential for depression and stronger anxiety still exists, and it is possible that in the long run, various depressive moods are a major threat as many aspects of the ecological crisis become more difficult in the coming decades.

If the educator is motivated and engages in self-reflection, they can develop skills to help students even further by providing opportunities for discussing climate emotions. Most uses of the Mind Map exercise fall into this category. Self-reflection is needed beforehand, so that the educator is not (too) surprised by what kinds of feelings the responses of others may engender or surface in them.

A third possible step would be to apply skills in facilitating embodied methods of exploring emotions. There is a wide array of possible methods here, ranging from walking outside to participatory dance. These kinds of methods require a bit more skill, but many educators have already gathered such skills and used them in relation to other subject matters.

BODILY MOVEMENT AND DISCUSSIONS WITH THE SPECTRUM LINE

The Spectrum Line of Climate Emotions exercise is an example of Step 3 methodology (additional detail for this activity is available online at the Existential Toolkit for Climate Justice Educators website). It is based on simple movement along an invisible spectrum line, one emotion word at a time. Pair discussions are followed by short group discussions. Bodily movement and the removal of rigid positional structures (such as sitting on chairs in rows and lines) enable much more feeling to emerge. At the same time, the facilitator can keep the situation safe by setting wise rules and a compassionate atmosphere. Pedagogically, the fluctuation between pair discussions and group discussions brings many benefits.

People are able to learn much both from others and from their own responses in such settings. If all goes well, ordinary psychological defenses may be removed and people connect with deep feelings. At the same time, there is a difference between the Spectrum Line and the more intensive emotional methodologies developed for example by Joanna Macy and studied by Jo Hamilton.[13] The Spectrum Line activity offers participants possibilities to choose the level of intimacy that they wish to engage with.

Many justice issues have come up during the Spectrum Line sessions. People connect with their climate grief and rage: specifically, the moral anger evoked by all the injustices of petro-patriarchal economic and political orders. Participants also wrestle with feelings of guilt and inadequacy, sometimes also with species shame. If time and the context allow, I have included discussion and movement about feelings of pride, which are very tricky also for many activists. It is very unjust if people who really try to make an impact still feel inadequate, as the quote in the beginning of this chapter showed. Emotional methods can help people practice pleasure activism[14] and feel self-efficacy and group efficacy.

CRITICAL OBSERVATIONS

While emotional methodologies like the ones introduced in this chapter can profoundly help both justice causes and psychological resilience, the use of these methods includes many challenges. The context and the role of the facilitator always need critical evaluation.

In many contexts, it has to be taken seriously into account that people may have profound traumas which affect their ability for emotional work. The methods discussed here are relatively easier to use in places like Finland, where the living conditions are relatively peaceful. When working with students who have suffered from many injustices, there is a need for extra carefulness. However, even in privileged contexts there will be participants who have traumas and vulnerabilities.[15]

The role of the facilitator is a tricky one. On one hand, the facilitator needs to be prepared enough, especially when using more intensive methods such as the Spectrum Line. Even with the Mind Map, one has to make sure that one's own opinions and biases do not shape the session in an unconstructive manner. Yet on the other hand, there is great need for methods like this, and one cannot ever be perfect, either. It is a fine line between too little skill and a problematic withdrawal. At best, these kinds of methods could be practiced either with trusted peers or even as part of formal education for educators.

A word of caution is needed also in relation to structural change. While it is true that there have been many advances in Finland regarding emotional methods in climate education, it is also true that in many ways the work has just begun. In a relatively large educational project called "Toivoa ja toimintaa" (Hope and Action), teachers and students have been trained with the Mind Map, with generally positive results.[16] But those participating have been volunteers. When taken into average school settings, many teachers have prejudices about climate emotions and emotional methods. For example, although there is a wide discussion about climate anxiety and its variations in Finland, many teachers say they wonder where the students with climate anxiety are, because they have never noticed one. In other words, they are unable to observe the emotions of students, because of cultural alexithymia and psychological defenses; yet surveys and studies show that a great number of Finnish students self-identify with some kind of climate anxiety.[17]

> Engaging with dark emotions can move participants toward a space where counterpart emotions are also present: a space of meaningfulness, joy and connection through shared work for the planet, courage in the face of finitude, and healthy pride for making an effort.

In 2021, I took part in conducting a global research project that showed that 16- to 25-year-olds worldwide experience many difficult climate emotions.[18] Of the 10,000 respondents, 56 percent thought that "humanity is doomed." The Pew Research Center's 2021 international survey also showed great climate concern, and after the most recent IPCC report, the media was full of concerned comments about the climate crisis. The existential pressure of the crisis is already very high and will probably continue to grow. A further step in working with climate emotions is to deepen the encounter with even more fundamental and existential feelings; I have observed that varieties of existential anxiety are interwoven with forms of climate anxiety and dread.[19] Climate justice educators need further skills in engaging with existential issues such as feelings of meaninglessness, deliberations related to freedom and responsibility, fear of death, and guilt. Engaging with dark emotions can move participants toward a space where counterpart emotions are also present: a space of meaningfulness, joy and connection through shared work for the planet, courage in the face of finitude, and healthy pride for making an effort. The full scale of emotions accompanies the shared mission for climate justice.

NOTES

1. Lertzman, *Environmental Melancholia*; Pihkala, "Environmental Education After Sustainability"; Ray, "Coming of Age at the End of the World."
2. Pekkarinen and Tuukkanen, *Maapallon tulevaisuus ja lasten oikeudet*.
3. Pihkala, "Eco-Anxiety and Environmental Education."
4. Jaakkola et al., "Holistic Effects of Climate Change."
5. Pihkala, "Climate Grief"; Tschakert et al., "One Thousand Ways to Experience Loss."
6. Greenspan, *Healing Through the Dark Emotions*; Lomas, *Positive Power of Negative Emotions*.
7. Pihkala, *Päin helvettiä?*; Santaoja, "Ilmastoahdistuksesta toimintaan."
8. Kretz, "Emotional Solidarity."

9. Pihkala, "Environmental Education After Sustainability"; Pihkala, "Eco-anxiety, Tragedy, and Hope; Pihkala, "Anxiety and the Ecological Crisis"; Pihkala, "Cost of Bearing Witness."

10. Pihkala, "Toward A Taxonomy of Climate Emotions"; Pihkala, "Process of Eco-Anxiety."

11. Ray, *Field Guide to Climate Anxiety*; Ojala, "Facing Anxiety."

12. Pihkala, "Eco-Anxiety and Environmental Education."

13. Hamilton, "Emotional Methodologies for Climate Change Engagement."

14. brown, *Pleasure Activism*; Ray, *Field Guide to Climate Anxiety*.

15. Pihkala, "Cost of Bearing Witness."

16. See the supplement in Pihkala, "Eco-Anxiety and Environmental Education."

17. Hyry, *Kansalaiskysely ilmastonmuutoksesta ja tunteista*.

18. Hickman et al., "Climate Anxiety in Children."

19. Pihkala, "Eco-anxiety, Tragedy, and Hope"; Budziszewska and Jonsson, "From Climate Anxiety to Climate Action."

REFERENCES

brown, adrienne maree. *Pleasure Activism: The Politics of Feeling Good*. Chico, CA: AK Press, 2019.

Budziszewska, Magdalena, and Sofia Elisabet Jonsson. "From Climate Anxiety to Climate Action: An Existential Perspective on Climate Change Concerns within Psychotherapy." *Journal of Humanistic Psychology* (2021). Ahead of print. https://doi.org/10.1177/00221678211993243.

"An Existential Toolkit for Climate Justice Educators." https://www.existentialtoolkit.com.

Greenspan, Miriam. *Healing through the Dark Emotions: The Wisdom of Grief, Fear, and Despair*. Boulder, CO: Shambhala, 2004.

Hamilton, Jo. "Emotional Methodologies for Climate Change Engagement: Towards an Understanding of Emotion in Civil Society Organisation (CSO)-Public Engagements in the UK." PhD dissertation, University of Reading, 2020.

Hickman, Caroline, Elizabeth Marks, Panu Pihkala, Susan Clayton, R. Eric Lewandowski, Elouise E. Mayall, Britt Wray, Catriona Mellor, and Lise van Susteren. "Climate Anxiety in Children and Young People and Their Beliefs about Government Responses to Climate Change: A Global Survey." *The Lancet Planetary Health* 5, no. 12 (2021): e863–73. https://doi.org/10.1016/S2542-5196(21)00278-3.

Hyry, Jaakko. *Kansalaiskysely ilmastonmuutoksesta ja tunteista* [National Survey on Climate Change and Emotions]. Helsinki: Sitra, the Finnish Innovation Fund, 2019.

Jaakkola, Jouni J. K., Suvi Juntunen, and Klemetti Näkkäläjärvi. "The Holistic Effects of Climate Change on the Culture, Well-Being, and Health of the Saami, the Only Indigenous People in the European Union." *Current Environmental Health Reports* 5, no. 4 (2018): 401–17. https://doi.org/10.1007/s40572-018-0211-2.

Kretz, Lisa. "Emotional Solidarity: Ecological Emotional Outlaws Mourning Environmental Loss and Empowering Positive Change." In *Mourning Nature: Hope at the Heart of Ecological Loss and Grief*, edited by Ashlee Cunsolo Willox and Karen Landman, 258–91. Montreal: McGill-Queen's University Press, 2017.

Lertzman, Renée Aron. *Environmental Melancholia: Psychoanalytic Dimensions of Engagement*. New York: Routledge, 2015.

Lomas, Tim. *The Positive Power of Negative Emotions: How Harnessing Your Darker Feelings Can Help You See a Brighter Dawn*. London: Piatkus, 2016.

Ojala, Maria. "Facing Anxiety in Climate Change Education: From Therapeutic Practice to Hopeful Transgressive Learning." *Canadian Journal of Environmental Education* 21 (2016): 41–56.

Pekkarinen, Elina, and Terhi Tuukkanen, eds. *Maapallon tulevaisuus ja lasten oikeudet*. Lapsiasiavaltuutetun toimiston julkaisuja 2020:4. [Helsinki]: Lapsiasiavaltuutetun toimisto, 2020.

Pew Research Center. "In Response to Climate Change, Citizens in Advanced Economies Are Willing to Alter How They Live and Work." 2021. https://www.pewresearch.org/global/2021/09/14/in-response-to-climate-change-citizens-in-advanced-economies-are-willing-to-alter-how-they-live-and-work.

Pihkala, Panu. "Toward a Taxonomy of Climate Emotions." *Frontiers in Climate* 3 (2022). https://www.frontiersin.org/article/10.3389/fclim.2021.738154.

———. "The Process of Eco-Anxiety: A Narrative Review and a New Proposal." *Sustainability* 14, no. 24 (2022). https://doi.org/10.3390/su142416628.

———. "Anxiety and the Ecological Crisis: An Analysis of Eco-Anxiety and Climate Anxiety." *Sustainability* 12, no. 19 (2020). https://doi.org/10.3390/su12197836.

———. "Climate Grief: How We Mourn a Changing Planet." BBC website, Climate Emotions Series, 2020.

———. "The Cost of Bearing Witness to the Environmental Crisis: Vicarious Traumatization and Dealing with Secondary Traumatic Stress among Environmental Researchers." *Social Epistemology* 34, no. 1 (2020): 86–100. https://doi.org/10.1080/02691728.2019.1681560.

———. "Eco-Anxiety and Environmental Education." *Sustainability* 12, no. 23 (2020). https://doi.org/10.3390/su122310149.

———. "Eco-Anxiety, Tragedy, and Hope: Psychological and Spiritual Dimensions of Climate Change." *Zygon* 53, no. 2 (2018): 545–69.

———. "Environmental Education After Sustainability: Hope in the Midst of Tragedy." *Global Discourse* 7, no. 1 (2017): 109–27.

———. *Päin helvettiä? Ympäristöahdistus ja toivo* [Eco-Anxiety and Hope]. Helsinki: Kirjapaja, 2017.

Ray, Sarah Jaquette. *A Field Guide to Climate Anxiety: How to Keep Your Cool on a Warming Planet*. Oakland: University of California Press, 2020.

———. "Coming of Age at the End of the World: The Affective Arc of Undergraduate Environmental Studies Curricula." In *Affective Ecocriticism: Emotion, Embodiment, Environment*, edited by Kyle A. Bladow and Jennifer Ladino, 299–319. Lincoln: University of Nebraska Press, 2018.

Santaoja, Minna. "Ilmastoahdistuksesta toimintaan—Mahdollisten tulevaisuuksien jäljillä." *Alue ja ympäristö* 47, no. 1 (2018): 112–16. https://doi.org/10.30663/ay.69980.

Tschakert, P., N. R. Ellis, C. Anderson, A. Kelly, and J. Obeng. "One Thousand Ways to Experience Loss: A Systematic Analysis of Climate-Related Intangible Harm from around the World." *Global Environmental Change* 55 (2019): 58–72. https://doi.org/10.1016/j.gloenvcha.2018.11.006.

FIFTEEN

Building Somatic Awareness to Respond to Climate-Related Trauma

EMILY (EM) WRIGHT

This chapter introduces somatics, or body-centered practices, as an approach to healing the stress and trauma related to climate change as it manifests in the body. Through a reading, activities, and reflection prompts, the chapter supports learners to become more aware of personal stress and learn somatic strategies to help find regulation and balance.

Climate change stresses every body, but not all bodies are identical. As the body's first responder to stress, the nervous system automatically engages to keep us safe from threats before we fully process what is happening. Like icebergs shaped by water and air, each nervous system is shaped by numerous forces over time and space, including genetics, culture, environment, systemic conditions, oppression, and trauma.[1] As a result, different people can have very different responses to the same stress.

In response to climate change, hope, despair, anger, grief, and anxiety are among the spectrum of emotions people experience. This diversity of responses may be partially explained by differences in nervous systems across individuals and communities. Another contributing factor is the unequal and inequitable distribution of climate impacts, risks, vulnerabilities, and resilience resources across geographies, demographics, and abilities,[2] which in turn, shape nervous systems.

Trauma occurs when a threat to a person's safety is "too much, too soon, or too fast" for them to successfully overcome.[3] Their attempted safety response is thwarted, creating a pattern in their nervous system that they are more likely to repeat in response to future threats of any kind. Trauma can occur from a single event, a set of conditions, or recurring experiences. Furthermore, research has shown that trauma can be passed from one generation to the next.[4] Climate-related trauma can include:

Primary trauma from directly experiencing severe weather or disaster (e.g., wildfires), living with daily stressors from this existential threat (e.g., higher cooling bills), and *pre-trauma* from fear of future catastrophe.[5]

Secondary trauma from witnessing and participating in systems, willingly or otherwise, that harm ecosystems and other species.

Historical trauma from centuries of humans' separation from and power over nature, dating back to the rise of Judeo-Christian and Cartesian belief systems, and burgeoning through colonization, capitalism, and land privatization.[6]

Like other types of trauma in our society, climate trauma is unequally distributed. Black and Indigenous communities and other communities of color have experienced disproportionate impacts from climate change and continue to do so, including displacement, loss of culturally important lands and species, financial setbacks, and physical and mental health impacts.[7] Moreover, trauma and stress from climate change layer onto existing traumas from living within systems of oppression, adverse childhood and adult experiences, and intergenerational trauma.[8]

STRESS AND SAFETY RESPONSES

One step toward healing climate trauma is to learn about one's own stress responses. To that end, Figure 15.1 presents a conceptual map of climate stress responses to guide self-reflection. The map is not a diagnostic tool, but rather a framework to support meaning-making and future research. The map is not comprehensive of all experiences.

All stress responses are valid—they are neurobiological tools to re-secure safety, dignity, and belonging. A climate trauma framework invites individuals into compassionate curiosity about their own responses and validates and builds empathy with other people's diverse responses.

> A climate trauma framework invites individuals into compassionate curiosity about their own responses and validates and builds empathy with other people's diverse responses.

INCREASING OPTIONS

When trauma has occurred, responses can feel outsized for the situation, automatic without choice, or inhibiting. Healing climate trauma involves increasing options for responding to stress. One approach with demonstrated success in healing trauma is the use of body-centered practices in the field of somatics.[9] *Somatics* describes numerous practices based in the holistic mind-body-spirit connection that increase awareness of physical sensations in the body as signals of emotions, needs, boundaries, and other embodied information. With cultural lineages around the world, diverse groups have practiced somatics for generations.

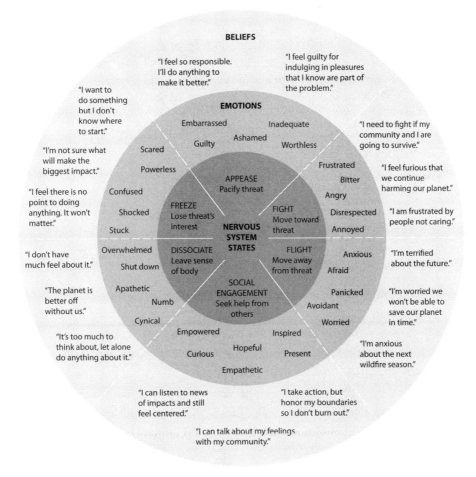

FIGURE 15.1 Climate Stress Responses Map.

Today, somatics is increasingly used to support healing and liberation for historically oppressed and marginalized groups.[10]

Somatic practices can help us understand the neurobiological mechanisms underlying our emotional and embodied experiences of climate disruption. They can heal traumas and expand our capacity to "read" our physical sensations, return to a sense of regulation, and mindfully respond to climate stress in a way that affirms our values, vision, purpose, and well-being.

> Somatic practices can help us understand the neurobiological mechanisms underlying our emotional and embodied experiences of climate disruption.

Building Somatic Awareness   135

ACTIVITIES

The following activities are offered to help participants: (1) cultivate awareness of physical sensations and emotions, (2) begin learning their responses to climate-related stress, and (3) take small steps to support their well-being and expand their choices.[11]

Before You Begin

These activities may bring up strong emotions and sensations, especially if participants have experienced trauma. Give participants sufficient notice of these potential outcomes and permission to adapt as needed or to opt out. Instruct participants to stop doing the activity if it feels too intense. When strong emotions do arise, invite and validate them, suggest the self-support techniques in Activity #2, and direct participants to appropriate resources at your institution for therapeutic support.

Ways to Support Participants

- Do the activities in a place where participants feel safe and supported. Depending on the group size, composition, and physical setting, these activities may be best suited to do individually at home, together as a group, or both.
- Be clear that participants can change their consent at any time during the activities and provide options for doing so.
- Give ample time after the activity for participants to process in the way they prefer (e.g., be alone, go outside, move their body, discuss with a trusted person).

For Each Activity

- If in a group setting, the educator can read the instructions aloud. If completed alone, the learner should read instructions for the entire activity before beginning.
- Reflection prompts can be adapted for a journaling or essay assignment.
- If in a group setting, invite participants to share in pairs about their experience by taking turns listening without interrupting, analyzing, or responding, but simply witnessing with curiosity and care.

INSTRUCTIONS FOR ACTIVITY #1—BODY MAPPING

In this practice, you will observe and document physical sensations and emotions. This can help increase awareness of the body and build capacity to identify emotions. This can be a daily at-home practice and/or a group activity before, during, and/or after a class as a way of checking in.

Materials

Comfortable chair or space to sit or lay down

Paper

Writing or drawing utensils

Instructions

- Draw an outline of a body (e.g., gingerbread person) on a piece of paper. Don't worry about being realistic; a simple outline will do.
- Sit comfortably or lay down. You're welcome to close your eyes. Take three to five deep breaths in and exhale slowly.
- Continue breathing and slowly scan your body from your head to toes, and notice what you feel. Notice temperature (e.g., heat, coolness), pressure (e.g., tightness, slackness), movement (e.g., pulsing, tingling), and any lack of sensation.
- If any place catches your attention, stay there for a few breaths. Notice any emotions present. Notice any stories, colors, or images that come to mind.
- Continue scanning until you've reached your toes. Then take another three to five deep breaths. When you're ready, open your eyes.
- On your gingerbread person, draw what you felt in your body—colors, shapes, words, etc. Don't worry about how it looks or being complete—just capture what you remember.

Reflection Questions

- What did you notice?
- What stood out to you about your experience?
- What might this tell you about your relationship to your body?

INSTRUCTIONS FOR ACTIVITY #2—LEARNING YOUR PATTERNS

In this practice, you will be invited to experience minor activation to learn your climate stress responses. Only engage in this practice if you feel relatively grounded and well-supported, and when you have time afterwards to let any emotions or activation settle.

Materials

Comfortable chair or space to sit or lay down

Instructions

- Sit comfortably, lay down, or stand. You're welcome to close your eyes. Take three to five deep breaths in and exhale slowly.
- Focus on the places where your body contacts the chair or floor. Feel gravity pull you toward earth. Keep breathing deeply.
- Call to mind a memory or image related to climate change that is *only mildly* activating—not intense or traumatizing, but that stirs something within you. Bring that memory or image to mind, noticing the sounds, sights, smells, sensations.
- Turn your attention to your body. Scan your body and notice sensations you feel (temperature, pressure, movement).
- If you feel stronger sensations in a particular place, pause for a few breaths and focus there. You can place one or both hands on that spot. Notice any mood or emotion.
- Release the memory. Take three to five deep breaths in and sigh out.
- Consider using self-support techniques:
    + Go or look outside. Take in the colors, textures, light, and shadows.
    + Place hands on your chest, belly, and/or forehead. Breathe and feel the support from your hands.
    + Imagine a place, person, or pet that helps you feel safe. Absorb the feeling of that safety.
    + Move your body by:
        * Rocking back and forth
        * With your hips and legs stable, twisting your torso from right to left with your arms swinging freely
        * Humming to feel the vibration in your throat and chest

Reflection Questions

- What emotions did you feel? Is there a story or belief underneath them? Do any beliefs or emotions contradict each other?
- Does your response seem familiar? Are there other experiences in your past when you remember responding similarly?
- Can you find the emotion(s) and belief(s) you experienced on the Climate Stress Responses map? If so, draw lines to connect them to the nervous system states.
- What does this experience tell you about how climate change impacts you?
- How might you practice self-support when you feel climate stress?

OUTCOMES

Participants who have experienced these somatic practices have expressed how they are simple, accessible tools for them to come back to a sense of grounding in the present moment and open up to more choices and alternative ways of being that they could not imagine or did not consider possible before. At the very least, these practices can assist students in expanding their range of resilience, or ability to respond to difficult content with greater skill. The vast majority of us (and our students) hold some degree of trauma within us, so having at least some trauma-informed pedagogical tools in our toolkits makes us more effective—and arguably more ethical—educators, for ourselves and for our students. At best, integrating some awareness of the body can even become a daily practice of resistance to the disconnecting tendencies between mind and body present in academia and higher education, which are increasingly understood as contributing to the climate crisis in the first place.

NOTES

1. Van der Kolk, *Body Keeps the Score.*
2. Reidmiller et al., *Impacts, Risks, and Adaptation.*
3. Menakem, *My Grandmother's Hands,* 14.
4. DeAngelis, "Legacy of Trauma."
5. Richardson, "Affects of the Catastrophe to Come"; Woodbury, "New Taxonomy of Trauma"; Kaplan, *Climate Trauma.*
6. Alberro, "Humanity and Nature;" Schultz, "Inclusion with Nature"; Vining et al., "Distinction between Humans and Nature."
7. Howell and Elliott, "Damages Done"; Jesdale et al., "Racial/Ethnic Distribution."
8. Felitti et al., "Relationship of Childhood Abuse and Household Dysfunction."
9. Levine, *Waking the Tiger*; Brom et al., "Somatic Experiencing"; Haines, *Politics of Trauma.*
10. Menakem, *My Grandmother's Hands*; Haines, *Politics of Trauma.* For examples, see the work of the organizations Holistic Resistance and Black Organizing for Leadership and Dignity.
11. These activities drew inspiration from the work of Resmaa Menakem as well as generative somatics, which supports somatic transformation in social and climate justice movements.

REFERENCES

Alberro, Heather. "Humanity and Nature Are Not Separate." *The Conversation,* September 17, 2019.

Brom, Danny, Yaffa Stokar, Cathy Lawi, Vered Nuriel-Porat, Yuval Ziv, Karen Lerner, and Gina Ross. "Somatic Experiencing for Posttraumatic Stress Disorder: A Randomized Controlled Outcome Study." *Journal of Traumatic Stress* 30, no. 3 (2017): 304–12.

DeAngelis, Toni. "The Legacy of Trauma." *Monitor on Psychology* 50, no. 2 (2019).

Felitti, Vincent, Robert F. Anda, Dale Nordenberg, David F. Williamson, Alison M. Spitz, Valerie Edwards, Mary P. Koss, and James S. Marks. "Relationship of Childhood Abuse and Household Dysfunction to Many of the Leading Causes of Death in Adults." *American Journal of Preventive Medicine* 14, no. 4 (1998): 245–58.

Haines, Staci. *The Politics of Trauma: Somatics, Healing, and Social Justice.* Berkeley: North Atlantic Books, 2019.

Howell, Junia, and James Elliott. "Damages Done: The Longitudinal Impacts of Natural Hazards on Wealth Inequality in the United States." *Social Problems* 66, no. 3 (2019): 448–67.

Jesdale, Bill, Rachel Morello-Frosch, and Lara Cushing. "The Racial/Ethnic Distribution of Heat Risk–Related Land Cover in Relation to Residential Segregation." *Environmental Health Perspectives* 121, no. 7 (2013): 811–17.

Kaplan, Ann. *Climate Trauma: Foreseeing the Future in Dystopian Film and Fiction*. New Brunswick: Rutgers University Press, 2016.

Levine, Peter A. *Waking the Tiger: Healing Trauma*. Berkeley: North Atlantic Books, 1997.

Menakem, Resmaa. *My Grandmother's Hands: Racialized Trauma and the Pathway to Mending Our Hearts and Bodies*. Las Vegas: Central Recovery Press, 2017.

Reidmiller, David, Christopher W. Avery, David R. Easterling, Kenneth E. Kunkel, Kristin Lewis, Thomas K. Maycock, and Brooke C. Stewart, eds. *Impacts, Risks, and Adaptation in the United States: Fourth National Climate Assessment, Volume II*. Washington, DC: U.S. Global Change Research Program, 2018.

Richardson, Micheal. "Climate Trauma, or the Affects of the Catastrophe to Come." *Environmental Humanities* 10, no. 1 (2018): 1–19.

Schultz, P. Wesley. "Inclusion with Nature: The Psychology Of Human-Nature Relations." In *Psychology of Sustainable Development*, edited by Peter Schmuck and Wesley P. Schultz, 61–78. Boston: Springer, 2002.

Van der Kolk, Bessel. *The Body Keeps the Score*. New York: Penguin Books, 2014.

Vining, Joanne, Melinda Merrick, and Emily Price. "The Distinction between Humans and Nature." *Human Ecology Review* 15, no. 1 (2008): 1–11.

Woodbury, Zhiwa. "Climate Trauma: Toward a New Taxonomy of Trauma." *Ecopsychology* 11, no. 1 (2019): 1–8.

SIXTEEN

Using Poetry to Resist Alienation in the Climate Change Classroom

MAGDALENA MĄCZYŃSKA

Poetry is not a luxury. It is a vital necessity of our existence. It forms the quality of the light within which we predicate our hopes and dreams toward survival and change, first made into language, then into idea, then into more tangible action.

    AUDRE LORDE

Poetry is a call to action, and it also is action.

    JUAN FELIPE HERRERA

This chapter offers a customizable lesson plan for incorporating poetry into climate education across disciplines, focusing on "Ode to Dirt" by Sharon Olds.

Although societal awareness of the climate crisis continues to grow, climate change is still frequently framed as happening at a remove: not *now* but *later*; not *here* but *out there*—in the Arctic wilderness or, conversely, the slums and war zones of "disaster porn." For college students, this spatial and temporal distancing may be experienced as emotional alienation, compounded by academia's hyper-intellectualized, hyper-specialized spaces. How can climate educators resist these intersecting alienations, and encourage students to inhabit their physical and emotional *here and now*? I propose that poetry provides a pathway to overcoming alienation in climate change courses (or in any field of study) by helping students and professors build community, make space for feelings, and develop habits of attention and connection. Too often treated as a frivolous add-on disconnected from on-the-ground struggles, poetry, as Audre Lorde and Juan Felipe Herrera remind us, is not a "luxury" but an inspiration for transformative action—and a form of action in its own right.

    Poems are a versatile pedagogical resource. Their most straightforward application in the climate change classroom is to provide an engaging, intimate introduction to course themes: inter-species kinship in "Characteristics of Life" by Camille T. Dungy; species extinction in "The Big Picture" by Ellen Bass; food sustainability and ecosystem health in "Notes from a Climate Victory Garden" by Louise Maher-Johnson; the human cost of intensified storms in "Man on the TV Say" by Patricia Smith; alternative models of leadership in "Calling All Grand Mothers" by Alice Walker; or "For Those Who Would Govern" by Joy Harjo.[1] And so on. But poems can do much more than introduce or supplement content. Unburdened by the pragmatic rules of everyday language usage, poetry invites readers to make unexpected connections between the sensory and the intellectual, the local and the cosmic, the human and the larger-than-human. This freedom to traverse category boundaries opens unexpected venues of conversation, inviting students to connect with course material in personal, idiosyncratic ways. Poetry adds a much-needed affective dimension to academic discussions of climate-related subjects. Sharing a poem at the beginning of a thematic segment signals acknowledgement of learners' emotional responses to course material and establishes such responses as a legitimate subject of classroom discussion. Finally, working with poetry can help students and professors build community through the practices of deep listening, emotional expression, and creative play.

Community and affective engagement are—as great educators have always understood, and the neuroscience of learning confirms—fundamental to successful learning.[2] They are especially important for addressing the educational and social alienation of students who have been historically excluded from academic spaces (and the spaces of eco-activism). This is not to say that incorporating poems into lesson plans will automatically create an affect-conscious, just, and equitable learning environment for every student. Like any aspect of schooling, a poetry lesson can become a tool of alienation. But it can, when used thoughtfully, become one step in a larger pedagogical commitment to fostering connection and honoring emotion beyond academia's conventional rigidities and exclusions.

I chose Sharon Olds's "Ode to Dirt" as my centerpiece because this poem addresses the problem of alienation directly by calling out human failure to perceive the networks of interdependencies that sustain us. In extolling humble "dirt," Olds invites us to overcome habitual inattentiveness and to honor our material kinship with the larger-than-human world. Thus, "Ode to Dirt" pushes against what environmental philosopher Val Plumwood calls *backgrounding*—systemic inattention to "inferiorized" elements within culturally established hierarchies.[3] Plumwood names *backgrounding* as one of many strategies through which relationships of dominance are maintained in Eurocentric colonial cultures; other such strategies include the *hyper-separation* of fluid categories into stark binary opposites (culture/nature, male/female, white/Black) and the *instrumentalizing* reduction of one side of the binary to the role of exploitable resource.[4] Plumwood's schema is especially useful for conversations about climate change because it links the backgrounding of human groups with the backgrounding of non-human life, integrating considerations of environmental stewardship with questions of decolonization and social justice.

Below, I offer a series of exercises and prompts centered around "Ode to Dirt." A recording of the poem read by the poet is available on YouTube via *The On Being Project* channel; the full *On Being* episode featuring Sharon Olds in conversation with Krista Tippett can be found in the podcast archive (at https://onbeing.org). What follows is not a fixed lesson plan (although it might be used as such), but rather a collection of modular activities that can be combined into longer segments to suit the needs and goals of each instructor and class. Most activities can be adapted, with some modification, to other poems.

INTRODUCING THE POEM

If time allows, you can start by sharing the poem's title, and then ask students for guesses/predictions/free-writes about its content and connection to your class. This should lay the ground for a more engaged conversation later in the class. For a real-time, emotionally resonant experience of the poem, make sure it is read out loud (you can read yourself, use a recording, or ask for volunteers). I like inviting students to start and stop reading whenever they wish, without raising hands or waiting for permission to proceed: one person starts from the top and goes on as long as they choose; when they stop, someone else picks up where the previous reader left off. This method requires tolerance for long pauses but yields a nicely idiosyncratic experience of the text. You can follow up by inviting students to "freestyle" by reading aloud any verses/phrases they choose, in any order, to create a new version of the poem particular to their group. (In very large classrooms, students can do this simultaneously, for a cacophony of verse!) I suggest reading the poem at least twice, ideally with different techniques/speakers, to allow the words and music to sink in.

INITIAL RESPONSES

A fun low-stakes activity that allows students to map the poem's thematic and affective patterns is making an "erasure poem" out of the original text. This is accomplished by deleting selected words/verses/stanzas with a dark marker or white-out to reveal an abridged, authorial version of the text. As each student's "erasure poem" will be different, the pieces serve as great starting points for small-group or all-class discussions, reflections, and in-depth responses.

FOCUSED RESPONSES

Once students are ready to explore the poem in a more systematic way, they can map out thematic patterns relevant to the class: food systems and sustainability; continuities/kinships between humans and non-human lives; the epistemological problem of inattention; and so on. This activity also lends itself to group work, with each group exploring one theme. Invite students to add their own prior knowledge (personal and academic) to these mappings, so that they may begin building bridges between the poem and their lived experiences.

CREATIVE MANIPULATION

Students can use the poem as a starting point for their own creative exploration. For example, they can compose "odes" to undervalued (human or non-human) entities identified within their own material and social environments. What kinships are we ignoring? Whose labor remains invisible? The creative pieces don't have to be poems. Letters, diary entries, Twitter threads, collages, and internet memes are all great options (it's best to give students the freedom to choose their mode and medium). As a follow-up, you can ask the class to brainstorm ideas for how to overcome the various kinds of inattentiveness/alienation identified in the previous activity. This discussion might lead to further creative projects focused on awareness-raising and communication.

REFLECTION

Although reflection comes last on this list of activities, it is anything but an afterthought. Time permitting, reflection should be integrated as often as possible, to allow space for students to process their responses, make connections to prior experiences, build knowledge networks, and develop metacognition (reflection about *how*, rather than just *what*, they are learning). Lack of metacognitive skills is one of the most important barriers faced by students with limited access to academic capital. Consequently, metacognitive reflection (which demystifies the processes of learning and knowledge production) is a powerful tool for advancing educational justice.[5] Invite students to reflect on their responses, gut reactions, resistances, and thought processes. Allow them to consider the knowledge they had brought into your classroom and to reflect on how these knowledges were acquired. If you can, make time for values-affirmations: prompts that invite students to identify their values and to think about how course material might intersect with or support those values.[6] Incorporating reflection throughout your course will allow students to develop a deeper relationship with themselves as learners and human beings—and thus lessen their alienation and deepen their connection to the course content.

Poetry unlocks new ways of perceiving the familiar. It opens oblique entries into complex subjects. It invites the integration of thought and feeling. It nourishes attention and vulnerability. College students often see poetry as intimidating or irrelevant—but helping them experience poems as living documents applicable to contemporary lives can yield surprising moments of dialogue, openness, and creativity. Making room for poems like "Ode to Dirt" (especially in academic fields that don't typically accommodate emotion or play) is one way to provide an affectively engaged, community-centered learning experience that counters the multiple alienations experienced by students in a climate change classroom.

Ode to Dirt

Dear dirt, I am sorry I slighted you,
I thought that you were only the background
for the leading characters—the plants
and animals and human animals.
It's as if I had loved only the stars
and not the sky which gave them space
in which to shine. Subtle, various,
sensitive, you are the skin of our terrain,
you're our democracy. When I understood
I had never honored you as a living
equal, I was ashamed of myself,
as if I had not recognized
a character who looked so different from me,
but now I can see us all, made of the
same basic materials—
cousins of that first exploding from nothing—
in our intricate equation together. O dirt,
help us find ways to serve your life,
you who have brought us forth, and fed us,
and who at the end will take us in
and rotate with us, and wobble, and orbit.

"Ode to Dirt," p. 96, from *Odes* by Sharon Olds, compilation copyright © 2016 by Sharon Olds. Used by permission of Alfred A. Knopf, an imprint of the Knopf Doubleday Publishing Group, a division of Penguin Random House LLC. All rights reserved.

NOTES

1. Johnson and Wilkinson, eds., *All We Can Save*.
2. Cavanagh, *Spark of Learning*.
3. Plumwood, "Decolonizing Relationships with Nature," 56.
4. Ibid., 54–59.
5. McGuire, *Teach Students How to Learn*.
6. Lang, *Distracted*, 103–6.

REFERENCES

Cavanagh, Sarah Rose. *The Spark of Learning. Energizing the College Classroom with the Science of Emotion*. Morgantown: West Virginia University Press, 2016.
Johnson, Ayana Elizabeth, and Katharine K. Wilkinson, eds. *All We Can Save: Truth, Courage, and Solutions for the Climate Crisis*. New York: One World, 2020.
Lang, James M. *Distracted: Why Students Can't Focus and What You Can Do about It*. New York: Basic Books, 2020.

McGuire, Saudra Yancy. *Teach Students How to Learn: Strategies You Can Incorporate into Any Course to Improve Student Metacognition, Study Skills, and Motivation.* Sterling, VA: Stylus Publishing, 2015.

Plumwood, Val. "Decolonizing Relationships with Nature." In *Decolonizing Nature: Strategies for Conservation in a Postcolonial Era,* edited by William M. Adams and Martin Mulligan, 51–78. Sterling, VA: Earthscan, 2003.

SEVENTEEN

Prompts for Feeling-Thinking-Doing

Somatic Speculation for Climate Justice

SARAH KANOUSE

> This chapter describes a set of artist-authored prompts for embodied, affective, and imaginative political-ecological inquiry. The exercises are freely available online and can be used in an undergraduate seminar or adult education context as warm-ups, reflections, or homework.

"We" sat in a misshapen circle, squashed into available shade on a sweltering afternoon. "We"—a group of non-Native and mostly White artists and writers—were discussing Leanne Simpson's argument that Native people should strategically withdraw from the "academic-industrial complex"—and its valuation of the written over the oral, word over action, credential over experience—and instead work for "a resurgence of Indigenous intellectual systems and a reclamation of the context within which those systems operate."[1] "We"—my partner and I—self-organized this seminar because institutional spaces do not cultivate convivial, multisensory, and material inquiry into how to live justly with the land. While the group could agree with Leanne Simpson's diagnosis of the toxic "academic-industrial complex," we were less sure about what to do with it. As settlers, we were not Simpson's primary audience, nor should we be. Yet her comments on the alienation of academia from relational, embodied, and affective knowledge-making resonated deeply. Mindful of the dual risks of extractive appropriation and performative moves to innocence, we asked how we might work through settler-colonial epistemes without further legitimizing their authority and reifying their affects?

> Institutional spaces do not cultivate convivial, multisensory, and material inquiry into how to live justly with the land.

This question guided an ongoing research and pedagogy project entitled "Over the Levee, Under the Plow: An Experiential Curriculum" (OLUP). Coordinated by myself and Ryan Griffis, OLUP is an evolving collection of books, prompts, and objects designed to foster embodied and imaginative political-ecological inquiry. Pushing back on the mind/body and theory/practice dualisms that still dominate Euro-American educational environments, the centerpiece of this curriculum is a deck of cards prompting individual and collective investigation into and beyond the settler-colonial landscape. The deck comprises four artist-developed suits of cards, each suit oriented around a particular eco-social concept: "Beyond Property" (by Sarah Kanouse, on relationships with land), "Amongst Relatives" (by Corinne Teed, on relationships with other species), "After Extraction" (by Ryan Griffis, on stretching conventional temporal and geographical perception), and "Shaped by Rivers" (by Heather Parrish, on evolving relationships with rivers in a climate-changed world). These cards offer experientially grounded, often poetic points of entry into complex questions of environmental ethics and affects. Their prompts variously recommend movement, sketching, multisensory observation, writing, mapping, and further research as ways of locating oneself amidst the institutional or ideological forces contributing to environmental crisis. Although each suit complements an artist's book originally produced for the 2019 Haus der Kulturen der Welt program "Mississippi: An Anthropocene River," the cards are available as a free digital download (see links to resources, at end of chapter). Classroom educators are encouraged to browse, draw on, and adapt the cards' exercises for discussion prompts, as experiential ways to introduce abstract concepts, or as suggested "warm-ups" for more traditional forms of research into climate justice topics.

> These cards offer experientially grounded, often poetic points of entry into complex questions of environmental ethics and affects.

Using varied strategies, the cards obliquely answer the question prompted by Simpson's text: how to work through the settler-colonial dimensions of the North American environment in ways that go beyond purely intellectual or academic critique, that make space for uncomfortable feelings, and that find new ways of relating to the more-than-human world outside the binaries of nature/culture, pure/spoiled, and conserved/consumed offered by Euro-American traditions. The cards emphasize the irreducibly social nature of embodied critical inquiry and offer points of entry grounded in personal experience but offering a change of scale or flipped perspective. For example, one of Parrish's cards prompts:

Go with a companion to a water-side place.

Ask each other the following questions and listen:
Do you have stories in your life that involve flooding? Where were you?

What age?

Who was with you?

What do you remember smelling, hearing, sensing on your skin and feeling emotionally before the flood?

During and directly after?

Months after?

Years after?

Who else in your life shares this memory?

Where do your memories of the experience converge and diverge?

Speak the story to the waterway, understanding that all water is connected.

Listen to the water in return. What do you hear?

Each suit of cards also asks the user to tune into or change habituated bodily responses. Some of these cards are influenced by generative somatics, a school of movement practice that recognizes that the traumas of systemic racism are registered in how all people inhabit their bodies and thus seeks to bring liberatory values into individual and collective embodiment. These exercises bear in mind the differently privileged or centered positions from which users might come to this process. For example, the last stage in a card on the somatics of private property asks the user to:

think about your own ancestral relationships to property. Go back to your family of origin as many generations as you know or can imagine. Recognize that distant ancestors might position you in different and conflicting ways. Choose whichever relationship to property you feel prepared to take on in this moment. Notice where it manifests in your body: how it changes your breath, where it introduces tension, weight, direction, and, perhaps, even pain. If it feels right, allow your body to move into these sensations, assuming a series of positions that hold and respond to them. Then, allow the tensions to melt with your breath as you assume a posture of dignity again.

Other cards address the embeddedness of the user's body in larger ecological cycles through prompts inviting meditation and speculation, as in this exercise from Ryan Griffis:

Drink your glass of water and imagine that your body *is* the watershed. Imagine that the glass is the initial source of water in the area—rainfall, a spring, an open hydrant. As the water passes through your mouth and throat, it encounters plants and animals, and collects materials from the surface of the land (sediment, agricultural chemicals and waste, industrial effluent, debris). Some of these contribute more materials to the water, and some of them filter the water. As the water passes further into your body, it continues the water cycle journey, joining groundwater, aquifers, oceans, lakes, and weather systems. Eventually, it may again become the water you, or someone else, drinks.

Finally, the deck includes specific prompts that draw on the insights of Indigenous philosophies. Through respectful citation and prior acknowledgment of the subject position of a card's author and user, these exercises seek to avoid the very practice of extracting Indigenous ecological knowledge for the benefit of the academic-industrial complex that Simpson warns against in "Land as Pedagogy." Corinne Teed, whose cards accompany the field guide she made with Dakota and Ho-Chunk interlocutors, carefully positions herself at the outset of her suit of cards and generously cites the work of others:

> The cards were influenced by Lakota, Dakota, and Anishinaabe writers including Mary Sissip Geniusz, Kim Tallbear, and Nick Estes. They were written on unceded Dakota lands in so-called Minneapolis, surrounded by the homelands of the Dakota and Anishinaabe people. They were also inspired by the poet CAConrad's somatic rituals. Conrad's rituals aim to create embodied presence, which serves as research fodder for their future poems.

In many ways, the OLUP curriculum extends calls stretching back decades for environmental education to attend to phenomenological experience, embodiment, and the practices of everyday life. For example, Phillip Payne argued in 1997 for an environmental pedagogy built on the recognition that "the locus of understanding, explanation and praxis for the 'environment' should be 'in here, with me and you' rather than 'out there, somewhere, to be found, identified, studied and solved.'"[2] More recently Sarah Jaquette Ray has used similar language in the context of climate education, suggesting that "people with privilege can be asking 'Who am I?' and 'How am I connected to all of this?'"[3] While cautioning about the "unbearable whiteness of climate anxiety," Ray has also called on educators to consider the "emotional arc" of the environmental studies curriculum and to hold space for the discomfort, resistance, and unhappiness that arise in examining the political roots and implications of ecological crisis.[4] It is not surprising that the arts—as one of the few publicly sanctioned spaces to grapple with subjectivity and emotion—have become favored means for educators to address the affective implications of climate change and climate (in)justice. Moreover, artists have in recent years reinterpreted pedagogical forms—lectures, readers, seminars, and assignments—to explore research as a practice at once critical and imaginative. While the OLUP cards may be uncommon among artists' projects in actively pursuing application in the environmental studies classroom, they will almost certainly not be the last.

> The arts—as one of the few publicly sanctioned spaces to grapple with subjectivity and emotion—have become favored means for educators to address the affective implications of climate change and climate (in)justice.

RESOURCES

The OLUP cards, as well as the artists' books that preceded them, are available in print, as digital downloads, and online. They also have been assembled into a limited-edition curriculum box for exhibitions and as gifts to the organizations and individuals whose work informed their creation. If you use these cards in the classroom, we ask that you consider—like an artist—how to contextualize and sequence the activities. How are you as an instructor positioned in relation to the prompts? How can you be upfront about that positioning with your students? Where in the "emotional arc" of your curriculum does the prompt work best—and how might it bend this arc toward climate justice?

Use the prompts online: https://regionalrelationships.org/olup.

Download the books and OLUP cards as PDFs from the Haus der Kulturen der Welt's Anthropocene Curriculum website: https://bit.ly/AC_OLUP. Although the project has since grown, the Anthropocene Curriculum initiative commissioned the original artists' books and continues to host these materials publicly.

NOTES

1. Simpson, "Land as Pedagogy," 22.
2. Payne, "Embodiment and Environmental Education," 133.
3. Ray, "Climate Anxiety Is an Overwhelmingly White Phenomenon."
4. Ray, "Coming of Age at the End of the World," 299.

REFERENCES

Griffis, Ryan, and Sarah Kanouse, eds. *Over the Levee, Under the Plow: An Experiential Curriculum*, 2021. https://regionalrelationships.org/olup.
Payne, Phillip. "Embodiment and Environmental Education." *Environmental Education Research* 3, no. 2 (1997): 133–53. http://dx.doi.org/10.1080/1350462970030203.
Ray, Sarah Jaquette. "Climate Anxiety Is an Overwhelmingly White Phenomenon." *Scientific American*, March 21, 2021. https://www.scientificamerican.com/article/the-unbearable-whiteness-of-climate-anxiety.
———. "Coming of Age at the End of the World." In *Affective Ecocriticism: Emotion, Embodiment, Environment*, edited by Kyle Bladow and Jennifer Ladino, 299–319. Lincoln: University of Nebraska Press, 2018.
Simpson, Leanne Betasamosake. "Land as Pedagogy." *Decolonization: Indigeneity, Education & Society* 3, no. 3 (2014): 1–25. https://jps.library.utoronto.ca/index.php/des/article/view/22170/17985.

PART FOUR

FUTURITY, NARRATIVE, AND THE IMAGINATION

Visualizing What We Desire

EIGHTEEN

The Tool of Imagination

DOREEN STABINSKY and KATRINE OESTERBY

This chapter discusses the possibilities inherent in the idea that the future is plural. We introduce two tools for imagining different climate futures, one comparing the stories provided in the IPCC assessments with other ways to write the future, and another a simple exercise to pause and to journal.

We make this contribution as a professor-student pair. Katrine was a student and then teaching assistant in Doreen's Climate Justice course. We are intrigued by the possibilities and potentialities that appear when we reframe the future as plural. In this chapter we describe some of the journey we took to get here and two exercises for the imagination that we developed along the way. What would it mean, if imagination of futures was part of our existential toolkit?

WHERE IT STARTED

The original Climate Justice course design was informed and inspired by Doreen's work over the years with developing-country governments and nongovernmental organizations who participate in the United Nations Framework Convention on Climate Change (UNFCCC) negotiations. Their framings of climate justice in terms of climate debt and how to determine what is a fair share of the remaining carbon budget[1] became definitional foundations for the course.[2]

Such a frame requires paying attention to emissions and budgets, which is understandably a very numbers-centric exercise. We look at graphs from the Intergovernmental Panel on Climate Change (IPCC) and learn how to interpret them.[3] It is a very sterile way to describe the magnitude of the challenge that we face, because the graphs can clinically show that—according to the numbers—we are fast using up the remaining budget to keep temperature rise below 1.5°C of warming above pre-industrial levels. After the IPCC 1.5 report was released in 2018,[4] that numerical challenge became translated in popular media conversations as "we have twelve years left."[5]

At the time, it seemed a very logical, rational, fact-based story that could help students understand developing-country demands for the fair sharing of the remaining carbon budget. Surely the conclusion that things are not really going so well, making urgent action absolutely necessary, is based on objective facts. The IPCC 1.5 report is framed to contrast impacts at 1.5°C of warming with impacts at 2°C. One way to visualize that difference is by looking at coral reefs: the world's reefs don't stand a chance on either trajectory—a 99 percent loss is estimated at 2°C of warming and only just a bit less than that at 1.5°C, which are temperature thresholds that at current rates of warming we will pass in the next 10 (1.5°C) to 25 or 30 (2°C) years. One recent term, this story sent students over a metaphorical cliff.

Students do a lot of emotional work to be able to engage in climate-related classes, organizations, and conversations. This particular way of explaining the current climate predicament provoked intense feelings of grief and anxiety; it became obvious that a different story needed to be told. It also became obvious that, fundamentally, the IPCC description of the current climate situation is just a story, based on a rather narrowly defined set of assumptions that the class could interrogate and unpack. The students could explore how this IPCC-carbon budget frame was just one way of laying out a narrative about the future. An examination of how it was produced could provide necessary context for them to explore other narratives and, most importantly, to create space to craft their own narratives of the future, writing themselves into those stories.

INSPIRATION FROM INDIGENOUS EXPERIENCES, FUTURES STUDIES, AND CLAY

Where might we find other narratives to think with or alongside into the future?

Indigenous peoples have lived through centuries of the brutal and deliberate destruction of their lifeworlds. That lived experience puts current messages of climate urgency into perspective. One reaction to the graphs of the IPCC and the gut-wrenching images of the latest wildfire infernos is horror at expected temperature rise over the coming decades. Indigenous peoples note that their worlds have already been on fire for five hundred years.[6] Indigenous stories illuminate the more-than-human (ontologies) and teach noncolonial ways of knowing about the world around us (epistemologies), helping to pull off the blinders of modernity and Eurocentrism to notice that many different worlds exist and deserve to exist, and to recognize that part of the work of climate justice is actually to ensure these many beings and knowings can exist and flourish.[7] The Zapatistas dream of a world where many worlds may fit (*un mundo donde quepan muchos mundos*).[8] Part of the necessary work of climate justice is to share that dream and make space for a multiplicity of present and future worlds.

We learn from Indigenous peoples that many worlds exist side by side. Futurists teach that the future is also plural. Futurists help us recognize that we bring the future into being through processes of collective imagining. The future is not already written, no matter what the IPCC says about the exhaustion of carbon budgets. Learning to dream of what might lie ahead, in the plural, and learning to imagine collectively are essential tools of the future world-making in which we are active participants. Stuart Candy thinks it is important that we consciously take on this task: "our collective future as a species depends on using our capacity to imagine worlds together."[9] In posing the question, "Whose future is this?" he admonishes us to not accept "the future that has been imagined for you and imposed upon you," but rather urges us to "cultivate our own capacity for telling our own stories."[10] Keri Facer, a scholar of education futures and anticipation studies, reminds us that "our responsibility as educators in these times . . . is to support our students to think with hope and with rigour about the sorts of futures that are being made today; and to enable them to care for, imagine and make liveable futures in collective dialogue with others whose futures are also at stake."[11] If the future is yet unwritten, then we must all be authors, working together on the manuscripts.

In *The Memory We Could Be*, in a chapter entitled "Hope, a Horizon," author Daniel Voskoboynik tells us that "all systems have leverage points that can rearrange possibilities." "Hope is not about obsessive optimism" but rather is "a commitment to dream and affirm that change is possible." He reminds us that "the world is made of clay, disguised as granite."[12] Participation in the making of future worlds requires that we take on roles as molders and shapers of what is to come. Maybe imagining the world as clay can liberate us from stories that others are telling about "the" future and help us envision our roles in creating many possible futures. Yet it bears noticing that in working with clay, you quickly learn that you can't hurry it. The time of rest for the clay is just as crucial for the creation of your imaginations as the molding of them. Bayo Akomolafe teaches us that "the times are urgent, let us slow down."[13]

TWO OFFERINGS FOR THE EXISTENTIAL TOOLKIT: A CLASS MODULE AND A JOURNALING EXERCISE

A Short Module for Climate Justice: Models, Scenarios, and Storylines for the Future

IPCC models employ an enormous set of assumptions about the future: what sorts of technologies might exist, what is the annual rate of economic growth (as measured through gross domestic product), what policies are in place to affect the global price of carbon. The IPCC Fifth Assessment Report details five different storylines about how the future might unfold, called shared socioeconomic pathways (SSPs), to help readers imagine various warming trajectories.

In the module offered here, which unfolds over three class meetings, students first look at the SSPs and discuss climate models and how they work.[14] Students are then introduced to a simpler set of scenarios developed by the Tellus Institute to contemplate alternative futures. Tellus's "taxonomy of the future" encompasses six alternative future worlds—bifurcation endpoints of three diverging pathways towards the future: conventional world, barbarization, and the great transition.[15] These sorts of scenarios—whether SSPs or the Tellus worlds—are common tools of futurists and foresight.

In advance of a class discussion on the Tellus scenarios, students read about all six endpoints and indicate to the educator/moderator in advance which most resemble the future that they imagine. In class they are put into small groups, mixing together students who chose different future endpoints. They discuss the inflection points and bifurcations.[16] How might we get to particular futures—what happens to push societies in one direction or another? Through a focus on inflection points, the idea is to draw students' attention to what Candy calls "seeds of change"—current conditions "that could be really transformative if they were to grow," to prompt them to think more deeply about how the plurality of futures might come about.[17]

The readings and materials for the discussion during the third and final class session in this short module include both inspiring visions for, and forceful rejections of, worlds that are not yet here.[18]

A Journaling Exercise: Reminding Ourselves to Slow Down

Clay teaches us that a crucial part of the work of molding our imaginations is remembering to take time to slow down. This exercise is a breathing moment or conversation starter for reflections on what students need as they grow up in the climate crisis. The exercise can work in a classroom setting as well as informal settings where students collaborate on projects on climate change.

The aim of the journaling exercise is to help create a space where thoughts and questions about the mental health aspect of the climate crisis can be shared and where students might find ways to articulate the support they need. This exercise does not require the facilitator to be a trained therapist; rather, it invites students to take their mental health seriously, to not pathologize it, to find they are not alone, and to imagine themselves as agents of their own mental health.

With this activity, there should be no rush. Students sit in quiet with their notebooks and are given a menu of prompts that they can reflect upon as they wish. There is no rule for having to answer all the prompts; they can freely pick the ones that resonate with them.

The prompts include:

How does the climate crisis impact your life and your decisions?

Which emotions come with climate change?

Which kinds of futures do you imagine and hope for?

What is something good you do for yourself or you would like to do for yourself, when you do climate work?

What kind of support do you think you might need?

Students are rarely given a free moment to reflect and journal. To some, it even feels like a moment of restoration and a reminder that it is alright to slow down. The exercise also removes focus from climate issues and creates space to reflect on which kinds of futures the students hope for and imagine, and how they wish to take care of themselves as they work for it. After fifteen minutes of quietly engaging with the questions, the conversation is opened for any reflections or thoughts that have arisen. For educators, this conversation can be a view into what it feels like to grow up in the climate crisis and what it feels like to be part of a generation which is constantly told to fix it.

IMAGINATION AS AN EXISTENTIAL TOOL

Climate education can move beyond facts and calls of urgency. Giving students pen and paper and permission to craft their own stories about what lies ahead can help them imagine many possible futures. Can we make a commitment to imagination and take a stance against the language of inevitability?[19] (See Friederici in this volume for more on this interrogation of the pitfalls of inevitability.) And can we move climate education to all the places where we find the climate crisis—out there and inside ourselves? Can we recognize imagination as a superpower to cultivate and an existential tool as we make our way?[20]

NOTES

1. The greenhouse gas carbon dioxide accumulates and remains in the atmosphere for hundreds to thousands of years. The "carbon budget" is the amount of carbon dioxide that might still be emitted by humans and safely remain below a warming threshold of 1.5°C.
2. Adow, "Climate Debt"; Climate Equity Reference Project, "Calculator"; Plurinational State of Bolivia, "Call"; Stilwell, "Climate Debt—A Primer."
3. For example, see figure SPM-10 of IPCC AR5.https://archive.ipcc.ch/report/graphics/index.php?t=Assessment%20Reports&r=AR5%20-%20WG1&f=SPM
4. IPCC, *Global Warming of 1.5°C*.
5. Thunberg, "Our House Is on Fire."
6. Whyte, "Indigenous Peoples and Climate Justice"; Whyte, "Too Late."
7. Berman, "Political Ontology."
8. Escobar, *Designs for the Pluriverse*.
9. Candy, "Three Dimensions of Foresight."
10. Candy, "Whose Future Is This?"
11. Facer, "Storytelling in Troubled Times."
12. Voskoboynik. *The Memory We Could Be*.
13. Akomolafe, " Times Are Urgent."
14. Carbon Brief, "Q&A"; Carbon Brief, "Explainer."
15. Tellus Institute, "Taxonomy."
16. Farmer et al., "Sensitive Intervention Points"; Meadows, "Leverage Points."

17. Candy, "Three Dimensions of Foresight."
18. Thanki, "Fuck Your Apocalypse"; *The Intercept,* "A Message from the Future."
19. Facer, "All Our Futures."
20. Solnit, "Ten Ways to Confront."

REFERENCES

Adow, Mohamed. "The Climate Debt: What the West Owes the Rest." *Foreign Affairs,* May/June 2020. https://www.foreignaffairs.com/articles/world/2020-04-13/climate-debt.

Akomolafe, Bayo. "The Times Are Urgent, Let Us Slow Down." Filmed April 26, 2019. Victoria, BC, Canada. Video, 1:49:26, https://www.youtube.com/watch?v=9qWaWGHNvy0.

Berman, Anders. "The Political Ontology of Climate Change: Moral Meteorology, Climate Justice, and the Coloniality of Reality in the Bolivian Andes." *Journal of Political Ecology* 24, no. 1 (2017): 921–30. https://doi.org/10.2458/v24i1.20974.

Candy, Stuart. "Whose Future Is This?" Filmed January 27, 2014 at TEDx Christchurch. https://www.youtube.com/watch?v=YxgVxu2mdZI.

———. "Three Dimensions of Foresight." April 23, 2020. https://futuryst.blogspot.com/2020/04/three-dimensions-of-foresight.html.

Carbon Brief. "Explainer: How 'Shared Socioeconomic Pathways' Explore Future Climate Change." April 19, 2018. https://www.carbonbrief.org/explainer-how-shared-socioeconomic-pathways-explore-future-climate-change.

———. "Q&A: How Do Climate Models Work?" January 15, 2018. https://www.carbonbrief.org/qa-how-do-climate-models-work.

Climate Equity Reference Project. "Climate Equity Reference Calculator." Accessed January 15, 2022. https://calculator.climateequityreference.org.

Escobar, Arturo. *Designs for the Pluriverse: Radical Interdependence, Autonomy, and the Making of Worlds.* Durham, NC: Duke University Press, 2018.

Facer, Keri. "All Our Futures? Climate Change, Democracy and Missing Public Spaces." Social Futures Conference, Turku, Finland, June 2019. https://futuresconference2019.files.wordpress.com/2019/08/keri_facer_csf2019.pdf.

———. "Storytelling in Troubled Times: What Is the Role for Educators in the Deep Crises of the 21st Century?" *Literacy* 53, no. 1 (2019): 3–13.

Farmer, J. D., C. Hepburn, M. C. Ives, T. Hale, T. Wetzer, P. Mealy, R. Rafaty, S. Srivastav, and R. Way. "Sensitive Intervention Points in the Post-Carbon Transition." *Science* 364 (2019): 132–34.

The Intercept. "A Message from the Future with Alexandria Ocasio-Cortez." April 17, 2019. Video, 7:35. https://www.youtube.com/watch?v=d9uTH0iprVQ.

Intergovernmental Panel on Climate Change (IPCC). *Global Warming of 1.5°C: An IPCC Special Report.* Edited by V. Masson-Delmotte, P. Zhai, H.-O. Pörtner, D. Roberts, J. Skea, P. R. Shukla, A. Pirani, et al. Cambridge: Cambridge University Press, 2018. https://doi.org/10.1017/9781009157940.

Meadows, Donella. "Leverage Points: Places to Intervene in a System." The Sustainability Institute, 1999. https://donellameadows.org/archives/leverage-points-places-to-intervene-in-a-system.

Plurinational State of Bolivia. "Call for Urgent and Equitable Action to Halt the Climate Collapse and Restore Balance with Mother Earth." June 16, 2021. https://observatorioccdbolivia.files.wordpress.com/2021/07/submission-colectiva-16.06.2021-final-en.pdf.

Solnit, Rebecca. "Ten Ways to Confront the Climate Crisis Without Losing Hope." *The Guardian,* November 18, 2021. https://www.theguardian.com/environment/2021/nov

/18/ten-ways-confront-climate-crisis-without-losing-hope-rebecca-solnit-reconstruction-after-covid.

Stilwell, Matthew. "Climate Debt—A Primer." *What Next?* 3 (2012): 41–46.

Tellus Institute. "Taxonomy of the Future." Accessed January 15, 2022. https://www.tellus.org/integrated-scenarios/taxonomy-of-the-future.

Thanki, Nathan. "Fuck Your Apocalypse: Between Denial and Despair, a Better Climate Change Story." July 11, 2017. https://worldat1c.org/fuck-your-apocalypse-c82696b533d9.

Thunberg, Greta. "Our House Is on Fire." *The Guardian,* January 25, 2019. https://www.theguardian.com/environment/2019/jan/25/our-house-is-on-fire-greta-thunberg16-urges-leaders-to-act-on-climate.

Voskoboynik, Daniel Macmillen. *The Memory We Could Be.* Gabriola Island, BC, Canada: New Society Publishers, 2018.

Whyte, Kyle. "Too Late for Indigenous Climate Justice: Ecological and Relational Tipping Points." *WIREs Climate Change* (2020). https://doi.org/10.1002/wcc.603.

Whyte, Kyle Powys. "Indigenous Peoples and Climate Justice." April 18, 2018. Video, 14:43. https://www.youtube.com/watch?v=7YPvsOCUhI8.

NINETEEN

Overcoming the Tragic

PETER FRIEDERICI

This chapter critiques common narrative frames often used in discussing climate change, and suggests that comic, locally focused, and radically inclusionary frames can better serve students by allowing them to experience agency and radical hope.

What sorts of stories ought we tell ourselves and our students about climate change now that the hour is so advanced? We've known for more than three decades that the growth orientation of dominant economic systems is altering the Earth's climate, yet in all that time the facts, the knowledge, and the stories tied to mainstream narratives have hardly nudged the trajectories of relentless expansion, consumption, and emissions. The human capacity for story no doubt grew in large part out of a need to share warnings of danger, but in the case of climate change it has fallen badly short, at least in the case of the most broadly accessible narratives. The objective facts of climate change are bad enough. But what makes the situation worse is that most of the narratives in which they are couched are structured—whether deliberately or not—in a manner that discourages action.

There are numerous reasons for this, some directly linked to deliberately crafted false narratives coined in the interests of power and greed,[1] some linked to how we function psychologically,[2] and some that I have recently explored from the perspective of narratology and science communication.[3] But the direness of our climate change future calls us to action as well as to understanding. We need to create irresistible new stories that transcend science, policy, and individual-level action. Informed by long traditions of storytelling from minority cultures, these stories can present new visions for how humans can live together on the planet in a climate-changed future—a future that will be characterized by immeasurable loss and grief, but also by new possibilities of community and of living within limits.

As educators, it is not our task to dictate these stories. Rather, it is our job to provide our students with tools that will allow them to incubate, articulate, and embody manifold new climate change stories over the years and decades to come. This essay is not intended to prescribe how to do that; a key element of what's needed is polyphonic-voiced storytelling whose components and takeaway messages should not be dictated by any one individual, and indeed cannot be predicted in advance at all. Rather, my intent is to provide a framework for thinking about how to re-envision climate change storytelling in ways that encourage hope and action, with a focus on key elements of framing that we as educators can share with our students.

THE TRAGIC NARRATIVES OF CLIMATE CHANGE

It has become almost impossible to avoid climate change as a topic in the linked worlds of public policy and the media, and increasingly often in personal experience too. People are faced with an ongoing barrage of bad news from today coupled with well-publicized projections that things are likely to become far worse still in the not-so-far-off future. As a result, they are increasingly unable to ignore the problem. Rather, they are regularly forced to choose what to believe about climate change and what to do about it—to make this challenging set of circumstances into a cohesive story.

Outright denial of the importance of climate change is increasingly restricted to a shrinking minority in the U.S.—albeit a minority that is heavily overrepresented in the inner circles of political and economic power. But exposure to news about climate change can result in other common forms of diversionary belief that admit the problem while denying its scale or the scale of the needed societal responses; for example, responses such as "techno-optimism," individualism, market fundamentalism, and "green-growthism" do not amount to explicit denial of the problem.[4] Rather, these ideological channels allow their adherents to believe that societies need not respond to the climate crisis with great sacrifice or any large-scale reordering of societal priorities. Such responses relegate climate change to the stature of a policy issue or (as former Exxon CEO and Secretary of State Rex Tillerson once claimed) "an engineering problem."[5] Like allied responses such as religious end-times beliefs and capitalist opportunism,[6] they don't provide any motivation for a broad mobilization around climate change.

Nor do they provide any succor to the increasing numbers of people who believe that climate change represents a profound crisis or emergency, even an existential threat to modern civilization. To them, geo-engineering proposals and the ramping up of renewable energy are akin to placing small band-aids on a suppurating wound. This community of belief includes an increasingly large segment of younger people.[7] If such people are not subject to the sorts of denial that are increasingly concentrated in older demographics, they are instead subject to other climate change responses that carry their own dangers, such as anxiety, depression, and paralysis.

Renee Lertzmann labeled this set of responses "environmental melancholia" in her 2015 book of that title; the Australian philosopher Glenn Albrecht, too, has literally worked to find words for the linked experiences of current loss and future dread.[8] Panu Pihkala has posited that "[m]any people in fact care too *much*, not too little, and as a result they resort to psychological and social defenses."[9] He suggests that many onlookers respond to the increasing signs of climate breakdown with symptoms of "terror management," meaning that they become preoccupied with fears of death and suffering—whether their own or in the people or places they love. Psychologically, this can lead people to a defense of the status quo as a means of effecting (illusory) stability. The vast scope of climate change can also lead to feelings of "pseudo-inefficacy,"[10] or the sense that because individuals cannot effect change on a sufficiently large scale they have little reason to attempt to do so on a small scale.

> Only by recognizing tragedy as the common theme of numerous, often conflicting narratives that all discourage action can we hope to find new storylines that encourage engagement, community, and courage itself.

Collectively, these psychological responses to climate change often lead in predictable directions, which can be thought of as discrete narratives erected to make sense of difficult news. Along with the varied facets of denial, these avenues include the manifold expressions of anxiety, depression, and other forms of paralysis[11] that have become unfortunately common among young people.[12] Another is Romantic fatalism, a sense that the world's future is doomed, that humanity is headed for a climate apocalypse that cannot be averted. Sarah Jaquette Ray has pointed out that this sort of fatalism is a textbook example of a tragic narrative, one that "makes us look for villains and heroes, ignore gray areas, and reject imperfect but meaningful proposals for progress."[13] But I want to suggest here that tragedy in fact links many of these prevalent climate change narratives, even those that contradict one another (e.g., "there is no problem"; "we'll solve this with green energy"; "the problem is so vast that we are all doomed"). Only by recognizing tragedy as the common theme of *numerous, often conflicting narratives that all discourage action* can we hope to find new storylines that encourage engagement, community, and courage itself.

The literature scholar (and naturalist) Joseph Meeker suggested five decades ago that tragedy in literature grows from the same seedbed of belief that has led us into the ecological crisis, especially in its focus on individualism and on a profound divide between humans and the rest of nature. It is a critique reminiscent of the emphasis on the collective and interdependence that is espoused by Indigenous and environmental justice communities. Though he was writing prior to widespread knowledge of climate change, Meeker's summary resonates today. "The tragic view," he wrote, "assumes that man exists in a state of conflict with powers that are greater than he is"; tragic literature undertakes "to demonstrate that man is equal or superior to his conflict."[14]

Tragic literature is also characterized by what Frank Kermode dubbed "the sense of an ending," meaning that its narratives achieve cohesive meaning only through the existence of a conclusion that defines and stitches together the story—an ending that is foreordained by the narrative's author, even if readers only discover its meaning when they get to it. Prevailing climate change narratives feature numerous such predetermined endings:

- Some ingenious technological innovation will save us from climate change
- The momentum of destructive modern economies cannot be resisted
- Climate change is pushing us toward end times or apocalypse as defined in religious tradition (or, increasingly, Hollywood storytelling)
- We are all doomed and might as well despair—or try to divert our attention as best we can

The common strand of tragedy that links all these narratives is the sense of inevitability that has characterized tragedy since the time of the ancient Greeks: once the story starts running, its course cannot be altered. No wonder young people are anxious.

ALTERNATIVES TO TRAGEDY

But tragedy is not the only game in town, or ought not be, and it is in explicitly looking for counter-traditions that we can find stories for grappling with climate change that do not embed predetermined outcomes. By guiding our students in the creation of such stories, we can help them to discover their power of agency; we can help them realize that today's young people do not need to follow preset narratives, however powerful they may seem to us today.

As we look for alternatives to the tragedy inherent in climate change narratives, a promising one is tragedy's ancient foil: comedy. Meeker argues exactly this, finding in comedy a kinship with the myriad unpredictable ways in which biological evolution has allowed life to persist and flourish in the face of innumerable environmental challenges. Like evolution, he writes, comedy is "concerned with muddling through, not with progress or perfection."[15] Comedy and evolution, Meeker suggests, both prioritize the survival of the whole over big, abstract ideals. They play roles similar to what Nicole Seymour called "bad environmentalisms" in her 2018 book of that title, which focuses on how irony, playfulness, and irreverence "are particularly suited to addressing" the eco-horrors of the twenty-first century.[16] It's important to note here that *comedy* does not always equate to *humor*: rather, it centers on practices of breaking down barriers, disrupting hierarchies, and finding unexpected new means of advancing the story, just as evolution can result in adaptations that could not have been predicted in advance. However, "funny" comedy does have great potential. Max Boykoff and Beth Osnes describe instituting a sequence of climate change-themed comedy performances and competitions among students at the University of Colorado, and found that students reported higher levels of efficacy and engagement as a result of their participation.[17]

A tragic sensibility can readily encourage stoicism, and here too we can profit by looking for stoicism's opposite, namely an openness to emotion. Authors such as Elin Kelsey have called for the forthright expression of climate change-related emotions as vital to the process of overcoming denial, anxiety, despair, and inertia. The Good Grief Network has developed a sequence of interventions (modeled on the 12-step recovery process pioneered by Alcoholics Anonymous) intended to allow participants to find agency and renewed meaning by plumbing climate change emotions. In the classroom, too, we can guide students in exploring emotion as a necessary processing mechanism for dealing with climate change turmoil and grief, and for finding productive new ways of moving forward through story (see also chapters by Bryant and Davenport in this volume).

Because much of the challenge of dealing with climate change is related to the vast spatial and temporal scales involved, we can also help students do the work of meaning-making at a more manageable level. One way to do this is to explicitly look for local expressions of and efforts against climate change. Students can, for example, gather oral histories centering on environmental change, or participate in university-level or local-level climate planning processes (see chapters by Trott, Holmes, and Frame et al. in this volume). Such practices can serve as remarkably effective bridges between students and those outside academia,[18] as well as between lived experience and the more theoretical predictions of climate science.[19] They can also provide students with experiences that can help ground them in a particular place or region, which has been suggested as an important step in effective climate change adaptation and activism[20] as well as in psychological health at a stressful time.[21]

It's also helpful to remember that dominant climate change narratives tend to be Western-centric and male-centric. Amitav Ghosh has explicitly linked climate change to Western storytelling traditions, and suggests that non-Western traditions may provide fertile ground for exploring alternative responses to climate change. Kyle Whyte, a member of the Citizen Potawatomi Nation, has contrasted how climate change has been expressed both in Indigenous dystopia stories and in Western fantasies of progress. Ursula Le Guin has called for feminist-centered storytelling that is more expansive and inclusive than the traditions that have prevailed in the male-dominated publishing and media industries. Excavating the assumptions that underlie particular lines of narrative can help students to perceive that there are alternative means of storytelling, relationship, and action that are positive and generative rather than reductive (see additional chapters on storytelling by Babatunde, Gray, and Richards in this volume).

Julia Corbett has suggested that a needed element in climate change storytelling is the explicit acknowledgement that humanity is moving through a "liminal" period, one in which our entire species is moving from one worldview or fundamental set of values to another—though many Indigenous or land-based groups of people already have values closely allied to where we need to go.[22] To succeed in that process, we need to acknowledge that this transition is underway—a recognition that should come as a relief to young people infused with the knowledge that the old narratives no longer make sense.

We also need to open the doorways to what philosopher Kathleen Dean Moore calls the "radical imaginaries" of climate change,[23] acts of reinvention that are based more on building on positive, existing traditions than on reacting to reductive mainstream framings. Such stories have long arisen out of fantasy and other visionary storytelling means of imagining the future,[24] often through linkages of old and new like those sampled in Walidah Imarisha and adrienne marie brown's anthology *Octavia's Brood*. These pathways offer fertile ground for the imaginations of young people.

> We can best prepare our students for their uncertain future by emphasizing that it is exactly in what we cannot foresee that hope lies.

Western education has often centered on the idea that it is the educators at the front of the classroom who have the answers. But among the most important lessons we can convey is that there is no "answer" to climate change, no one approach that will save us. Rather, we should inspire our students to put their faith in emergent phenomena, in "unexpected collaborations and combinations, in hot compost piles," as Donna Haraway puts it,[25] or in the ability to adapt to the unexpected that Moore has dubbed "presilience,"[26] with its echo of Gerald Vizenor's earlier call for Indigenous "survivance."[27] We can best prepare our students for their uncertain future by emphasizing that it is exactly in what we cannot foresee that hope lies.

NOTES

1. Summarized in Cook et al., *America Misled*.
2. For example, see Stoknes, *What We Think About*.
3. Friederici, *Beyond Climate Breakdown*.
4. Petersen et al., "Reconceptualizing Climate Change Denial."
5. Daily, "Exxon CEO Calls Climate Change Engineering Problem."
6. Friederici, *Beyond Climate Breakdown*, 72–76.
7. Pew Research Center, "Climate Change Still Seen as the Top Global Threat."
8. Albrecht, *Earth Emotions*.
9. Pihkala, "Eco-Anxiety, Tragedy, and Hope," 548.
10. Västfjäll et al., "Pseudoinefficacy and the Arithmetic of Compassion."
11. Reviewed in Ashlee Cunsolo et al., "Ecological Grief and Anxiety."
12. Kelsey, *Hope Matters*, 31–32.
13. Ray, *Field Guide to Climate Anxiety*, 81.
14. Meeker, *Comedy of Survival*, 22.
15. Ibid., 26.
16. Seymour, *Bad Environmentalism*, 234.
17. Boykoff and Osnes, "Laughing Matter?"
18. Friederici, "Introduction."
19. Friederici, "Private Memories of Public Precipitation."
20. Devine-Wright, "Think Global, Act Local?"

21. Green, "New Understanding and Appreciation."
22. Corbett, *Communicating the Climate Crisis*, 192.
23. Moore, *Great Tide Rising*, 261.
24. Oziewicz, "Fantasy for the Anthropocene."
25. Haraway, *Staying with the Trouble*, 4.
26. Moore, *Great Tide Rising*, 213.
27. Vizenor, *Manifest Manners*.

REFERENCES

Albrecht, Glenn A. *Earth Emotions: New Words for a New World*. Ithaca, NY: Cornell University Press, 2019.

Boykoff, Max, and Beth Osnes. "A Laughing Matter? Confronting Climate Change through Humor." *Political Geography* 68 (2019): 154–63.

Cook, John, Geoffrey Supran, Stephan Lewandowsky, Naomi Oreskes, and Ed Maibach. *America Misled: How the Fossil Fuel Industry Deliberately Misled Americans about Climate Change*. Washington, DC: George Mason University Center for Climate Change Communication, 2019.

Corbett, Julia B. *Communicating the Climate Crisis: New Directions for Facing What Lies Ahead*. Lanham, MD: Lexington Books, 2021.

Cunsolo, Ashlee, Sherilee Harper, Kelton Minor, Katie Hayes, Kimberly Williams, and Courtney Howard. "Ecological Grief and Anxiety: The Start of a Healthy Response to Climate Change?" *The Lancet Planetary Health* 4 (July 2020): e261–e263.

Daily, Matt. "Exxon CEO Calls Climate Change Engineering Problem." Reuters, June 27, 2012. https://www.reuters.com/article/us-exxon-climate-idUSBRE85Q1C820120627.

Devine-Wright, Patrick. "Think Global, Act Local? The Relevance of Place Attachments and Place Identities in a Climate Changed World." *Global Environmental Change* 23 (2013): 61–69.

Friederici, Peter. *Beyond Climate Breakdown: Envisioning New Stories of Radical Hope*. Cambridge, MA: MIT Press, 2022.

———. "Introduction." In *What Has Passed and What Remains: Oral Histories of Northern Arizona's Changing Landscapes*, edited by Peter Friederici, 1–9. Tucson: University of Arizona Press, 2010.

———. "Private Memories of Public Precipitation." In *The Land Speaks: New Voices at the Intersection of Oral and Environmental History*, edited by Debbie Lee and Kathryn Newfont, 35–47. New York: Oxford University Press, 2017.

Ghosh, Amitav. *The Great Derangement: Climate Change and the Unthinkable*. Chicago: University of Chicago Press, 2016.

Green, Amanda S. "'A New Understanding and Appreciation for the Marvel of Growing Things': Exploring the College Farm's Contribution to Transformative Learning." *Food, Culture & Society* (2021). https://doi.org/10.1080/15528014.2021.1883920.

Haraway, Donna. *Staying with the Trouble*. Durham, NC: Duke University Press, 2016.

Imarisha, Walidah, and adrienne marie brown. *Octavia's Brood: Science Fiction Stories from Social Justice Movements*. Chico, CA: AK Press, 2015.

Kelsey, Elin. *Hope Matters: Why Changing the Way We Think Is Critical to Solving the Environmental Crisis*. Vancouver, BC, Canada: Greystone Books, 2020.

Kermode, Frank. *The Sense of an Ending: Studies in the Theory of Fiction*. New York: Oxford University Press, 1967.

Le Guin, Ursula K. "The Carrier-Bag Theory of Fiction." In *Dancing at the Edge of the World*, 165–70. New York: Grove Press, 1989.

Leiserowitz, Anthony, Connie Roser-Renouf, Jennifer Marlon, and Edward Maibach. "Global Warming's Six Americas: A Review and Recommendations for Climate Change Communication." *Behavioral Sciences* 42 (2021): 97–103.

Lertzmann, Renee. *Environmental Melancholia: Psychoanalytic Dimensions of Engagement*. London: Routledge, 2015.

Meeker, Joseph W. *The Comedy of Survival: Studies in Literary Ecology*. New York: Charles Scribner's Sons, 1974.

Moore, Kathleen Dean. *Great Tide Rising: Towards Clarity and Moral Courage in a Time of Planetary Change*. Berkeley: Counterpoint, 2016.

Oziewicz, Marek. "Fantasy for the Anthropocene: On the Ecocidal Unconscious, Planetarianism, and Imagination of Biocentric Futures." In *Fantasy and Myth in the Anthropocene: Imagining Futures and Dreaming Hope in Literature and Media*, edited by Marek Oziewicz, Brian Attebery, and Tereza Dědinová, 58–69. London: Bloomsbury Academic, 2022.

Petersen, Brian, Diana Stuart, and Brian Gunderson. "Reconceptualizing Climate Change Denial." *Human Ecology Review* 25, no. 2 (2019): 117–41.

Pew Research Center. "Climate Change Still Seen as the Top Global Threat, but Cyberattacks a Rising Concern." February 2019. https://www.pewresearch.org/global/2019/02/10/climate-change-still-seen-as-the-top-global-threat-but-cyberattacks-a-rising-concern/.

Pihkala, Panu. "Eco-Anxiety, Tragedy, and Hope: Psychological and Spiritual Dimensions of Climate Change." *Zygon* 53, no. 2 (June 2018): 545–69.

Ray, Sarah Jaquette. *A Field Guide to Climate Anxiety: How to Keep Your Cool on a Warming Planet*. Oakland: University of California Press, 2020.

Seymour, Nicole. *Bad Environmentalism: Irony and Irreverence in the Ecological Age*. Minneapolis: University of Minnesota Press, 2018.

Stoknes, Per Espen. *What We Think about When We Try Not to Think about Climate Change*. White River Junction, VT: Chelsea Green, 2015.

Västfjäll, Daniel, Paul Slovic, and Marcus Mayorga. "Pseudoinefficacy and the Arithmetic of Compassion." In *Numbers and Nerves: Information, Emotion, and Meaning in a World of Data*, edited by Scott Slovic and Paul Slovic, 42–52. Corvallis: Oregon State University Press, 2015.

Vizenor, Gerald. *Manifest Manners: Narratives on Postindian Survivance*. Lincoln: University of Nebraska Press, 1999.

Whyte, Kyle P. "Indigenous Science (Fiction) for the Anthropocene: Ancestral Dystopias and Fantasies of Climate Change Crises." *Environment and Planning E: Nature and Space* 1, no. 1–2 (2018): 224–42.

TWENTY

Practicing Speculative Futures

APRIL ANSON

When did the future switch from being a promise to being a threat?

CHUCK PALAHNIUK, *Invisible Monsters*

The story we tell ourselves about environmental crises, the story of humanity's place on the earth . . . determines how we understand how we got here, where we might like to be headed, and what we need to do.

HEATHER DAVIS AND ZOE TODD, "On the Importance of a Date"

Imagination is one of the spoils of colonization, which in many ways is who gets to decide the future for a given geography.

ADRIENNE MAREE BROWN, *Emergent Strategy*

What's past is prologue.

WILLIAM SHAKESPEARE, *The Tempest*

This chapter describes a module that centers grief and storytelling in imagining climate change. Students practice speculative climate futurism by imagining how an issue they care about reshapes the world of 2050. Course framing and assignments can be adapted to suit upper division majors, general education undergraduate classrooms, or be revised for high school students.

The four provocations in the epigraphs frame a course that I first taught as "Apocalypse and the Anthropocene" to upper division English majors at the University of Pennsylvania and then revised for a general education course entitled "The Future" at San Diego State University (SDSU). In each, we responded to our favorite of the four above quotations. An overwhelming majority of both graduating Ivy League students and predominantly first generation SDSU freshman chose Palahniuk. Despite divergent institutional contexts and backgrounds, students seemed similarly scared, uncertain, and, above all, angry as they echoed Palahniuk's desire to know when, exactly, the future transformed from a word connoting hope to one importing terror.

> We are simultaneously tired and tenacious in our climate alarm.

Across these classes, students expressed emotional realities and lived experiences that transcend political, social, economic, and educational divides. Their thoughtful vulnerabilities, their insightful interrogations, their always-in-progress-heartbreaks remind me that I cannot "meet students where they are at" if I assume that to be a stable, shared, or even definable, place. Instead, we inhabit dynamic, shifting intimacies with the charged and often confusingly complex considerations of climate chaos. When students convey both hope and nihilism in the same breath, they remind me how much emotional and imaginative labor every single person is already doing in the climate classroom—far more than we often imagine in our simultaneously tired and tenacious climate alarm. And yet, many who have come before us expressed similar fears in the face of fickle futures. Joining Palahniuk and others, our shared trepidation about the future testifies to Antonio's reminder in *The Tempest*—that "the past is prologue."

COURSE DESCRIPTION

If the future is what we make it, and we are, indeed, to make it, we must determine what from the past deserves repeating and what demands revolt. We must decide what worlds we should work to create and which ones we need to end. In other words, our futurist questions demand we reckon with stories of apocalypse. In this course, we think with North American novels, film, poetry, and video games that detail surviving apocalypses, predict current ones, or imagine future ends-of-worlds, near and far. We approach the essayistic prose of Amitav Ghosh and Mary Heglar and fiction from the likes of N. K. Jemisin and Louise Erdrich, all with the recognition that "Art is not neutral. It either upholds or disrupts the status quo, advancing or regressing justice. We are living now inside the imagination of people who thought economic disparity and environmental destruction were acceptable costs for their power. It is our right and responsibility to write ourselves into the future. All organizing is science fiction. If you are shaping the future, you are a futurist."[1] As futurists, then, we must wrestle with how stories of the end-of-the-world help us imagine more just worlds or reinvigorate ecofascist ideologies. We discuss how apocalyptic stories can represent and contest the deadly legacies of genocide, slavery, colonialism, and capitalism now manifesting in changing climates and other unequal distributions of promises and threats. Our class investigates the ways works of art attempt to render these complex and perhaps overwhelming concepts comprehensible so that we may envision and enact just futures. Still, as much as a single sentence can, and often does, transform our own orientations toward the future, it does not change our headlines as quickly. When we are weighted with new nuances, the news can seem even more daunting than before. This class does not shy away from that phenomenon, but sees that space as the exact place that cultivates our ability to navigate the realities of climate change. In other words, grief is not to be moved through, but to be greeted, affirmed, engaged.[2] It offers a complex site of searingly responsive theories about the world.

> Grief is not to be moved through, but to be greeted, affirmed, engaged. It offers a complex site of searingly responsive theories about the world.

WEEKLY ASSIGNMENT

This weekly response assignment is meant to be invitational and affirmative, to honor students' affective responses to trying topics without the worry of doing it "correctly." Their responses make take any visual, written, or other artistic form students choose, so long as they: (1) respond to one prompt given in that week's lecture, (2) connect to the week's readings, (3) cite one direct quotation with full citation format, and then (4) relate to other issues, examples, or questions that come to their mind. For each week's readings, students can: write a standard 400-word response; compile a playlist and explain the connections; make "agitprop" posters; write or record voicemails to the future; sketch or record a podcast; write lesson plans; make a video about a key concept; compile or draw four landscapes that they saw when reading; screenshot four parts of the readings and add emojis, captions, links, or questions; sketch a map of the central concepts; cast the play or film version; investigate who has cited that piece of writing; imagine the future that the piece is bringing into being; or another form of response that suits them.

FINAL PROJECT

The final project asks students to move from responder to practitioner. We begin discussions of the final project by acknowledging the difficult work that students are already doing. Having engaged with issues of climate change, racial injustice, colonialism, genocide, fascism, and other horrors, they have—in mere months—become capable of holding the immense complexities needed to build better worlds. Guided by Walidah Imarisha's reminder that "Whenever we try to envision a world without war, without violence, without prisons, without capitalism, we are engaging in speculative fiction," our final project invites students to their own version of this speculative fiction practice.[3] This is an exercise of hope in the face of fear. It regards emotional experiences like grief and loss, anxiety and anger, determination and courage, desperation, or hope-filled feelings as vital theoretical tools for dreaming better worlds into being.[4] Our final project is a public exercise in this practice, asking students to revise a piece of writing from the term into a formal contribution to an open-access, student-designed speculative fiction digital magazine.

Final Project Description

In this class, you have read and made your own sense of challenging ideas, inspiring words, terrifying histories, and a diversity of possible futures. Still, it often takes intentional reflection to recognize our growth through such complexities. In that spirit, revisit your reflections from the term and find one or two that still powerfully affect you. Ask yourself, What world was I hoping would exist? Then, imagine your ideas today did change the world by 2050. What world would that 2050 be? Your final project will practice an answer to that question. Specifically, you will create a single page of a magazine as if it were the year 2050. You will revise ideas from your weekly responses into a single-page contribution to our speculative fiction digital magazine, *Speculate 2050*. To create your magazine page, you must not only think about what world you would like to bring into being, but also acknowledge the ways this is necessarily not a "perfect" future, an idea we have learned to view with suspicion. Try to begin from the assumption of imperfection. Start with the understanding that imagining the futures we want to live in requires revision. This beginning helps us practice what we've learned this term—that we must adjust our present to build towards the worlds we desire. *Speculate 2050* will be open access so you can share it with friends, family, and even list it on your résumé. It will be graded according to the rubric below, which corresponds to these requirements:

Your 1-page pdf Magazine entry should:

- Elaborate and refine ideas from the course to suit the year 2050. Question, revise, and expand your early vision.
- Include a striking visual component. If you wish to produce audio such as a podcast, include the transcription with visuals; if you wish to produce a video, include screenshots. If you use any image other than your own, include full citation.
- Include a direct quotation from the course readings with full citation.
- Be drafted prior to the class workshop date.

Your 1-page Reflection essay should:

- Reflect on your final project as well as the whole term. Discuss how your initial response changed, grew, or failed in the face of the year 2050. What does your magazine page represent, for you, about the future? What in the class sparked, bored, inspired, terrified, or changed your futures?
- Be formatted in Times New Roman, 12-point font, double-spaced, 1-inch margins.

TABLE 20.1 Final Project Grading Rubric

	Minimal (2)	Basic (3)	Proficient (4)	Advanced (5)
Content	The project is lacking most requirements. There are many gaps in information presented. Unclear what concepts from class are represented in the pages.	Lacks many required elements and identifiable class concepts. Information is presented in an unorganized fashion and may be hard to understand.	Project meets requirements and includes identifiable concepts from class; however, it is not well-organized or easily understood. May lack some required elements.	The project includes all requirements and clear, nuanced engagement with class concepts. The project is clear, concise, and over and above requirements.
Revision Quality	Revisions do not reflect any change in time-scale or are incomplete. Multiple identifiable errors, lack of editing, and/or the final remains the same as the original.	Revisions are evident, but the work feels rushed and/or incomplete.	Revisions completed with evidence of the time and thought given to the changes made.	The revisions exceed all expectations, evidence diligent and sustained efforts.
Citation	No direct quote; images used without citation.	Direct quote does not come from course readings; images unclearly cited.	Direct quote and/or images are included but missing information in citation.	Direct quote and images include full citation.
Voice	Students did not speak/write in a tone that could be understood by all students. Information too hard to understand.	Many gaps in flow of information, or voice at times unclear or inappropriate.	Minimal gaps in flow of information. Clear voice, original messaging.	Clear and appropriate voice. Information was easy to understand and showed subtlety and nuance.
Reflection	Reflection does not answer required questions.	Reflection is missing some required questions.	Reflection answers all required questions but has multiple errors.	Reflection answers all required questions and contains minimal (no more than 2) errors.

LOGISTICS

We spend the last two weeks sharing in-progress ideas, and swapping design and formatting strategies from programs like PowerPoint as well as analog examples like photographs of physical collages. Each method produces powerful results. I then compile all the contributions into a single pdf and upload it to Calameo, which allows me to embed the magazine into a WordPress site for no cost. I have assigned both 2-page and 1-page entries based on class size, examples of which you can see on the Existential Toolkit website.

CONCLUDING ACTIVITY

We end the class with Ayanna Elizabeth Johnson's Venn diagram from *How to Save a Planet*, which asks, "What are you good at?" "What brings you joy?" and "What is the work that needs doing?" Johnson suggests that where our answers intersect indicates work we should do. How beautiful that each of us answers differently, how encouraging that those answers can and should be revised, and how capable we already are of those changes. Together, these assignments emphasize that revision itself is the most radical, brave, and necessary skill we can cultivate. The futures and presents of climate chaos demand we tenaciously commit to justice while holding our speculations about how to get there as lightly, inclusively, and compassionately as we can.

> The futures and presents of climate chaos demand we tenaciously commit to justice while holding our speculations about how to get there as lightly, inclusively, and compassionately as we can.

NOTES

Epigraphs: Palahniuk, *Invisible Monsters*, 103; Davis and Todd, "On the Importance of a Date," 763–64; brown, *Emergent Strategy*, 163; Shakespeare, *The Tempest*, Act 2, Scene 1, line 986.
  1. brown, *Emergent Strategy*, 197.
  2. Wray, "Why Activism."
  3. Imarisha, "Introduction," 3.
  4. Ray, *Field Guide to Climate Anxiety*; Consulo et al., "Ecological Grief and Anxiety."

REFERENCES

brown, adrienne maree. *Emergent Strategy: Shaping Change, Changing Worlds*. Chico, CA: AK Press, 2017.
Consulo, Ashlee, Serilee L. Harper, Kelton Minor, Katie Hayes, Kimberly G. Williams, and Courtney Howard. "Ecological Grief and Anxiety: The Start of a Healthy Response to Climate Change?" *The Lancet Planetary Health* 4 (July 2020). https://doi.org/10.1016/S2542-5196(20)30144-3.
Davis, Heather, and Zoe Todd. "On the Importance of a Date, or Decolonizing the Anthropocene." *ACME: An International Journal for Critical Geographers* 16, no. 4 (2017): 761–80.

How to Save a Planet. Ayanna Elizabeth Johnson's framework for figuring out what you should do about the climate crisis. All We Can Save Project, Twitter, March 19, 2021. https://twitter.com/allwecansave/status/1372939917861474308.

Imarisha, Walidah. "Introduction." In *Octavia's Brood: Science Fiction Stories from Social Justice Movements,* edited by Walidah Imarisha and adrienne maree brown, 3–6. Chico, CA: AK Press, 2015.

Palahniuk, Chuck. *Invisible Monsters.* New York: W. W. Norton, 1999.

Ray, Sarah Jaquette. *A Field Guide to Climate Anxiety: How to Keep Your Cool on a Warming Planet.* Oakland: University of California Press, 2020.

Shakespeare, William. *The Tempest.* Edited by Alden T. Vaughan and Virginia Mason Vaughan. London: Bloomsbury, 2011.

Wray, Britt. "Why Activism Isn't *Really* the Cure for Eco-Anxiety and Eco-Grief." *Gen Dread,* August 5, 2020. https://gendread.substack.com/p/why-activism-isnt-really-the-cure.

TWENTY-ONE

Cultivating Radical Imagination through Storytelling

SUMMER GRAY

This chapter presents storytelling as a tool for cultural recovery, empathetic connection, radical transformation, and self-reflection. It includes three guided activities that can be used together or independently to engage with and cultivate narratives that embrace a wider sense of possibility.

Students encounter stories everywhere they look, from social media to the classroom, often without fully recognizing them as such. This set of activities provides an opportunity for young adults to engage with and practice critical thinking and affective literacy, enabling them to see storytelling as a means of cultural recovery, empathetic connection, radical transformation, self-reflection, and also as a source of potential harm, influenced by worldviews that sometimes limit their view of what is possible. The activities detailed below engage with stories of climate injustice, racial injustice, trauma, and empathy to explore how narratives impact our desire and ability to respond effectively to the challenges of our time, including the climate crisis. Students are invited to explore stories that cultivate a wider sense of possibility and a more intersectional view of climate justice.

THEORETICAL FRAMING

Storytelling is fundamental to how humans connect and engage with the world around them. Because stories are so embedded in social life, it is easy to mistake them for reality.

The Three Pillars of Storytelling

1. *Stories are social constructions embedded in relationships of power.* Stories are selective representations that mirror the demands, sentiments, and interests of their tellers, whether consciously or not.[1] They necessarily build upon assumptions and power dynamics, and they often relay prejudices and untruths.[2] This includes stories about the environment.[3] The stories we tell about ourselves, our communities, and our collective destinies have a profound impact on our beliefs and actions.

> The stories we tell about ourselves, our communities, and our collective destinies have a profound impact on our beliefs and actions.

2. *Stories have the potential to reinforce or challenge dominant narratives.* While stories can reinforce political divides, they can also transcend those same divides.[4] Stories can also challenge dominant narratives and contribute to cultural recovery, defined by Jennifer Machiorlatti as "the recuperation and regeneration of personal and community identity, language, cultural practice, and responsibility to/with the ecosphere."[5]

3. *Stories have the emotive power to transport us into other worlds and states of being.* As Walidah Imarsha and adrienne maree brown write, "Whenever we try to envision a world without war, without violence, without prisons, without capitalism, we are engaging in speculative fiction."[6] In communications studies, transportation is described as "the process of becoming fully engaged in a story, an integrative melding of attention, imagery, and feelings."[7] Theories of transportation involve understanding how stories impact, or transport the listener, by evoking empathy, an emotion that is widely considered to be distinct from more limited sentiments of pity and sympathy.[8] The emotive capacity of storytelling links the wider field of human affect.

> "Whenever we try to envision a world without war, without violence, without prisons, without capitalism, we are engaging in speculative fiction."—Walidah Imarsha and adrienne maree brown

ACTIVITIES AND ASSIGNMENTS

Activity 1: Three Stories of Our Time

This first activity, adapted from the University of California Student Leadership Institute for Climate Resilience, can be done in one hour. Students start with a positionality prompt and then read an excerpt from "Three Stories of Our Time" by Joanna Macy and Chris Johnstone.[9] Macy and Johnstone define three stories commonly held in the world today: (1) "Business as Usual," which assumes that there is little that can be done about the way we live; (2) "The Great Unraveling," which highlights the sixth great extinction of species and the collapse of our biological, ecological and social systems, due to the impacts of activities within the first story; and (3) "The Great Turning," which involves a just transition from an unsustainable model to one that respects and supports the entire web of life. The key takeaway is that we live in all three stories and have the power to shift our perspective.

Instructions

- *Describe the purpose of this activity*—to identify and challenge dominant narratives about the current state of the world that frame our perceptions and attitudes towards the climate crisis. By developing a critical sensitivity to the way in which emotive storytelling is utilized in daily communication and in the construction of common beliefs, students can start to learn emotional agency.
- *Note that not everyone experiences the climate crisis in the same way*. Describe and define positionality for the students: "The notion that personal values, views, and location in time and space influence how one understands the world. In this context, gender, race, class, and other aspects of identities are indicators of social and spatial positions and are not fixed, given qualities. Positions act on the knowledge a person has about things, both material and abstract."[10]
- *Create a positionality mind map.* Have students create a simple mind map based on the following prompts: What conditions (personal, societal, national, or global) inform my positionality? How do the conditions of my positionality shape how I experience the climate crisis? Allow students to share and discuss key features of their mind map.
- *Introduce "Three Stories of Our Time" and have the students read each story, one at a time.* For each story, ask students to reflect on what happens when we believe and live into each narrative. What kinds of feelings, emotions, thoughts, or actions emerge? At the end of the reading, have students reflect on which story they are currently living in. Which story do they want to live in more fully? Ask students to reflect back on their mind map and identify one area in their life where they can change their story.

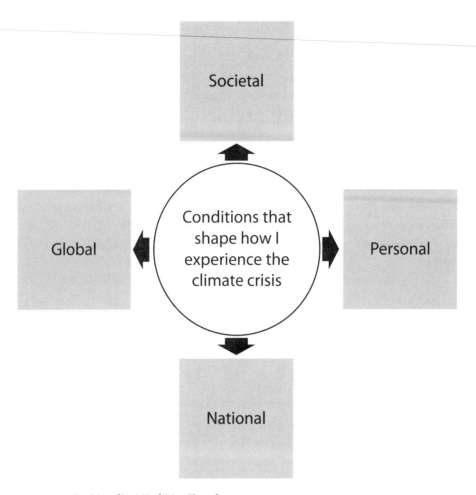

FIGURE 21.1 Positionality Mind Map Template.

- *Optional homework.* Have students select a set of news articles, social media posts, or Instagram photos and videos and use the three stories as a lens for doing a critical read of how the media frames the challenges and opportunities of the current moment.

Activity 2: Stories that Transport Us into Radically Different Worlds

The second activity involves reading an Afrofuturist novel that evokes strong feelings of narrative transportation. Students read Octavia Butler's *Parable of the Sower*, which follows the story of Lauren Olamina, who must survive her way through a brutal dystopian landscape to sow the seeds of a utopian future.[11] Butler's portrayal of hyper-empathy syndrome, which causes Lauren to feel both the pain and pleasure of those around her, draws the reader in and forges a deep and lasting connection—transporting students into a new world of feeling.

Pre-Reading Instructions

- *Describe the purpose of this activity*—to heighten students' radical imagination while deepening their sense of connection, belonging, and interdependence with the web of life. To create further opportunities to reflect on and apply the previous "Three Stories of Our Time" activity.
- *Introduce Octavia Butler and the meta-story of her work.* Butler was one of the first African American science fiction novelists to gain worldwide recognition. She was born in Pasadena, California, and was raised by her widowed mother. Her dystopian novels explore themes of racial injustice, climate change, women's rights, and political disparity. She passed away in 2006 and never got to see *Parable* rise to fame. The book became a best seller only recently in the context of the COVID-19 pandemic and the rise of the Black Lives Matter movement. Now, her books inspire community organizers and activists to make real and lasting change in the world.
- *Provide a trigger warning and relevant campus resources.* A trigger warning is absolutely necessary when assigning this text. There are depictions of violence and rape. Be prepared by identifying mental health resources and support networks on campus and list those on the syllabus or course website.
- *Set up reading group options for students who want them.* Some students will need space to reflect with others in the class on some of the more difficult scenes in the book.
- *Offer a pre-reading activity to get students engaged and interested.* Have students explore the first chapter of the "Octavia's Parables" podcast by Toshi Reagon and adrienne maree brown.[12]

Part 1

- *Assign the first 150 pages of the book.* Encourage students to meet with their reading groups.
- *Discuss affect and hyper-empathy.* Provide classroom discussion time or an online forum for students to share their thoughts on the book so far. Ask them to think about what kinds of emotions stand out and which of the "Three Stories of Our Time" best describes the narrative. Start a discussion about why they think hyper-empathy is significant to the story.

Part 2

- *Assign the second half of the book.* Depending on how much time is available, this can take an additional 1–2 weeks. Encourage students to continue to meet with their reading groups.

- *Discuss contemporary social and environmental linkages.* In class, or on their own, have students explore the webinar "Octavia Tried to Tell Us: Parable for Today's Pandemic."[13] Ask the students to think about *Parable of the Sower* in relation to the COVID-19 pandemic or the current climate crisis.

Reflection

- Have students write a short reflection paper. How does Octavia Butler help us to envision other worlds while challenging dominant narratives?

Activity 3: Telling Your Story

The third activity turns inward and invites students to create a short story that they want to embrace and live into. Students work on these stories individually and then share them with others in the class during a workshop.

Instructions

- *Describe the purpose of this activity*—(1) To use storytelling as a form of self-reflection and community engagement; (2) To begin to foster a sense of purpose and a deeper commitment to social and environmental justice.
- *Invite the students to create a story.* This can either be open ended or more structured using one of the prompts featured in the book *Active Hope* (e.g., "Letter to the Future"; "A Letter from Gaia"; or "Letter from the Seventh Generation").[14] Allow students a week or more to work on this.
- *Workshop.* Set up a workshop (digitally or in person) and have students read and provide supportive feedback on the work of their peers.
- *Creative meta-reflection.* Have students reflect on their story journey. What stood out? Did they learn anything new about themselves? Ask them to write out their thoughts in a poem or illustrate a map of their journey.

NOTES

1. Bridger, "Community Imagery and the Built Environment."
2. Moezzi et al., "Using Stories, Narratives, and Storytelling."
3. Cronon, "Place for Stories."
4. Gustafson et al., "Personal Stories."
5. Machiorlatti, "Ecocinema, Ecojustice, and Indigenous Worldviews," 63.
6. Imarsha and brown, *Octavia's Brood*.
7. Green and Brock, "Role of Transportation," 701.
8. Wald et al., "How Does Personalization in News Stories Influence Intentions?"
9. Macy and Johnstone, *Active Hope*, 13–34 (available for free as an e-book at some university libraries).
10. Sánchez, "Positionality."
11. Butler, *Parable of the Sower* (available for free as an e-book at some university libraries).

12. Reagon and brown, "Octavia's Parables."
13. Due and Coleman, "Octavia Tried to Tell Us."
14. Macy and Johnstone, *Active Hope*, 117, 158, 159.

REFERENCES

Bridger, Jeffrey C. "Community Imagery and the Built Environment." *Sociological Quarterly* 37, no. 3 (1996): 353–74. https://doi.org/10.1111/j.1533-8525.1996.tb00743.x.

Butler, Octavia. *Parable of the Sower*. New York: Warner Books, 1993.

Cronon, William. "A Place for Stories: Nature, History, and Narrative." *Journal of American History* 78, no. 4 (1992): 1347–76. https://doi.org/10.2307/2079346.

Due, Tananarive, and A. Monica Coleman. "Octavia Tried to Tell Us: Parable for Today's Pandemic with Monica A. Coleman and Tananarive Due." Filmed May 2020 on Zoom. Video, 1:24:49. https://www.youtube.com/watch?v=GyLsqaFjl44.

Green, Melanie C., and Timothy C. Brock. "The Role of Transportation in the Persuasiveness of Public Narratives." *Journal of Personality and Social Psychology* 79, no. 5 (2000): 701–21. https://doi.org/10.1037/0022-3514.79.5.701.

Gustafson, Abel, Matthew T. Ballew, Matthew H. Goldberg, Matthew J. Cutler, Seth A. Rosenthal, and Anthony Leiserowitz. "Personal Stories Can Shift Climate Change Beliefs and Risk Perceptions: The Mediating Role of Emotion." *Communication Reports* 33, no. 3 (2020): 121–35. https://doi.org/10.1080/08934215.2020.1799049.

Imarsha, Walidah, and adrienne maree brown. *Octavia's Brood: Science Fiction Stories from Social Justice Movements*. Oakland, CA: AK Press, 2015.

Machiorlatti, Jennifer. "Ecocinema, Ecojustice, and Indigenous Worldviews: Native and First Nations Media as Cultural Recovery." In *Framing the World*, edited by Paula Wiloquet-Maricondi, 62–80. Charlottesville: University of Virginia Press, 2010.

Macy, Joanna, and Chris Johnstone. *Active Hope: How to Face the Mess We're in without Going Crazy*. Novato, CA: New World Library, 2012.

Moezzi, Mithra, Kathryn B. Janda, and Sea Rotmann. "Using Stories, Narratives, and Storytelling in Energy and Climate Change Research." *Energy Research & Social Science* 31 (2017): 1–10. https://doi.org/https://doi.org/10.1016/j.erss.2017.06.034.

Reagon, Toshi, and adrienne maree brown. "Octavia's Parables." In *Parable of the Sower: Chapter 1*, produced by Kat Aaron, June 22, 2020, 37 min. https://www.readingoctavia.com/.

Sánchez, Luis. "Positionality." In *Encyclopedia of Geography*, edited by Barney Warf, 258. Thousand Oaks, CA: Sage, 2010.

Wald, Dara M., Erik W. Johnston, Ned Wellman, and John Harlow. "How Does Personalization in News Stories Influence Intentions to Help with Drought? Assessing the Influence of State Empathy and Its Antecedents." *Frontiers in Communication* 5, no. 111 (2021). https://doi.org/10.3389/fcomm.2020.588978.

PART FIVE

UNSETTLING PEDAGOGIES

Discomfort and Difficult Knowledge

TWENTY-TWO

Critical Journalism, Creative Activism, and a Pedagogy of Discomfort

KIMBERLY SKYE RICHARDS

This chapter describes how to create a fictional newspaper containing stories of the news we *want* to see in the world and positions the assignment in the context of climate arts activism. It also offers a reflection on the pedagogical challenges of deconstructing cherished colonial investments and facilitating creative activism in the classroom.

I strive to support Indigenous-led movements of water, mountain, and land protection against extractive industrial development in my teaching, research, and artistic practice. In 2020, when I was a Lecturer at the University of the Fraser Valley in Abbotsford, British Columbia, I designed a class called "Extractivism in Canadian Literature, Film, and Performance" that engaged novels, plays, songs, and films that examine resource extraction and its impacts on frontline communities. Prompted by Naomi Klein's question—"Why is it so hard for Canadian political leaders . . . to design climate policies that are guided by climate science?"—we discussed how the answer exists not simply in Canada's economic dependence on staple commodities (fish, fur, lumber, wheat, minerals, and fossil fuels), but also in the founding myths of the nation. We analyzed stories that challenged the myth created by European explorers that there were abundant resources available on Canadian land for the taking and that depicted the social and environmental consequences of that history: Fred Stenson's *Who by Fire*, Marie Clements's *Burning Vision*, Cherie Dimaline's *The Marrow Thieves*, and Elle-Maija Tailfeathers's *Bloodlands*.

I anticipated difficult emotions would manifest in the classroom as we confronted coloniality. For some of my students, deconstructing cherished Canadian myths and examining the roots of climate change might feel like an existential attack. Another group of students would have personally experienced racialized violence, so this material would be charged. The violence perpetrated by non-Indigenous people on Indigenous communities and lands is "difficult knowledge"; for non-Indigenous students in my class, coming to understand that one's privilege is the result of the dispossession of Indigenous communities is cognitively, psychologically, emotionally, and politically destabilizing.[1] Moreover, I suspected members of my students' families worked in forestry, fishing, agriculture, mining, or oil and gas, thereby making the critiques of extractive industries and projects deeply personal. I worried about alienating these students and how their feelings of embarrassment, anger, shame, and guilt might shut down conversation, or provoke reactions that could harm students with lived experiences of environmental racism resulting from these very projects.

Emotions like anger, fear, denial, and guilt are defensive mechanisms that can make it more difficult to address colonial thought among white students. Like white discomfort and fragility, settler discomfort (or unsettlement) can be productively engaged through "pedagogies of discomfort" which emphasize the need for both the educator and students to move outside of their comfort zones, "the inscribed cultural and emotional terrains that we occupy less by choice and more by virtue of hegemony."[2] A pedagogy of discomfort is a political endeavor whereby students are encouraged to question assumptions and privilege, pay attention to emotion, and work for collective change to foster social justice. Judith Walker and Carolina Palacios argue that a pedagogy of discomfort is a pedagogy for the privileged.[3] They explain the role of the educator is not necessarily to cause discomfort, but to allow for discomfort, to be okay with discomfort, to invite students into a place for discomfort. Yet they point out there can be a fine line between discomfort that moves and discomfort that forestalls. A pedagogy of discomfort can quickly become a pedagogy of despair:

> Discomfort can quash a pedagogical moment when: (1) It produces overwhelming guilt, which results in defensiveness and shutting down. (2) It produces shame and self-consciousness so that a student is overcome with a fear of "getting it wrong," making things worse, or committing some sort of social justice faux pas so they stay silent; or . . . (3) [A student feels] an overwhelming sense of hopelessness and despair in the face of learning about so much injustice and suffering in the world.[4]

While our readings helped us to understand the impetus for extractivism in Canada, and the past and present realities of injustice and violence involved, I wanted to engage the "productive tension between denunciation and annunciation, and at the nexus of love-indignation-hope" where a pedagogy of discomfort can become a pedagogy of hope.[5] Further, I wanted to design an assignment that would simultaneously deliver this pedagogy of discomfort for my privileged students without falling into the trap of centering their experience as dominant in the classroom, attending only to their journeys. Thus, the assignment I designed also lifts up the knowledge of my non-white students, which directed us in the co-creation of new forms of consciousness about extraction, social change, and futurity.

I created an assignment in which we would co-create a newspaper containing stories of the news we wanted to see in the world, but even the discussion of who "we" were in this class required some critical reflection. Some of these stories were more desirable than others, depending on students' positionality, to be sure, but we collectively agreed we would like to hear stories of local iterations of the repatriation of Indigenous land; decolonized governance structures with Indigenous leaders at the helm; missing Indigenous women and two-spirit people found safe; the death of pipelines; re-allocation of funding towards solar panels and remediation of impacted ecosystems; progressive policies to reduce greenhouse gas emissions; breakthroughs in alternative energy research and uptake; reparations for laborers harmed by the lack of workplace safety, etcetera. These news stories—whether humorous or serious in tone—would tell accounts of demands being met and progressive policies being implemented, thanks to popular pressure. Students researched Indigenous, decolonial, and anti-racist futures being proposed and fought for by movement leaders, identified the political and structural barriers to making those futures possible, and crafted stories about their undoing.

This assignment was inspired by two examples of creative activism. The first is a letter from the future penned by poets Rita Wong and Hiromi Goto, and published in the *Vancouver Observer*, detailing how the Site C Dam—a massive hydroelectric generating station on the Peace River in northwestern British Columbia—was almost constructed but eventually called off. In a letter to her granddaughter one hundred years in the future, a grandmother recounts how the revaluing of clean flowing waters and intact ecosystems, cultural knowledge, and sacred sites led to a shift in thinking that, in turn, yielded abundant produce, "green jobs," and improved relations between Indigenous and non-Indigenous people. The letter counters the narratives of extractive development circulating in relation to the project and illustrates the imaginative work of "writer activists" who "convert into image and narrative the disasters that are slow moving and long in the making, . . . into stories dramatic enough to rouse public sentiment and warrant political intervention."[6]

A second source of inspiration was the secret creation and surprise distribution of three radical prank newspapers detailed by Larry Bogad in *Tactical Performance: The Theory and Practice of Serious Play*. In 2008 and 2009, Bogad and collaborators (including Jacques Servin of the notorious Yes Men) wrote, designed, and distributed fantasy versions of the *New York Times, International Herald Tribune,* and *New York Post* to spread ideas about what would be possible, "not if those in power suddenly did the right thing, but if masses of people mobilized and pressured them to do so."[7] While the efficacy of Bogad's papers lay in their distribution and the brief suspension of disbelief experienced by their readers, I suspected that the experience of writing such stories, and dreaming these victories, would also be elating for white and BIPOC students alike. The resulting assignment, "Special Edition," would transform students into writer activists animated by fantastic visions of societal transformation. It would invite them to contribute to the unmaking of the colonial world as "implicated subjects," not merely victims of, or bystanders to, the harms associated with colonial culture, but proximate to, and bearing some responsibilities for, these systems of injustice.[8] Doing this requires courage and risk-taking. This, I think, is the productive zone within a pedagogy of discomfort and a pedagogy of hope.

Instructions

1. Determine the scope and intended audience for your newspaper. Are you going to create a mock issue of an existing newspaper, or a fictional one? Be specific about the projected date of publication; projecting further into the future allows you to imagine the world you want to see in ten, twenty, one hundred years, while selecting a closer date heightens the disparity between the news announced and the status quo.

2. Introduce the assignment and the issues you want to focus on (i.e., the legacies of colonialism in Canada as they impact Indigenous life, reconciliation, and the environment). Brainstorm segments of the newspaper. I required each student to write one local, provincial, national, or international story and to make one other contribution (e.g., a sports story, film review, obituary, weather report, comic strip, advice column). Brainstorm potential topics and storylines. Allow students to sign up based on interests.

3. Students are likely to feel overwhelmed if they think they need to come up with solutions to colonialism and climate change themselves. Use this line of tactical questioning: What are you frustrated/angry/afraid of? Why do you feel so strongly about this issue? Who is doing work on this issue in the face of their own anger, sadness, and despair? What is the most efficacious action given the problem? Direct them to information about Indigenous leaders, social movements, progressive think tanks, their own ancestors and elders, and researchers who have policy recommendations, demands, answers, and steps for action. Paolo Freire instructs, "One of the tasks of the progressive educator, through a serious, correct political analysis, is to unveil opportunities for hope, no matter what the obstacles may be."[9] This assignment becomes part of a pedagogy of hope when students amplify these voices and their visions of more just, livable, and sustainable futures. Be prepared for differing views about what angers students most and what they think the best solutions are. Use this conversation as part of the learning process about how epistemology shapes knowledge production and power. Illuminate the structures at play in their identities, and even bring in notions of empathy and audience to help students expand their communication toolkit.

4. Allow time for ideas to gestate and be refined. Determining what constitutes the "news you/we want to see in the world" requires a lot of reflection, learning, and unlearning. Budget extra time for office hours, if possible. It may be helpful to add additional layers of scaffolding in which students "pitch" their ideas and receive feedback. It may also be helpful to add a research component so that

students detail the science, policy, history, Indigenous, or anti-colonial knowledge they are engaging.

5. Once the ideas for the stories are determined, turn to focusing on the genre. Study newspapers and invite student and community-based journalists as special guests to discuss the tone and style of journalistic writing. Have students develop a draft to share. Establish systems for peer review. Allow students time to revise and refine their story. Do not underestimate the time it takes to learn this genre of writing, and the lessons layered within it about audience, appeals, epistemologies, and knowledge production. This project is challenging conceptually and formally. Provide instruction and support at both levels.[10]

6. Determine how the special edition will be shared, and with whom. Consider collaborations with your school newspaper or other local publications. This project is labor-intensive.[11] If you are intending to print and distribute it, you may need specific software or support with layout and graphics, as well as a larger budget for printing and distribution plan.

7. Read and discuss. Invite students to examine their spheres of influence—where they may have access to power to effect change—in relation to the stories covered in your "Special Edition."

NOTES

1. Britzman, *Lost Subjects, Contested Objects*; Zembylas, "Theorizing 'Difficult Knowledge'"; Bryan, "Affective Pedagogies."
2. Zembylas, "Affect, Race, and White Discomfort"; Boler and Zembylas, "Discomforting Truths," 108.
3. Walker and Palacios, "Pedagogy of Emotion," 178.
4. Ibid., 184.
5. Ibid., 187.
6. Nixon, *Slow Violence*, 3.
7. Bogad, *Tactical Performance*, 233.
8. Rothberg, *Implicated Subject*.
9. Friere, *Pedagogy of Hope*, 9.
10. An earlier assignment in my course focused on "media literacy" to build students' capacity to critically read and interpret news stories.
11. I will also note that this assignment requires an innovative strategy for assessment. How do you assess the cultivation of a radical imagination focused on transforming what is to what could be?

REFERENCES

Bogad, Larry. *Tactical Performance: On the Theory and Practice of Serious Play*. New York: Routledge, 2016.

Boler, Megan, and Michalinos Zembylas. "Discomforting Truths: The Emotional Terrain of Understanding Difference." In *Pedagogies of Difference: Rethinking Education for Social Justice*, edited by Peter Pericles Trifonas, 115–38. London: Routledge, 2002.

Britzman, Deborah. *Lost Subjects, Contested Objects: Toward a Psychoanalytic Inquiry of Learning.* Albany: State University of New York Press, 1998.

Bryan, Audrey. "Affective Pedagogies: Foregrounding Emotion in Climate Change Education." *Policy and Practice: A Development Education Review* 30 (2020): 8–30.

Goto, Hiromi, and Rita Wong. "Opinion: A Letter from the Future of Site C." *Vancouver Observer,* November 30, 2017.

Friere, Paolo. *Pedagogy of Hope: Reliving Pedagogy of the Oppressed.* Translated by Robert R. Barr. New York: Continuum, 1992.

Klein, Naomi. "Canada's Founding Myths Hold Us Back from Addressing Climate Change." *Globe and Mail,* September 23, 2016.

Nixon, Rob. *Slow Violence and the Environmentalism of the Poor.* Cambridge, MA: Harvard University Press, 2011.

Rothberg, Michael. *The Implicated Subject: Beyond Victims and Perpetrators.* Palo Alto: Stanford University Press, 2019.

Walker, Judith, and Carolina Palacios. "A Pedagogy of Emotion in Teaching about Social Movement Learning." *Teaching in Higher Education* 21, no. 2 (2016): 175–90.

Zembylas, Michalinos. "Affect, Race, and White Discomfort in Schooling: Decolonial Strategies for 'Pedagogies of Discomfort.'" *Ethics and Education* 13, no. 1 (2018): 86–104.

———. "Theorizing 'Difficult Knowledge' in the Aftermath of the 'Affective Turn': Implications for Curriculum and Pedagogy in Handling Traumatic Representations." *Curriculum Inquiry* 44, no. 3 (2014): 390–412.

TWENTY-THREE

Why Worry? The Utility of Fear for Climate Justice

JENNIFER LADINO

This chapter explores variants of fear and asks readers to reconsider the capacity of worry to spark concern about the climate crisis among majority white, rural communities, and to help identify "objects of care" that may catalyze action. The chapter recounts two brief pedagogical experiences to illustrate ways to invite vulnerability in our classrooms and mobilize emotions for justice.

"When I was younger I could play outside any time I wanted to. Now, going outside can be dangerous."

So began what seemed like a lightbulb moment for one young white man in Coeur d'Alene, Idaho, in May of 2019. He recalled the smell and feel of the smoke from wildfires in the summer of 2017. He had been visiting the neighboring city of Spokane, Washington, he explained, headed to the Apple store. But the store was closed. All the stores were closed. The smoke was too thick; the air quality was too dangerous. His story ended with a moving admission and a striking shift from the past tense to the present: "I am really scared."

My co-facilitator, Environmental Science doctoral candidate Kayla Bordelon, and I were wrapping up the first of an Idaho Humanities Council-sponsored discussion series in which we visited Idaho towns to discuss the local impacts of climate change through the relatively safe space of fiction, specifically Barbara Kingsolver's *Flight Behavior*. Our events prompted dialogue through an audience-centered approach, inspired by National Park Service interpretive strategies, and we brought in scientific data about climate change, Yale Climate Opinion Maps, and quotes from the diverse cast of characters in the novel. We treated personal stories from participants as valuable data in understanding the climate crisis.

Incentivized by his high school teacher to attend the event for extra credit, the young man in Coeur d'Alene hadn't linked smoky summers to climate change until Kayla helped him connect those dots. It seemed like his answer to our question, "When did climate change become real to you, if it has?" was . . . today.

It's moments like this that motivate my teaching. I am a professor at the University of Idaho, a "land grab" university located on unceded Nimiipuu (Nez Perce), Palus (Palouse), and Schitsu'umsh (Coeur d'Alene) homelands. My university town of Moscow (pop. around 25,000), part of the Inland Pacific Northwest, is nestled in Idaho's panhandle, among the rolling hills of the Palouse, a few hours' drive from the Canadian border. Approximately 75 percent of UI's students are white, and an equal percentage hails from in-state. Many grew up in smaller towns than Moscow. The English majors I teach lean left, politically, and Moscow is a purplish-blue dot in an otherwise red state. (Coeur d'Alene is much redder.) Idaho has made the headlines for Ammon Bundy, anti-mask rallies, and legislation against critical race theory, abortion, and gender-affirming care for transgender youth. My state is also home to a disproportionately large number of climate change skeptics. Because my thinking about fear is rooted in this sociopolitical context, it might resonate more, or less, with educators in other places. When it comes to climate emotion pedagogy, a first step is to get to know the emotional weather in your own classrooms, including the causes and conditions of your students' emotional lives.

In this essay, I'll suggest that feeling "scared" is one way to spark concern about climate change among groups who have been privileged enough to afford skepticism or apathy. In classrooms and communities like mine—largely white and rural, and often right-leaning—confronting climate change with a healthy dose of fear might be useful. I started to wonder about fear, its variants, and its utility for climate justice when I saw footage of Greta Thunberg in 2019 fiercely telling world leaders "I want you to feel the fear I feel every day." Perhaps less intense forms of fear, specifically *worry,* can be motivating for a broad demographic, especially if we pinpoint the "objects of care"[1] the emotion of worry is directed toward.

Research on fear often dismisses it as an obstacle to overcome, usually by way of hope or another positive emotion. Articles with provocative titles like "Inspire Hope, Not Fear"[2] and "Fear Won't Do It"[3] are plentiful. Two of my UI sociology colleagues, Kristin Haltinner and Dilshani Sarathchandra, are among many who've suggested climate change skepticism is fueled by an "exaggerated ostrich effect" in which fear drives information aversion.[4] Sarah Jaquette Ray cites their 2018 article in one of her recent essays, suggesting we should couch climate science in terms of the "benefits of making change" rather than the "dangers of continuing with the status quo."[5] I agree that the "carrot of desire" is often more productive than the "sticks of guilt or fear,"[6] and I am very onboard with empowering students to work for change. But I see two problems with the widespread dismissal of fearful feelings. First, I don't think it's usually fear we are actually talking about. Fear is an immediate response to an "imminent threat"[7]—not what most of us are feeling day-to-day in relation to the climate crisis. Second, the way discrete emotions such as worry, dread, or anxiety are reduced to, or used interchangeably with, fear oversimplifies our affective lives. I have written elsewhere about differences between fear, anxiety, dread, worry, and "pandemic panic"[8] and suggested that some critiques of fear—that it's a "fight, flight, or freeze" response, that it's paralyzing or fundamentally conservative, that it prevents long-term engagement—might not be true of worry.[9] Indeed, worry could be an affective resource for tackling climate change.

Importantly for environmental educators, worry is not synonymous with the "chronic" feeling of eco-anxiety defined by the American Psychological Association.[10] In addition to Ray's important work on climate anxiety, other participants in the Climate Justice Educators network are expanding conceptions of anxiety and worry. Panu Pihkala has developed influential typologies of ecological anxiety.[11] Krista Hiser and Matthew Lynch's innovative research tracks what college students in Hawaii believe, feel, and do about climate change, as well as where their knowledge comes from.[12] "Worried" is among the more frequently mentioned responses to their team's question, How do you feel about all of this? Alan E. Stewart's Climate Change Worry Scale provides another new approach for measuring "proximal worry about climate change rather than social or global impacts."[13]

My interest in worry grew from a collaboration with the aforementioned colleagues, Drs. Haltinner and Sarathchandra, whose research focuses on understanding U.S. climate skepticism. These colleagues invited me to partner with them because of my background in affect studies, which helped us distinguish worry as a troubled or unsettled mental state that can fuel ongoing anxiety but that is not reducible to it. Our collaboration generated a definition of worry as "an affective state that directs anxious feelings toward particular objects of care without succumbing to the pitfalls of fear."[14] This understanding dovetails with research by Pihkala, for whom worry is not synonymous with eco-anxiety but can still fall under its umbrella, and with Susan Clayton and Bryan T. Karazsia's work on climate change anxiety, which suggests worry has a less serious impact on individuals.[15]

Starting with the premise that worry, unlike anxiety, takes an object, revealed different things in the Contexture data set—33 interviews and 1000 surveys with self-identified climate skeptics. It turns out that a significant number of self-identified skeptics express worry and dread about climate change. Many of these people point to specific "objects of care," a phrase Susie Wang et al. coined to refer to the "valued objects" that climate change threatens.[16] Climate change is not, itself, an object of care; rather, we care, and worry, about the people, places, and species that climate change stands to impact. In our data set, 15.7 percent of skeptics expressed having "quite a bit, very much, or an extreme amount" of worry about the impacts of climate change, including pollution, environmental damage, and species extinction. Another 11.5 percent expressed similar amounts of dread, which they connected to even more existential threats, such as environmental disasters and their consequences for their home regions, our own species, and the planet itself. Importantly, these concerns exist even though skeptics don't think of the threats in terms of anthropogenic climate change.

This research also supports what earlier studies suggested: that worry correlates strongly with environmental concern and policy support. Smith and Leiserowitz suggest that, provided they are "calibrated to be neither too mild nor too intense," appeals to worry "can motivate and promote . . . the kind of deliberative and iterative decision-making climate change requires."[17] Haltinner and Sarathchandra's more recent data updates and extends these findings by considering a unique group—climate change skeptics—and by adding specificity to understandings of what worry is as well as which policies skeptics tend to support: reducing pollution, preventing deforestation, and investing in renewable energy.[18] Interestingly, even some skeptics might be allies in mitigating the effects of climate change, an important reminder for those of us in classrooms and communities like my own.

A (VERY CONDENSED) TALE OF TWO CLI-FI COURSES

I remind my students frequently that the word *emotion* carries, etymologically, a connotation of movement and agitation; in a sense, emotions are always "anxious" or unsettling. But we are each unsettled by different triggers, and in different ways, depending on our positionality, including aspects of privilege. Many inhabitants of North Idaho are not yet registering the effects of climate change as intensely and urgently as people in other places. Many of my students have been relatively insulated from the climate anxiety others feel constantly. But that seems to be changing, perhaps due in part to the new summer normal of constant wildfires and apocalyptic smoke—conditions that feel, as the young man in Coeur d'Alene said, scary.

In the final weeks of my spring 2018 upper division climate fiction course (provocatively titled "Apocalypse Now: Disaster, Risk, and Ecocatastrophe in American Environmental Literature and Film"), I asked my students how they were feeling. We'd weathered some heavy subject matter: refugees, starvation, cannibalism, innovative forms of slavery, devastating fires and floods, and new drugs that make people orgasmic over watching other people burn. While I'd expected my students to fess up to ecological grief or climate anxiety, they told me they felt afraid, but they were glad about it. The fear these texts sparked was too rare, they said. It motivated them.

Two years later, in spring of 2021, I taught the same class, but now with pandemic-weary students. Not surprisingly, these students were not so eager to celebrate fear. Overwhelm, anxiety, despair, and sadness haunted our Zoom chats and Slack posts. Anticipating this, I had retooled the reading list to be less depressing. McCarthy was out; Kingsolver was in. With the help of a brilliant MFA student, Gianna Stoddard, who worked with me as a teaching assistant, we foregrounded fear and other emotions from day one. We opened with one of Jessica Creane's climate change games (see Creane's chapter in this volume) and asked four questions in the Zoom chat: What are your environmental fears? What are your environmental hopes? What are you willing to do for the environment? What are you willing to do for each other?[19]

We returned to our "fears and hopes" list—and the specific objects of care articulated there—repeatedly. Foregrounding these emotions was a great way to lead in to self-reflexivity around climate change and privilege, a process Gianna initiated in their first facilitation day. Inspired by sociologist Peter Kaufman's work on reflexivity,[20] they asked students to respond to two questions, and then to one another's answers, in an anonymous Google doc:

- What feelings come up when you think about climate change? How do you think these feelings have been shaped by your social positionality (race, class, age, gender, ability, sexuality, national identity, etc.)?
- What actions have you taken/do you take in your everyday life to combat climate change? What feelings or associations come up when you think about someone who does not perform similar actions?

Our students added "hometown" to the categories of positionality listed, an important reminder that identity is anchored in place—often, for Idaho students, a rural and/or remote home (see also Hall's chapter in this volume). Relatedly, some students noted the lack of access to resources like recycling due to our remote locations and/or political/ideological resistance to environmentalist practices. Although many students mentioned individual actions—like composting or opting out of "fast fashion"—they also noticed the tension between their felt lack of individual agency and the need to change our larger political and economic systems. A smattering of hope showed through, and several were pleasantly surprised at how much their classmates knew, and were doing, about climate change.

Gianna's activity laid a strong foundation for discussing Kingsolver's *Flight Behavior*. One poignant scene in the text features an environmental activist who descends on a small Appalachian town in order to ask residents to reduce their carbon footprints and sign a "sustainability pledge." The novel's feisty protagonist, as it turns out, doesn't eat out, doesn't buy new clothes, has never flown on an airplane. Her carbon footprint is tiny, and her relative lack of privilege is exposed in the conversation. Many of our students could relate to her rural, working-class perspective.

We started that class with an environmental privilege checklist adapted from Peggy McIntosh's classic white privilege "knapsack" exercise.[21] One brave Latinx student shared a story about being racially profiled in Southern California, and that opened up a conversation about how the checklist presumes, and risks recentering, a white reader. Interestingly, very few of my students found their experiences represented in the checklist, likely because of the class privilege it highlights. The list may presume a white audience, but its attempt to raise awareness of class privilege seems effective when that is, indeed, the audience involved. I'd like to do more next time by way of identifying and critiquing the objects of care embedded in the list—public parks, local produce, wilderness areas, and more—and comparing them to our own, perhaps by way of a "mind map" exercise, such as the "mind map of ecological emotions" (see Pihkala's chapter in this volume).

CONCLUDING THOUGHTS

The points I've been making about fear and climate privilege, as well as how worry might be harnessed for climate justice, might seem controversial. I want to be clear that I'm not suggesting we should aim to terrify our students. But, as the story about the young man in Coeur d'Alene suggests, fear can be a wake-up call. Fear might also be a catalyst for a longer-term, low-grade worry that—when tied to objects of care—can motivate action. Perhaps the key is to nurture the right kind of worry: "normal and adaptive" rather than the "persistent, problematic" kind, which feels "repetitive and uncontrollable" and leads to disengagement.[22] While both can feel stressful, "normal" worry "involves attention to a threat in the environment and provides resources for thinking and problem solving."[23] We educators can help provide such resources by connecting our own and our students' worries to specific objects of care.

As more Americans in particular express worry, concern, and alarm,[24] let's not dismiss, reduce, or pathologize these feelings but rather couple them with self-reflexivity, a sense of efficacy, and opportunities for collective action. I'm inclined to encourage students to better understand our affective lives in terms of a "spectra of micro-emotions"[25] fraught with cognitive and (I'd add) *affective* dissonance.[26] Providing occasions for "cognitive resonance that reduce emotional stress and anxiety"—small actions like composting that resonate with students' environmental beliefs—is a useful strategy for tempering stressful emotions.[27] The tools in the Climate Justice Educators Toolkit will be invaluable in offering additional resources for navigating dissonant feelings and working toward climate justice in a range of classrooms, including those with demographics similar to mine.

Finally, I'm sensitive to the fact that focusing on the affective lives and political agencies of my largely white students risks recentering whiteness, shoring up white fragility, or perpetuating a "white-savior industrial complex."[28] I am still learning how to mitigate these risks.[29] But we shouldn't avoid engaging the emotions of significant subsets of the population just because they are in positions of privilege. The "scared" young man in Coeur d'Alene, and other rural, white, and/or working-class students, are no less in need of support in understanding, legitimizing, and contextualizing their emotions. Indeed, inviting a broader demographic to identify their worries about the local impacts they can no longer ignore could help combat U.S. political polarization as we encourage climate skeptics and others to acknowledge threats to commonly held objects of care. Making space for white *men*, in particular, to express emotions about perceived environmental changes is also a step toward confronting toxic masculinity—including the pressure for rural white men to embrace a stubborn brand of "toughness" in the face of environmental and economic challenges—and toward dismantling stereotypes about emotions and gender, including that women are more emotionally attuned to "nature."

Our classrooms should be spaces for vulnerability and humility across identity categories, as well as places where we use intersectional approaches to probe the ways that emotions are politicized, foreclosed, presumed, or unevenly available to us, depending on who we are and how we identify. As Ray puts it: "we should approach the emotional landscape of climate change with a robust awareness of the cultural politics of emotion: how they shape power relations and change culture, and how culture and power mediate emotions."[30] Working with our students to be self-reflexive about our beliefs and forms of privilege, to listen to and collaborate with all sorts of people, and to recognize the complex cultural politics of emotion are important goals for all of us who want to move toward a more just future.

NOTES

1. Wang et al., "Emotions Predict Policy Support."
2. Ring, "Inspire Hope, Not Fear."
3. O'Neill and Nicholson-Cole, "Fear Won't Do It."
4. Haltinner and Sarathchandra, "Climate Change Skepticism."
5. Ray, "Climate Change Is Scary."
6. Ibid.
7. Stewart, "Psychometric Properties."
8. Ladino, "Who's Afraid of the Climate Crisis?"
9. Haltinner et al., "Feeling Skeptical."
10. Clayton et al., *Mental Health*.
11. Pihkala, "Anxiety."
12. Hiser and Lynch, "Worry and Hope."
13. Stewart, "Psychometric Properties."
14. Haltinner et al., "Feeling Skeptical."
15. Clayton and Karazsia, "Development and Validation."
16. Wang et al., "Emotions Predict Policy Support," 26.
17. Smith and Leiserowitz, "Role of Emotion," 945.
18. Haltinner and Sarathchandra, "Pro-Environmental Views of Climate Skeptics."
19. Creane, "Climate Change Games."
20. Kaufman, "Scribo Ergo Cogito."
21. McIntosh, "White Privilege." See also Crampton, "Race and Class Privilege."
22. Stewart, "Psychometric Properties."
23. Ibid.
24. Leiserowitz et al., *Climate Change in the American Mind*.
25. Hiser and Lynch, "Worry and Hope."
26. Ladino, *Memorials Matter*, 22–23.
27. Hiser and Lynch, "Worry and Hope."
28. Cole, "White-Savior Industrial Complex."
29. I'm grateful to Sarah Jaquette Ray for helping me think through these risks as well as the potential benefits to engaging white men and others in positions of privilege. Her input shaped the thinking in this paragraph in substantial ways. For more of her thoughts on anxiety, identity, and threat perception, see Ray, "Who Feels Climate Anxiety?"
30. Ibid.

REFERENCES

Cole, Teju. "The White-Savior Industrial Complex." *The Atlantic*, June 6, 2021. https://www.theatlantic.com/international/archive/2012/03/the-white-savior-industrial-complex/254843.

Clayton, S., and B. T. Karazsia. "Development and Validation of a Measure of Climate Change Anxiety." *Journal of Environmental Psychology*, no. 69 (2020): 1–11. https://doi.org/10.1016/j.jenvp.2020.101434.

Clayton, S., C. M. Manning, K. Krygsman, and M. Speiser. *Mental Health and Our Changing Climate: Impacts, Implications, and Guidance*. Washington, DC: American Psychological Association, and ecoAmerica, 2017.

Crampton, Liz. "Race and Class Privilege in the Environmental Movement." *Pachamama*, September 9, 2012. https://news.pachamama.org/news/race-and-class-privilege-in-the-environmental-movement.

Creane, Jessica. "Climate Change Games." Penn Program on the Environmental Humanities, May 2020. https://ppehlab.com/csds/climategames.

Haltinner, K., and D. Sarathchandra. "Climate Change Skepticism as a Psychological Coping Strategy." *Sociology Compass*, no. 12 (2018). https://doi.org/10.1111/soc4.12586.

———. "Pro-Environmental Views of Climate Skeptics." *Contexts* (2020): 36–41. https://doi.org/10.1177/1536504220902200

Haltinner, K., J. Ladino, and D. Sarathchandra. "Feeling Skeptical: Worry, Dread, and Support for Environmental Policy Among Climate Change Skeptics." *Emotion, Space, and Society*, no. 39 (2021). https://doi.org/10.1016/j.emospa.2021.100790.

Hiser, Krista K., and Matthew K. Lynch. "Worry and Hope: What College Students Know, Think, Feel, and Do about Climate Change." *Journal of Community Engagement and Scholarship* 13, no. 3 (2021): Article 7.

Kaufman, P. "Scribo Ergo Cogito: Reflexivity through Writing." *Teaching Sociology* 41, no. 1 (2013): 70–81. http://www.jstor.org/stable/41725581.

Ladino, Jennifer. *Memorials Matter: Emotion, Environment, and Public Memory at American Historical Sites*. Reno: University of Nevada Press, 2019.

———. "Who's Afraid of the Climate Crisis? Fear, Anxiety, Dread, and Pandemic Panic." *Edge Effects*, April 21, 2020. https://edgeeffects.net/whos-afraid-of-the-climate-crisis-fear-anxiety-dread-and-pandemic-panic.

Leiserowitz, A., E. Maibach, S. Rosenthal, J. Kotcher, P. Bergquist, M. Ballew, M. Goldberg, A. Gustafson, and X. Wang. *Climate Change in the American Mind*. New Haven: Yale Program on Climate Change Communication, 2020.

McIntosh, Peggy. "White Privilege: Unpacking the Invisible Knapsack." *Peace and Freedom Magazine*, July 1989, 10–12.

O'Neill, S., and S. Nicholson-Cole. "'Fear Won't Do It': Promoting Positive Engagement with Climate Change through Visual and Iconic Representations." *Science Communication* 30, no. 3 (2009): 355–79.

Pihkala, Panu. "Anxiety and the Ecological Crisis: An Analysis of Eco-Anxiety and Climate Anxiety." *Sustainability*, no. 12 (2020).

Ray, Sarah Jaquette. "Climate Change Is Scary: Here Are 7 Tools to Help You Keep Your Cool on a Warming Planet." *Resilience*, February 5, 2020. https://www.resilience.org/stories/2020-02-05/climate-change-is-scary-here-are-7-tools-to-help-you-keep-your-cool-on-a-warming-planet.

———. *A Field Guide to Climate Anxiety: How to Keep Your Cool on a Warming Planet*. Oakland: University of California Press, 2020.

———. "Who Feels Climate Anxiety?" *Cairo Review*, no. 43 (2021). https://www.thecairoreview.com/home-page/who-feels-climate-anxiety.

Ring, W. "Inspire Hope, Not Fear: Communicating Effectively about Climate Change and Health." *Annals of Global Health* 81, no. 3 (2015): 410–15.

Smith, N., and A. Leiserowitz. "The Role of Emotion in Global Warming Policy Support and Opposition." *Risk Analysis* 34, no. 5 (2014): 937–48.

Stewart, A. E. "Psychometric Properties of the Climate Change Worry Scale." *International Journal of Environmental Research and Public Health*, 18, no. 2 (2021): 494. https://doi.org/10.3390/ijerph18020494.

Wang, S., Z. Leviston, M. Hurlstone, C. Lawrence, and I. Walker. "Emotions Predict Policy Support: Why It Matters How People Feel about Climate Change." *Global Environmental Change*, no. 50 (2018): 25–40.

TWENTY-FOUR

The Social Ecology of Responsibility

Navigating the Epistemic and Affective Dimensions of the Climate Crisis

AUDREY BRYAN

This chapter presents a pedagogical framework and toolkit to explore the complex relationship between "ordinary" climate-related harms committed by individuals and those carried out by larger entities, including governments, industries, and corporations. It also explores the potential of art to illuminate these complex relationships and help learners navigate difficult emotions in relation to their own "implicatedness" in the climate crisis.

This chapter seeks to inform the development of educational resources that address the complex range of emotions learners experience when encountering knowledge that relates to their own involvement or implication in the climate crisis.[1] A central premise underpinning the chapter is the need for climate change education to pay attention to both the *epistemic* (knowledge-based) and *affective* (emotional) complexities of global warming. The affective dimensions of climate change discussed in this chapter relate primarily to individuals whose lifestyles are heavily dependent on fossil fuels, a majority—although not all of whom—are located in the Global North.[2] Individuals living in emissions-intensive societies can experience a range of psycho-affective responses when confronted with difficult knowledge about their involvement in the unfolding climate crisis, including varying levels of denial, discomfort, guilt, anxiety, ambivalence, apathy, embarrassment, grief, and despair.[3]

The chapter considers the potential for climate change art and environmental imagery to serve as a springboard for reflecting on—and working through—difficult emotions that arise when learners are confronted with knowledge about their own culpability for the climate crisis. It seeks to illuminate art as a pedagogical tool for generating productive questions about what we can do in the face of climate change harms and injustices. For illustrative purposes, I focus on a public art installation—*Pyramids of Garbage*—as an example of how artistic representations can at once promote critical reflection and understanding of learners' own implication in the climate crisis (and the complicated emotions this entails) while simultaneously inviting them to view themselves as active agents in its amelioration.[4] I present this illustration alongside a pedagogical model—the Social Ecology of Responsibly Framework (SERF)—which seeks to illuminate the complex interconnections between the micro world of everyday encounters and the operations and influence of institutions and the macro forces of economy, culture, media, and ideologies.

CLIMATE CHANGE ART AS A PEDAGOGICAL AND POLITICAL TOOL

Visual representation of the climate crisis has been identified as an important yet underexplored aspect of the politics of global warming.[5] The emergence of new visual artistic forms and photojournalistic images depicting various aspects of the ecological crisis provides a rich source of visual material that can prompt critical reflection about global warming and how best to respond to it. Whereas certain artistic genres have been criticized for their tendency to position viewers as mere *voyeurs* or *passive spectators*—rather than *active agents*—in the climate crisis, others encourage viewers to apprehend their role in the crisis and to reconfigure their relationship with the planet.[6]

> Artworks that implicate the viewer in the conditions being depicted serve as an "invitation to human agency" and therefore have the potential to be both pedagogically transformative and politically efficacious.

Yet the scale and severity of the climate-related harms that many individuals living in emissions-intensive societies are responsible for can bring about a range of negative psychoaffective responses or defense mechanisms, ranging from deep or outright forms of denial to feelings and processes that minimize the ecological crisis and their involvement in it.[7] Moreover, the sheer complexity of the dynamics that operate within and across different levels of the social system to produce multiple and mutually reinforcing climate-related harms and injustices can obscure these transformative possibilities. For these reasons, justice-oriented climate change education needs to pay attention to both the epistemic (knowledge-based) and affective complexities of the climate crisis (see Trott's essay in this volume for more on education's aversion to emotion as a justice issue). Pedagogical encounters with questions of personal and social responsibility for the ecological crisis must attend simultaneously to the affective responses that climate change knowledge can elicit as well as to the cognitive or epistemological complexity that is associated with attempting to illuminate complex chains of causal responsibility in consumer capitalist, emissions-intensive contexts (see Oziewicz's chapter on the politics of guilt in this volume).

The next section outlines a model that explicates the "symbiotic relationship" that exists between micro-level, routinized acts and practices carried out by individuals and macro-level "carbon crimes of the powerful."[8] It then briefly discusses Bahia Shehab's public art installation—*Pyramids of Garbage*—as an example of how climate change art can simultaneously promote critical reflection and understanding of learners' own implication in the climate crisis (and the complicated emotions this entails) while also inviting them to view themselves as active agents in its amelioration.

THE SOCIAL ECOLOGY OF RESPONSIBILITY FRAMEWORK (SERF)

SERF was designed to promote understanding of how individual acts and structural forces interact and reinforce one another in the context of global warming, so that individuals can better apprehend their own role in exacerbating and alleviating the climate crisis.[9] It is premised on the notion that all of those living in emissions-intensive societies bear at least some responsibility for the climate crisis, while acknowledging that this responsibility, and the trauma associated with ecological breakdown, are unevenly and unequally distributed, with those least responsible for the problem often bearing the brunt of the pain. While acknowledging the uneven distribution of responsibility as well as climate-related trauma and responsibility within emissions-intensive societies, the framework conceptualizes the climate crisis as a form of difficult knowledge for those whose lifestyles are heavily dependent on fossil fuels, a majority of whom are located in the Global North.[10] As such, it highlights the "ordinary harms" that those who lead fossil-fuel-soaked lifestyles routinely commit, which collectively have a substantial and deleterious impact on environmental problems (see Figure 24.1).[11] While not discounting differences of scale between individual actions and state-corporate practices, SERF transcends either/or approaches to climate responsibility which privilege either personal actions or corporate-state crimes. From the perspective of climate justice, this binaristic thinking is problematic because it fails to recognize or at least sufficiently acknowledge how individuals occupy different social-structural locations that position them as relatively more or less responsible for the harms that contribute to global warming.

As shown in Figure 24.1, the "implicated subject" stands at the center of a wider ecology of responsibility, comprising social actors, norms, institutions, ideologies, conditions, and processes involved in the production and intensification of the climate crisis.[12] The ecology comprises a series of nested and interrelated systems, ranging from the microsystem (the immediate contexts, relationships or organizations we interact with) to the macrosystem (societal and transnational structures, political-economic arrangements, cultural norms/values, etc.). Each element or "system" that makes up the wider ecology allows us to visualize how individual, ordinary acts that are carried out in everyday settings (the home, school, workplace, etc.) are influenced by, but also shape, increasingly distal forces and highlights

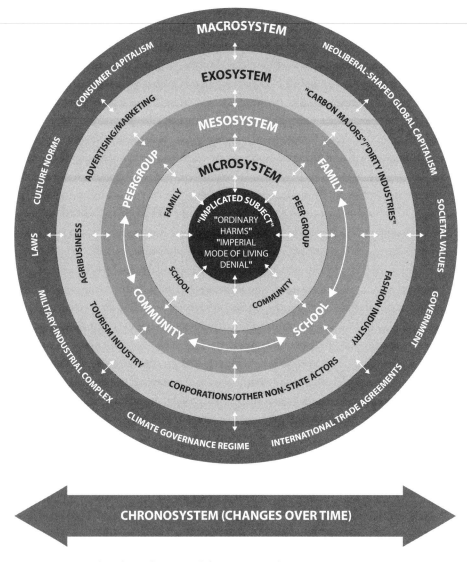

FIGURE 24.1 Social Ecology of Responsibility Framework (SERF).

the need for radical change within and across multiple scales and domains. In other words, it helps learners to come to a deeper understanding of the symbiotic relationship that exists between individuals and larger entities and encourages them to engage with the crisis from a position of non-innocence or self-implication. Furthermore, it enables them to develop a deeper understanding of themselves as agents who have the capacity to de-implicate themselves from the climate crisis by taking steps to mitigate the effects of global warming, and to appreciate the unevenness of humanity's responsibility for these ecological injustices.

Bahia Shehab's *Pyramids of Garbage* (see onsite photograph, Figure 24.2) maps the social ecology of responsibility in relation to patterns of over-consumption in consumer capitalist societies and provides a concrete example of how SERF can be used to inform the interconnections between individuals and different systems that contribute to global warming. SERF documents how the wider social, political, economic, and cultural environment *promotes* harmful consumerist philosophies and practices amongst individuals, while helping to explain the context within which individuals become such enthusiastic *practitioners* of these harms.[13] The framework depicts individuals as "implicated subjects" who form part of a wider infrastructure or *machinery* that produces ecological degradation and illuminates the need for radical change within and across multiple scales and domains.[14] This framework also allows for a consideration of those individuals and communities within emissions-intensive societies who proactively take measures to de-implicate themselves from the climate crisis and/or to dismantle the wider infrastructure which has created the problem in the first instance. The youth-led global climate movement inspired by young people such as Zeena Abdulkarim, Autumn Peltier, and Greta Thunberg has highlighted the vital role that individuals collectively play in putting pressure on world leaders to commit to leaving carbon reserves in the ground, and the radical steps that some are prepared to take to minimize the harms that are wreaking havoc on the planet.

CLIMATE CHANGE ART AS AN INVITATION TO AGENCY

One of the major challenges for climate change educators is to scaffold learners living in emissions-intensive societies to better comprehend their role in the existential threat of global warming, while at the same time not foreclosing possibilities for action and amelioration. A related challenge entails lifting up the efforts by those communities who resist global warming and other effects of the colonial-capitalist project, with which students may identify or belong to.

> Climate change education must tackle learners' negative affective reactions to difficult knowledge that implicates many of them in the conditions being taught.

Climate change art can scaffold and augment epistemic or knowledge-based learning encounters by at once prompting critical self-reflection on the myriad and profound ways that ordinary citizens, especially the most privileged, are implicated in the climate catastrophe while also providing opportunities for meaningful engagement with "messy" emotional responses—and indeed resistances—to the imagery such as loss, guilt, shame, disgust, anger, apathy, denial, etc. When these messy affective responses become the focus of individual and collective inquiry, reflection, and analysis, they serve as the basis for productive

FIGURE 24.2 *Pyramids of Garbage.* © Bahia Shehab.

questioning about what can be done in the face of profound climate change harms and injustices. In other words, in helping us to recognize some of the psychic/affective mechanisms that prevent us from taking effective action in relation to the crisis or from seeing our own implication, climate change art can serve as an invitation to respond differently to the crisis.[15]

PYRAMIDS OF GARBAGE

Bahia Shehab's *Pyramid of Garbage* installation in Cairo, Egypt, home to the great pyramids of Giza, offers a particularly striking and affecting illustration of our implication in the destruction of the planet. The enormity of the structure—which juxtaposes the eternity and majesty of the great pyramids of Giza with the ugliness and wastefulness of overconsumption—makes it difficult to deny the destructiveness of routine consumeristic practices and choices while simultaneously posing wider questions of collective responsibility.

In making us bear witness to the destruction of this majestic landscape and to global ecological decline more broadly, this arresting monument paves the way to ethical and political transformation. The negative affect viewers experience while interacting with the installation may derive from being asked to bear witness to the deleterious effects of consumer capitalistic ideologies which foster human greed and wastefulness.

Yet this negative affect that the artwork induces also serves as an "invitation to human agency" and therefore opens up vital possibilities for social transformation.[16] In other words, the negative affective weight of the monument serves as an invitation to respond otherwise to the ecological crisis by altering consumption practices that drive climate change as well as the wider ideologies and systems that endorse and sustain them. In the artist's own words:

As a species, we have built monuments that have defeated time. We have designed civilizations that dreamt of eternity. With climate change, this eternity is now challenged. Now is the time for us to rethink our legacy on this planet. Are we going to come together to build a sustainable future for all of us—or will our new legacy be pyramids of garbage?[17]

From this perspective, the grief that accompanies this challenged eternity can be a productive, rather than an immobilizing, emotion.[18] In asking us to bear witness to the magnitude and ugliness of ecological decline, and to consider "our legacy on this planet" by critically reflecting on—and working through—our epistemological and affective responses to this difficult knowledge, the possibility of mitigating the crisis becomes more palpable. For example, when affective responses to the difficult knowledge of climate change become a focus of inquiry this can serve as a basis for interrupting different levels and forms of climate change-related denial.[19]

The conceptual framework presented in this chapter comprises a social ecological model of responsibility that demonstrates the impossibility of disarticulating individual, private actions from state-corporate climate injustices and locates the "implicated subject" at the heart of this complex ecology.

Art has the potential to promote deeper awareness about how difficult climate-related knowledge is experienced, negotiated, and resisted and can arouse learners' ethical and political imaginations.[20] Most significantly, the artistic and epistemological tools outlined above can help to bring into conscious awareness the complex dynamics and ecology of implication, from which meaningful solutions to the climate crisis begin to reveal themselves.

NOTES

1. Andreotti, "Educational Challenges"; Rothberg, *Implicated Subject*.
2. Brand and Wissen, *Imperial Mode of Living*.
3. Garrett, "Learning to Tolerate"; Kelly and Kelly, "Becoming Vulnerable"; Seymour, *Bad Environmentalism*; Stein et al., "From 'Education for Sustainable Development'"; Van Kessel, "Teaching the Climate Crisis."
4. Shehab, *Pyramids of Garbage*.
5. O'Neill, "More Than Meets the Eye."
6. Nurmis, "Visual Climate Change Art 2005–2015."
7. Adams, *Ecological Crisis*.
8. Agnew, "Ordinary Acts," 95.
9. Bryan, "Pedagogy of the Implicated."
10. Brand and Wissen, *Imperial Mode of Living*.
11. Daggett, "Petro-masculinity," 29; Agnew, "Ordinary Acts."
12. Rothberg, *Implicated Subject*.
13. Agnew, "Ordinary Acts."

14. Rothberg, *Implicated Subject*.
15. Hariman and Luciates, *Public Image*, 93.
16. Ibid., 92.
17. Shehab, *Pyramids of Garbage*.
18. Atkinson, "Climate Grief"; Cunsolo and Landman, "Introduction."
19. Adams, *Ecological Crisis*; Bryan, "Pedagogy of the Implicated"; Garrett, "Learning to Tolerate."
20. Bryan, *Affective Pedagogies*.

REFERENCES

Adams, Matthew. *Ecological Crisis, Sustainability and the Psychosocial Subject*. New York: Palgrave Macmillan, 2016.

Agnew, Robert. "The Ordinary Acts That Contribute to Ecocide: A Criminological Analysis." In *Routledge International Handbook of Green Criminology*, edited by Avi Brisman and Nigel South, 58–72. New York: Routledge, 2013.

Andreotti, Vanessa. "The Educational Challenges of Imagining the World Differently." *Canadian Journal of Development Studies /Revue canadienne d'études du développement* 37, no. 1 (2016): 101–12. https://doi.org/10.1080/02255189.2016.1134456.

Atkinson, Jennifer. "Climate Grief: Our Greatest Ally?" *Resilience*, August 27, 2020. https://www.resilience.org/stories/2020-08-27/climate-grief-our-greatest-ally.

Brand, Ulrich, and Markus Wissen. *The Imperial Mode of Living. Everyday Life and the Ecological Crisis of Capitalism*. London: Verso, 2021.

Bryan, Audrey. "Affective Pedagogies: Foregrounding Emotion in Climate Change Education." *Policy and Practice: A Development Education Review* 30, no. 1 (2020): 8–30. https://www.developmenteducationreview.com/issue/issue-30/affective-pedagogies-foregrounding-emotion-climate-change-education.

———. "Pedagogy of the Implicated: Advancing a Social Ecology of Responsibility Framework to Promote Deeper Understanding of the Climate Crisis." *Pedagogy Culture and Society* 30, no. 3 (2022): 329–48. https://doi.org/10.1080/14681366.2021.1977979.

Craps, Stef. "Introduction: Ecological Grief." *American Imago* 77, no. 1 (2020): 1–7. https://doi.org/10.1353/aim.2020.0000.

Cunsolo, Ashlee, and Karen Landman. "Introduction: To Mourn beyond the Human." In *Mourning Nature: Hope at the Heart of Ecological Loss and Grief*, edited by Ashlee Cunsolo and Karen Landma, 3–26. Montreal: McGill-Queen's University Press, 2017.

Daggett, Cara. "Petro-masculinity: Fossil Fuels and Authoritarian Desire." *Millennium: Journal of International Studies* 47, no. 1 (2018): 25–44. https://doi.org/10.1177/0305829818775817.

Garrett, James. *Learning to Be in the World with Others: Difficult Knowledge and Social Studies Education*. New York: Peter Lang, 2017.

———. "Learning to Tolerate the Devastating Realities of Climate Crises." *Theory and Research in Social Education* 47, no. 4 (2019): 609–14. https://doi.org/10.1080/00933104.2019.1656989.

Hariman, Robert, and John Louis Luciates. *The Public Image: Photography and Civic Spectatorship*. Chicago: University of Chicago Press, 2016.

Kelly, Ute, and Rhys Kelly. "Becoming Vulnerable in the Era of Climate Change: Questions and Dilemmas for a Pedagogy of Vulnerability." In *Pedagogy of Vulnerability*, edited by Edward Brantmeier and Maria McKenna, 177–202. Charlotte, NC: Information Age Publishing, 2020.

Milner, Murray. *Freaks, Geeks, and Cool Kids: Teenagers in an Era of Consumerism, Standardized Tests, and Social Media.* New York: Routledge, 2015.

Nurmis, Joanna. "Visual Climate Change Art 2005–2015: Discourse and Practice." *Wiley Interdisciplinary Reviews* 7 (2016): 501–16. https://doi.org/10.1002/wcc.400.

O'Neill, Saffron. "More Than Meets the Eye: A Longitudinal Analysis of Climate Change Imagery in the Print Media. *Climatic Change* 163 (2020): 9–26. https://doi.org/10.1007/s10584-019-02504-8.

Pugh, Alison. *Longing and Belonging: Parents, Children and Consumer Culture.* Berkeley: University of California Press, 2009.

Rothberg, Michael. *The Implicated Subject: Beyond Victims and Perpetrators.* Palo Alto: Stanford University Press, 2009.

Seymour, Nicole. *Bad Environmentalism: Irony and Irreverence in the Ecological Age.* Minneapolis: University of Minneapolis Press, 2018.

Shehab, Bahia. *Pyramids of Garbage.* 2020. https://www.bahiashehab.com/public-installations/pyramids-of-garbage.

Stein, Sharon, Vanessa Andreotti, Rene Suša, Cash Ahenakew, and Tereza Čajková. "From 'Education for Sustainable Development' to 'Education for the End of the World as We Know It.'" *Educational Philosophy and Theory* 54, no. 3 (2022): 274–87. https://doi.org/10.1080/00131857.2020.1835646.

O'Neill, Saffron. "More Than Meets the Eye: A Longitudinal Analysis of Climate Change Imagery in the Print Media. *Climatic Change* 163 (2020): 9–26. https://doi.org/10.1007/s10584-019-02504-8.

Van Kessel, Cathryn. "Teaching the Climate Crisis: Existential Considerations." *Journal of Curriculum Studies Research* 2, no. 1 (2020): 129–45. https://doi.org/10.46303/jcsr.02.01.8.

White, Rob. "Ecocide and the Carbon Crimes of the Powerful." *University of Tasmania Law Review* 37, no. 2 (2018): 95–115.

TWENTY-FIVE

Beyond the Accountability Paradox

Climate Guilt and the Systemic Drivers of Climate Change

MAREK OZIEWICZ

This chapter considers climate guilt as a form of misremembering and suggests two activities that help students confront the accountability paradox which enables climate guilt as a substitute for constructive climate action.

In an episode of the HBO series *American Gods,* dead character Laura finds herself in purgatory. In a clichéd version of an afterlife experience, she is forced to watch a video projection of her own life. Laura is livid because she *knows* her life's story. As the film begins, she steps into the screen to fast-forward the projection. She tells the audience how she had facilitated her father's sex prowls; how she encouraged his promiscuity and enjoyed the pain it caused to others. But as she is made to *actually watch* the story, it turns out that she had misremembered everything. Laura was a victim. As she and the audience realize, her father's worst abuse was to make little Laura internalize the guilt for how he broke up their family.

This chapter invites you to imagine a scenario in which Laura's story is our story as individuals living in a climate-changed world—to imagine that we have misremembered what brought about climate disasters. It invites us to consider that we have accepted the guilt for ecocide that has been enabled not by our individual choices but by systemic drivers. The activities I share here are not meant to suggest that our individual choices are inconsequential. Quite the contrary: these choices are crucial for enabling a transition to an ecological civilization. That said, I have seen students crushed by climate guilt and accepting climate change as a kind of punishment for being implicated as its drivers. Not surprisingly, just what it means to be implicated can take various meanings: being born a human, living in the U.S., being white, being middle class, being part of the consumerist culture, eating meat, etc. Based on conversations in the classroom, I feel that *different degrees* of climate guilt afflict BIPOC and white students alike; they afflict students born into privilege and students from underprivileged backgrounds, even those who have a clear sense of being victims of the settler-colonial, white-supremacist, neoliberal, ecocidal system. Climate guilt, in other words, appears to be a generation-specific phenomenon that transcends identity politics and emerges as a first-order emotional response to learning about the systemic drivers of climate change. This reaction is made possible by the disconnect, embedded in today's (technocratic) climate change education, between the epistemic and affective learning (see Bryan's essay in this volume). While it is often linked to privilege, climate guilt is not an issue faced solely by predominantly white, privileged audiences. Climate guilt is an important factor in any considerations of climate justice (see essays in Part II of this volume). And it makes a difference between passivity and action. For these reasons, I consider addressing guilt to be among our key challenges as climate educators. Specifically, I propose that helping students overcome climate guilt requires conversations about the systemic drivers of climate change. It demands confronting the accountability paradox which lies at the heart of the identity narrative of our petronormative civilization.

> Helping students overcome climate guilt requires conversations about the systemic drivers of climate change.

Climate guilt—also called "eco-guilt"[1] and "green guilt"[2]—is one of the emotions on the spectrum of eco-anxiety[3] closely related to shame.[4] The sources of climate guilt are many: from acknowledging the realities of climate racism and injustice,[5] to guilt about not doing enough as a climate-conscious citizen, not knowing what to do, or guilt for having failed today's children by not stopping climate change in time.[6] In all its varieties, climate guilt is disempowering. On the individual level, climate guilt leads to fear and helplessness;[7] as a collective force, it contributes to a cultural trauma that maintains what Brulle and Norgaard call "climate change social inertia."[8] The paradox, Sarah Jaquette Ray notes, is that while climate guilt is clearly "ineffective for working toward [climate] justice" it is widely used "as an affective strategy in most environmental messages."[9] In other words, our students grow up in a world where the mainstream environmental narrative tells them they should feel guilty about the environment. This message, as I have argued elsewhere, is compounded by the fact that our students also happen to grow up in a culture dominated by dystopian narratives which project human-caused ecological collapse as inevitable (see also Friederici's essay in this volume).[10] The dystopian framework of impending biospheric collapse legitimizes the emotion of climate guilt for humanity's collective present-into-future crimes. Beyond dystopia, the guilt complex has also infiltrated the dominant story systems of Western civilization as part of its ecocidal unconscious.[11]

Here, then, is the accountability paradox. The more students learn about the ecocidal operations of neoliberal capitalism, the easier it is for them to "misremember" ecocide as nobody's fault (i.e., the "inevitable" price of technological progress) or as everybody's fault (i.e., responsibility is shared by everyone who participates in the system). This menu of two options is false, part of the neoliberal legitimizing discourse. Yet it tends to affect even those BIPOC and white students who realize that the demolition of the planet's life support systems is a continuation of the extractive logic "set in motion by the European colonization of much of the world,"[12] and is thus a fault of a particular group of people who have never been held accountable for their actions. In each case, the misremembering generates disempowering emotions. The "nobody's fault" approach induces the cognitive dissonance of feeling guilty that nobody appears to be held accountable for what science clearly identifies as a human-driven damage to the planet. In my experience, the sense of guilt is even more pronounced when students acknowledge their complicity in the ecocidal system, embracing the "everybody's fault" framework. Their climate guilt is then the expression of "doing my part." It builds on young people's sense of fairness and communicates their readiness to shoulder the blame for the sorry state of the planet even if they are unable to take any other constructive action. In fact—and perversely so—guilt *is* a powerful substitute for constructive climate action, all the more so because it appears to be a form of action too.[13]

> Climate guilt is a powerful substitute for constructive climate action, all the more so because it appears to be a form of action too.

I want to share two activities I have used to help my undergrads grapple with climate guilt. The first is an alien ultimatum questionnaire. I designed it to tease out their ideas about the systemic drivers of climate change, which I then put side by side with the students' articulations of climate guilt. The questionnaire opens by asking students to consider the following hypothetical story:

The Guests arrived in 20xx [I have always used the "immediate future" year]. It was like the Second Coming. Just not what anyone expected. Armed with evidence they had collected on humanity's reckless destruction of the biosphere, they came to salvage whatever life remained. Except for humans.

Unable to eliminate or ignore them, the Earth's leaders begged for another chance. The Guests considered. A highly advanced species who have long acted as gardeners of life in other galaxies, they unexpectedly gave us twelve months. Twelve months to reshape our global civilization: our political institutions, social organization, and dominant technologies. Unless we make it sustainable—environmentally, politically, and economically—in twelve months they will start draining the planet's water. They will also relocate all surviving life forms. Except for humans.

This scenario is then followed by the following questions. (1) Would we be able to change our ways? (2) Would governments, banks, and corporations be motivated to re-channel their resources into creating a sustainable global order? (3) Do you believe that we can save the planet and create a future for your grandchildren that's worth living for? (4) Are sustainable technologies, energy, and progress possible? and (5) What do we need to do to make them a reality? In questions 1 through 4 possible answers include Yes, No, and It's complicated, each followed by an optional prompt: In one sentence, explain your choice above.

Between 2018 and 2022 I collected 77 responses from students who consented to this research. One distinct trend I have seen is this: while students have a better grasp of the systemic drivers of climate change than they realize—they do recognize that the assault on the biosphere is a direct consequence of our current economic mode through which we organize the planet's resources, production, and distribution—they nevertheless tend to misattribute blame for this systemic outcome to people in general or human nature in particular. For example, 86.6 percent of students believe that sustainable progress, technologies, and growth are possible; 84.1 percent believe that "we can save the planet and create a future for your grandchildren that's worth living for." That said, only 40.6 percent contend that we would be able to change our ways, even in the face of an alien ultimatum. And 64.7 percent are doubtful whether a capitalist economy can ever be transformed to work for the planet. As one undergraduate explains, "I think that institutions themselves might resist the changes necessary simply because of the way they are designed" (2019). "Even in the event of the end of the world," another student says, "I think that the capitalist society that we have formed would still be looking for money and profit" (2020). Extrapolating from the position that "sustainability can simply not exist in capitalism" (2020), many students speculate that the systemic change would either be superficial or would be derailed altogether. "I think companies would invest in the creation of underground bunkers, water stockpiles and space travel to escape their fate. The resources to fix the planet would not be utilized for that purpose" (2018). "It is more likely that they would work to escape Earth and the Guests [i.e. aliens], using their monetary resources to build rockets and take resources to settle on Mars" (2019).

But the paradox remains. Despite articulating an understanding that aligns with evidence about neoliberal capitalism as the primary driver of climate change,[14] many students continue to blame themselves or human nature for what, in an earlier question, they had correctly attributed to the workings of larger systemic forces. Numerous statements that "people" are greedy, unwilling to change, power-driven, and "selfish beings that have created the habit of making themselves the top priority" (2020); or that "we" are a destructive species characterized by unlimited wants, unwarranted entitlement, and "idiotically wasting th[e] knowledge [that could save the planet]" (2021)—all strike me as versions of an implicit belief that the real blame for climate change lies with people and human nature rather than with the way our socioeconomic system is set up. While these claims do not come close to the more direct expressions of climate guilt from class discussions, forums, and essays, my sense is that climate guilt is actively sabotaging students' grasp of the systemic drivers of climate change. Like Laura in *American Gods*, students tend to accept the blame for what is *not* their fault (see Verlie's notion of "greenhouse gaslighting," in this volume). They seem to believe, as one undergrad puts it, that to stop climate change "we would need to completely change human behavior" because the real issue is that "people would still be people" (2019).

This vague cluster of beliefs that feed climate guilt may seem innocuous. But unless brought into the open and confronted, they fester and actively block visions of change. Students who believe that destruction is our "nature" tend to be resigned and unable to appreciate that a genuine social transformation is possible. My role is to help them understand that one of the key rhetorical strategies of the ecocidal status quo in which we live is to naturalize capitalism as a system of relations that reflect human nature.[15] I ask them to unpack what happens if one subscribes to this essentialist (neo-Hobbesian) view. Once you accept that the planetary devastation stemming from the operations of neoliberal capitalism is an expression of human nature, are you better positioned to challenge ecocide? Or to resign yourself to it? Who benefits from or is threatened by your attitude in each case? The essentialist view contends that we are competitive, destructive beings; that destroying the planet reflects who we are; and that we really have no choice. If destruction is our nature, we have no more power to control it than we have controlling our pancreas. But is this view of human nature correct? Or is it perhaps another "truth" sold in the same way as nineteenth century claims that colonialism was the white man's burden or that racial hierarchies were a part of the God-given natural order? Drawing on existing scholarship,[16] I help students realize that claims about capitalism as an expression of human nature are historically located, and quite recent too. They are constructs that can be challenged. As Ursula K. Le Guin famously pointed out: "We live in capitalism. Its power seems inescapable. So did the divine right of kings. Any human power can be resisted and changed by human beings. Resistance and change often begin in art, and very often in our art: the art of words."[17]

Another thing I aspire to help my students grasp is that claims about the naturalness of capitalism represent what Naomi Klein calls the "scapegoating [of] 'humanity' for the greed and corruption of a tiny elite."[18] This naturalization of capitalism is a foundational untruth of the neoliberal order, at once the central tenet and excuse of today's reigning market ideology. What makes this view especially dangerous is that to attribute the current ecocide to peculiarities of human nature renders it impossible to grasp neoliberalism as a real driver of climate change even when evidence is staring us in the face. Thus, no matter what my students may believe about human nature, I see it as my responsibility to help them understand the nature of the socioeconomic system in which they live. This shift toward thinking about systems comes with its own challenges—like questions about what an individual can do to oppose the market system, and an attendant unraveling of the framework of individualism itself—but at least it frees students from guilt that would otherwise cripple their agency to push for change.

> Much of the climate guilt students feel has been manufactured to redirect their attention away from demanding systemic change.

The misremembering of climate guilt does not just happen on its own. Thus, another activity I have used is to invite students into conversations about how and why much of the climate guilt they are urged to feel is manufactured. We may read fragments from Oreskes and Conway's *Merchants of Doubt* (2010) or watch the documentary (2014) to give students a sense of the propaganda campaigns funded by Big Oil since the 1980s; that said, we mostly focus on a more recent push to make the Big Pollute accountable.[19] Part of this new trend is a growing focus on "deflection"—a strategy of the Big Pollute that helps explain *how* the "misremembering" climate guilt is fed. As Michael Mann defines it, deflection refers to strategies adopted by the petronormative institutions—whose operations are the systemic drivers of climate change—to project the responsibility for the destruction onto individual consumers. In all its forms, deflection is aimed to blame individual consumers rather than corporate actors, emphasize individual responsibility over corporate culpability, personal change over systemic change, individual action over collective action, personal choice over government regulations. "The focus on the individual's role in solving climate change," Mann writes, has been "carefully nurtured by industry."[20] I challenge my students to consider how much of the climate guilt they may feel has been manufactured to redirect their attention away from demanding systemic change. For example, students are better empowered to grapple with climate guilt when they know that the notion of a personal carbon footprint was introduced by BP (formerly known as British Petroleum) to divide climate advocates by generating conflict, behavior-shaming, and deflecting the conversation away from systemic change.[21] They are better empowered to deal with climate guilt when they know that companies like Chevron had no qualms exploiting the 2020 Black Lives Matter protests to drive racial wedges between white and BIPOC activists[22] by creating different versions of divisive climate guilt for each group. No matter what evidence you use, students have a right to know that "the discourse of individualizing responsibility" for climate change[23] is a carefully manufactured rhetorical product designed to support what Geoffrey Supran and Naomi Oreskes call "the Fossil Fuel Savior frame": a narrative of the Big Pollute that presents the continued use of fossil fuels as inevitable and driven by individual choices.[24]

Back in 2011, describing "the environmentalism of the poor," Rob Nixon proposed a notion of slow violence, "a violence that occurs gradually and out of sight, a violence of delayed destruction that is dispersed across time and space, an attritional violence that is typically not viewed as violence at all."[25] In today's world, climate guilt has become a form of slow violence too, except that its primary targets are middle-class, Global North-residing, educated, and climate-concerned audiences. While BIPOC and white students experience climate guilt differently, they are all victims of a rhetoric meant to confuse them into misremembering who is responsible for climate change. Climate guilt is a form of framing that enables the ecocidal status quo to continue. If we are serious about climate justice and hope to unlock our students' soul force in the transition to an ecological civilization, we have to start by helping them confront climate guilt and break out of its insidious fetters.

NOTES

1. Pihkala, "Anxiety and the Ecological Crisis," 9.
2. Ray, *Field Guide to Climate Anxiety*, 114.
3. See Pihkala.
4. Hiser and Lynch, "Worry and Hope," 103.
5. Ray, *Field Guide to Climate Anxiety*, 114ff.
6. Norgaard, *Living in Denial*, 8.
7. See Norgaard, *Living in Denial*; Schneider-Mayerson, "Influence of Climate Fiction."
8. Brulle and Norgaard, "Avoiding Cultural Trauma," 901.
9. Ray, *Field Guide to Climate Anxiety*, 115.
10. Oziewicz, "Planetarianism NOW," 241–56.
11. Oziewicz, "Fantasy for the Anthropocene," 58–69.
12. Ghosh, *Nutmeg's Curse*, 167.
13. Ray, *Field Guide to Climate Anxiety*, 116–20.
14. See, for example, Naomi Klein, *This Changes Everything: Capitalism vs. the Climate* (New York: Simon and Schuster, 2014); Amitav Ghosh, *The Great Derangement: Climate Change and the Unthinkable* (Chicago: University of Chicago Press, 2016); Peter Joseph, *The New Human Values Movement: Reinventing the Economy to End Oppression* (Dallas: BenBella Books, 2017); Imre Szeman and Dominic Boyer, eds., *Energy Humanities: An Anthology* (Baltimore: John Hopkins University Press, 2017); Rupert Read and Samuel Alexander, *This Civilization Is Finished: Conversations on the End of Empire—and What Lies Beyond* (Melbourne: Simplicity Institute, 2019).
15. For more on this foundational lie, see Ghosh, *Nutmeg's Curse*; Jeremy Lent, *The Patterning Instinct* (New York: Prometheus Books, 2017); Dipesh Chakrabarty, *The Climate of History in a Planetary Age* (Chicago: University of Chicago Press, 2021); The Red Nation, *The Red Deal* (Common Notions, 2021); Tyson Yunkaporta, *Sand Talk* (New York: HarperOne, 2020).
16. See, for example, Eileen Crist, *Abundant Earth: Toward an Ecological Civilization* (Chicago: University of Chicago Press, 2019); and Jason Moore, ed., *Anthropocene or Capitalocene: Nature, History, and the Crisis of Capitalism* (Oakland, CA: PM Press, 2016).
17. Le Guin, "National Books Awards Acceptance Remarks."
18. Klein, *The (Burning) Case*, 252.
19. See, for example, Timperley, "Who Is Really to Blame?"
20. Mann, *New Climate War*, 63.
21. Ibid., 64.
22. Ibid., 96.
23. Supran and Oreskes, "Rhetoric and Frame Analysis," 698.
24. Ibid., 696.
25. Nixon, *Slow Violence*, 2.

REFERENCES

Brulle, Robert J. and Kari Marie Norgaard. "Avoiding Cultural Trauma: Climate Change and Social Inertia." *Environmental Politics* 28.5 (2019): 886–908. https://doi.org/10.1080/09644016.2018.1562138.

Ghosh, Amitav. *The Nutmeg's Curse: Parables for a Planet in Crisis*. Chicago: University of Chicago Press, 2021.

Hiser, Krista K., and Matthew K. Lynch. 2021. "Worry and Hope: What College Students Know, Think, Feel, and Do about Climate Change." *Journal of Community Engagement and Scholarship* 13, no. 3 (2021). https://digitalcommons.northgeorgia.edu/jces/vol13/iss3/7.

Klein, Naomi. *The (Burning) Case for a Green New Deal*. New York: Simon and Schuster, 2019.

Le Guin, Ursula K. "National Book Awards Acceptance Remarks." YouTube, November 19, 2014. https://www.youtube.com/watch?v=Et9Nf-rsALk.

Mann, Michael E. *The New Climate War: The Fight to Take Back Our Planet.* New York: Public Affairs, 2021.

Nixon, Rob. *Slow Violence and the Environmentalism and of the Poor.* Cambridge, MA: Harvard University Press, 2011.

Norgaard, Kari Marie. *Living in Denial: Climate Change, Emotions, and Everyday Life.* Cambridge, MA: MIT Press, 2011.

Oziewicz, Marek. "Fantasy for the Anthropocene: On the Ecocidal Unconscious, Planetarianism, and Imagination of Biocentric Futures." In *Fantasy and Myth in the Anthropocene: Imagining Futures and Dreaming Hope in Literature and Media,* edited by Marek Oziewicz, Brian Attebery, and Tereza Dědinová, 58–69. London: Bloomsbury Academic, 2022.

———. "Planetarianism NOW: On Anticipatory Imagination, Young People's Literature, and Hope for the Planet." In *Pedagogy in the Anthropocene: Re-wilding Education for a New Earth,* edited by Michael Paulsen, Shé Mackenzie Hawke, and Jan Jagodzinski, 241–56. London: Palgrave, 2022.

Oreskes, Naomi, and Erik Conway. *Merchants of Doubt.* New York: Bloomsbury Press, 2010.

Pihkala, Panu. "Anxiety and the Ecological Crisis: An Analysis of Eco-Anxiety and Climate Anxiety." *Sustainability* 12, no. 19 (2020). https://doi.org/10.3390/su12197836.

Ray, Sarah Jaquette. *A Field Guide to Climate Anxiety: How to Stay Cool on a Warming Planet.* Oakland: University of California Press, 2020.

Schneider-Mayerson, Matthew. "The Influence of Climate Fiction: An Empirical Survey of Readers." *Environmental Humanities* 10, no. 2 (November 2018): 473–500.

Supran, Geoffrey, and Naomi Oreskes. "Rhetoric and Frame Analysis of ExxonMobil's Climate Change Communications." *One Earth* 4 (2021): 696–719. https://doi.org/10.1016/j.oneear.2021.04.014.

Timperley, Jocelyn. "Who Is Really to Blame for Climate Change?" BBC Future, June 18, 2020. https://www.bbc.com/future/article/20200618-climate-change-who-is-to-blame-and-why-does-it-matter.

PART SIX

JOY AND RESILIENCE AS RESISTANCE

TWENTY-SIX

Joyful Climate Work

The Power of Play in a Time of Worry and Fear

CASEY MEEHAN

> This chapter asks readers to consider how leveraging their own sense of joy and playfulness in service to climate action can bring about a sense of personal renewal and regeneration for our communities and planet.

As I write this, I'm on a bluff overlooking a portion of Lake Superior, the largest of North America's Great Lakes. The water is a striking shade of blue, flecked with whitecaps from a westerly wind. Between me and the lake grows a mile of white pine, aspen, birch, and balsam forest sprinkled with fields of thimbleberry bushes. The songbirds have gone silent, replaced by the shrill calls of a family of four sharp-shinned hawks who recently moved into the neighborhood. My kids are out exploring a set of nearby trails with the family dog and will soon return to breathlessly report on all the wondrous things their young minds encountered.

Meanwhile, in events from the decidedly less natural world, the U.S. government passed a major infrastructure deal that pours billions of dollars into exciting opportunities that will help us mitigate and adapt to the climate crisis.

I open this essay on play and joy by way of this juxtaposition: the climate crisis is real; yet the world remains a joyful place full of abundance and breathtaking possibility.

I will be the first to admit that I tend toward the optimistic side of things, but the point here is not that everything is okay. Far from it. It's hard to escape the fact that the state of the planet, even in this idyllic little corner of the upper Midwest, is experiencing a climate in upheaval. But personally, I also know that dwelling on the angst of our situation blinds me to the beauty in the everyday world and all the things going right in humanity's response to the climate emergency. It's the awe and joy this engenders—not anxiety and fear—that pulls me out of bed each morning, ready to do my part. I know I'm not alone in this thinking.

Morgan Florsheim, a recent college graduate who is exhausted and demoralized from hearing professors and climate activists proclaim that the world her generation is taking over is quickly falling to ruins, wonders, "Where is the joy or satisfaction in fighting for a world that is already damned?"[1] Florsheim calls for a different narrative, "one that encourages self-care alongside activism, that works intentionally to foster excitement for the future we desire." The climate movement is often accused of pedaling "gloom and doom." My experiences researching, teaching, and communicating about the climate crisis to a wide range of audiences, including K-16 students, have led me to believe that a different narrative drives me and others like Florsheim. I call it joyful climate work.

JOYFUL CLIMATE WORK

Joyful climate work is a way of addressing the climate crisis that notices the delight, wonder, surprise, abundance, and playfulness inherent in our world and leveraging those characteristics to bring about a sense of renewal for ourselves and regeneration for our communities and planet.

This is not to say that joyful climate work is about toxic positivity, the "good vibes"-only mentality that requires optimism no matter how horrible the situation. Nor is joyful climate work meant to trivialize feelings of eco-anxiety, despair, anger, and shame some people experience. The work conducted by authors of this volume—and myriad other scholars and practitioners exploring the emotional impact of the climate crisis—gives credence to the importance of validating and attending to the unpleasant emotions we're feeling.

> Yes, we grieve; we fret; we anguish. But we also celebrate; we laugh; we take pleasure. In the human project, grief lives side-by-side with joy.

However, to *only* recognize the unpleasant emotions inherent in the climate crisis limits by at least half the spectrum of human experience. In danger of being sidelined is the joy, wonder, and delight that serve as important sources of human renewal, inspiration, and action. Through joyful climate work we remind ourselves to attend to the whole of the human spirit and psyche. Yes, we grieve, we fret, we anguish. But we also celebrate, we laugh, we take pleasure. In the human project, grief lives side-by-side with joy. "Joy is not found in the absence of pain and suffering," writes Imani Perry, "It exists through it."[2]

Given the pressing urgency of the climate crisis to all life on the planet, and especially given that marginalized populations already disproportionately suffer from the impacts, working through a lens of joy might seem like a privileged position to take. However, this ignores the rich traditions harnessing joy that marginalized communities have embraced for centuries. For instance, the Black Joy movement serves as a form of resistance reminding us that people are more than their setbacks and struggles.[3] Educator Tracey Michae'l Lewis-Giggets goes further, arguing that Black Joy is more than a tool for resistance: it is a means to develop resilience and renewal.[4] Celebrating, living fully, laughing . . . what might those working against a hegemony hellbent on burning up the planet learn from the Black Joy movement about resistance?

In her book *Joyful: The Surprising Power of Ordinary Things to Create Extraordinary Happiness,* designer Ingrid Fettell Lee writes that the drive toward joy is synonymous with the drive toward life, toward renewal.[5] Isn't this what our sustainability work is meant to bring about: nurturing healthy, regenerative systems so that all life can thrive?

Attending to joy isn't simply a theoretical exercise. Joyful climate work takes its cue from the field of positive psychology. Instead of fixating on deviance and maladaptive behavior and thought-processes, positive psychology studies the behaviors and emotions humans experience when they are flourishing. Empirical work exploring pleasant emotions has shaped what is known as the broaden-and-build theory.[6] Pleasant emotions *broaden* what we're willing to think and do. For instance, feeling joyful sparks one's urge to explore and consider alternate possibilities. This is the opposite of what happens with unpleasant emotions like fear, anger, and sadness that engender our fight, flight, and freeze responses. An expanded thought-action repertoire then *builds* one's physical, intellectual, social, and psychological resources. Importantly, according to the broaden-and-build theory, while pleasant emotions are temporary, the resources they build are durable. According to the broaden-and-build theory the benefits of pleasant emotions like joy can be drawn upon in the future to improve the odds of successful coping and survival even if one isn't feeling particularly joyful in the moment.

We manifest joy in our lives in myriad ways, but I'd like to consider for a moment play and playfulness as compelling though overlooked tools to instigate joyful climate work.

WHAT IS PLAY AND WHY SHOULD WE DO IT?

While there exists a lively debate about what, specifically, constitutes play, most attempts ascribe characteristics of play as voluntary, pleasurable, separate from the "real world" (through imagination or rules), and having inconsequential outcomes.

Play is a behavior. Playfulness, on the other hand, is a disposition. There is a social stigma attached to play, especially in the realm of work and adult life in general. Therefore, I've often found it easier to engage adults in discussions around their playfulness rather than play (see also Creane's chapter on play in this volume).

While the origins and purpose of play are still unsettled in the literature, play appears to be ubiquitous among animals. Anyone who has lived with a puppy or kitten or who has watched primates, dolphins, or other mammals can attest to their playfulness. Indeed, play research has grown this playful circle to include birds, fish, insects, and octopuses, among other life forms.

From a human perspective, we tend to view play and playfulness as the domain of children; however, this temporally bound perspective on play is inaccurate. Experts in education,[7] psychology,[8] and medicine[9] suggest that play is important across the human lifespan. Educator and performer Michael Rosen writes that "play is fundamental to our development as people, and more broadly, as a society and culture."[10]

> Why would we choose to play at a time like this? It turns out that a rich and growing body of literature offers empirical evidence for the extensive benefits of play.

Despite the seemingly innate drive to play, play is often viewed in secondary and college classrooms as a frivolous endeavor not suitable as a means for working on Serious Topics. And if the climate crisis is anything, it's a Serious Topic. Why would we choose to play at a time like this? It turns out that a rich and growing body of literature offers empirical evidence for the extensive benefits of play.

First, play nourishes mental well-being by activating the reward and social centers in the brain.[11] One of the foremost play scholars, the late Brian Sutton-Smith, stated that the opposite of play is not work, it's depression. Studies that scan the brain at play bear this out. During play the brain releases dopamine, a neurotransmitter associated with feelings of joy and linked to enhanced attention and creativity. Play also positively impacts our serotonin uptake centers. Dysfunctional serotonin centers are linked to depression, whereas high-functioning serotonin centers correlate with higher levels of optimism and altruism. As incidences of anxiety and depression are on the rise, play and playfulness offer a welcome antidote.

Second, play connects us to others. Physiologically, playful interactions trigger the release of oxytocin in the brain. Oxytocin, sometimes referred to as the "cuddle hormone," produces feelings of connection and trust. Psychologically, play is a vehicle for attunement—accepting people where they are without shame or judgment. Play naturally breeds attunement as we yield to others, making and reimagining rules so that all who want to can play equally regardless of ability, belief, or emotional state (see also chapters by Frame et al. and Holmes on community-building in this volume).

Still another benefit of play is that it affords us the space to explore new ways of acting and being in the world. As discussed above, play is seen as separate from the "real" world since players experiment with different rules, norms, and/or roles. In this imaginary world of play we can test and retest our understanding of a situation, learning about how tweaks to our alternate universe impact how we feel, how we perform, and how the "game" turns out. In the process, we spur creative solutions to problems that arise. Indeed, as the late play expert Bernie De Koven wrote, "we play our way to understanding."[12]

As we play, we come to realize that the process is more important than the outcome. The meaningful part of play doesn't live in the results but rather in how the play unfolds. Failure becomes perfectly acceptable, if not encouraged, as players apply what they have learned from past iterations. Given the scope of the challenges humanity faces in the climate crisis, we must not be afraid to try (and fail) with new ideas as we learn to (re)develop regenerative ways of being with each other and the planet.

Finally, play is a way of experiencing the world that requires us to be vulnerable. Anyone over a certain age understands that when we play publicly, we risk looking foolish, unprofessional, and unserious in front of our colleagues and friends. It takes courage to be vulnerable, yet courage, not hope, is exactly what renowned climate scientist Kate Marvel claims is needed to mentally handle the climate crisis. "We are inevitably sending our children to live on an unfamiliar planet," she writes. "We need courage, not hope. . . . Courage is the resolve to do well without the assurance of a happy ending."[13]

Play benefits our mental health, connects us to others, spurs creativity, gives us permission to fail, and gives us practice at being brave. But I suspect what attracts us to play is that it just feels good. When we play fully we feel renewed and ready to face the challenges we're up against, even if just bit by bit.

A MORE PLAYFUL PATH TO JOYFUL CLIMATE WORK

To be sure, play is not the panacea for helping ourselves and others reckon with the crisis we find ourselves in, and it isn't for everyone at all times. Some of us are grieving and can't or don't feel like playing. That's okay; we play when we're ready. When we are, how might we start down a more playful path?

What many people immediately jump to when they think of play, especially in the classroom, is games. This may partly be due to the recent popularity of "gamifying" curricula. Gamification uses elements of games (think tokens, badges, points, missions, storylines, etc.) to enhance users' experiences with what might be considered ordinary—or even dry—content. It's the spoonful of sugar that helps the medicine go down. Games are a subset of play to be sure, but what I'm describing here goes beyond the recent trend of gamification. The type of play we're interested in here is sparked by joy and a spirit of playfulness, not coercion. It is often more spontaneous, more free flowing, with rules that change with the circumstances and players who play because they find it intrinsically motivating.

A good first step to foster play in service of serious climate work is to nurture a playful attitude. Playfulness is the disposition that makes play possible. In *A Playful Path*, De Koven contends that "Playfulness . . . allows you to transform the very things that you take seriously into opportunities for shared laughter; the very things that make your heart heavy into things that make you rejoice. . . . It turns problems into puzzles, puzzles into invitations to wonder."[14] From a perspective of play in classrooms, playfulness is the tendency to (re)frame a situation to include possibilities for exploration, choice, and enjoyment.[15] The gateway to playfulness is wonder—a desire to explore the world and how it might work. How does your curriculum turn your course into an invitation to wonder—a space for students to ask, "I wonder what would happen if . . . ?" What opportunities do students have to improvise and create in your classroom?

In playful classrooms learners have choices about what and how they engage. This needn't be a free-for-all; rather, students might choose rules and roles that guide activities or assessments or have the choice to interact with a diversity of perspectives, materials, and texts. Playful experiences like running a UN Climate Change conference simulation using the free, award-winning C-Roads platform (available at https://www.climateinteractive.org) allows for students to negotiate, try different strategies, and learn about the diverse set of issues the climate crisis imposes on different regions of the world—all elements of choice used playfully.

Playful environments seek moments to manifest joy. Play and playfulness are not about taking life less seriously, but about taking *ourselves* less seriously. Where in our work and teaching do we build space for surprise, humor, celebration, laughter, and the time to connect with each other? This isn't time wasted: it is a recognition that we must nourish our whole selves when engaging with the emotionally fraught topics we teach.

Game designer Ian Bogust writes, "Play invites you to consider your surroundings as a vast domain of essentially limitless meaning and potential. . . . Playgrounds are places where we dig deep, where we mess things up and toss them asunder—ourselves included—in order to discover what else is possible."[16] I invite all of us to consider how we might treat our climate work more like a playground instead of a battleground.

So olly-olly oxen free! Let's come out of our hiding places and move forward in our response to the climate crisis together. Joyfully.

NOTES

1. Florsheim, "Don't Tell Me."
2. Perry, "Racism Is Terrible."
3. Ibid.
4. Lewis-Giggets, *Black Joy*.
5. Lee, *Joyful*, 296–98.
6. Fredrickson, "Role of Positive Emotions."
7. See Alison James and Chrissi Nerantzi, *The Power of Play in Higher Education: Creativity in Tertiary Learning* (Cham, Switzerland: Palgrave Macmillan, 2019).
8. See Stuart Brown and Christopher Vaughan, *Play: How it Shapes the Brain, Opens the Imagination, and Invigorates the Soul* (New York: Penguin, 2009).
9. See Anthony DeBenedet, *Playful Intelligence: The Power of Living Lightly in a Serious World* (Solana Beach, CA: Santa Monica Press, 2018).
10. Rosen, *Michael Rosen's Book of Play!*, 228.
11. Koeners and Francis, "Physiology of Play," 145.
12. De Koven, *Playful Path*, 247.
13. Marvel, "We Need Courage, not Hope."
14. De Koven, *Playful Path*, 31.
15. Mardell et al., "Towards a Pedagogy of Play."
16. Bogust, *Play Anything*, 219

REFERENCES

Bogust, Ian. *Play Anything: The Pleasure of Limits, the Uses of Boredom, and the Secret of Games.* New York: Basic Books, 2016.

De Koven, Bernard. *A Playful Path.* ETC Press, 2014.

Florsheim, Morgan. "Don't Tell Me to Despair about the Climate: Hope Is a Right We Must Protect." *Yes!,* June 15, 2021. https://www.yesmagazine.org/opinion/2021/06/15/climate-despair-hope.

Fredrickson, Barbara. "The Role of Positive Emotions in Positive Psychology. The Broaden-and-Build Theory of Positive Emotions." *American Psychologist* 56, no. 3 (2001): 218–26. https://doi.org/10.1037//0003-066x.56.3.218.

Koeners, Maarten, and Joseph Francis. "The Physiology of Play: Potential Relevance for Higher Education." *International Journal of Play* 9, no. 1 (2020): 143–59. https://doi.org/10.1080/21594937.2020.1720128.

Lee, Ingrid Fettell. *Joyful: The Surprising Power of Ordinary Things to Create Extraordinary Happiness.* New York: Little, Brown Spark, 2018.

Lewis-Giggetts, Tracey M. *Black Joy: Stories of Resistance, Resilience, and Restoration.* New York: Gallery Books, 2022.

Mardell, Ben, Daniel Wilson, Jen Ryan, Katie Ertel, Mara Krechevsky, and Megina Baker. "Towards a Pedagogy of Play." Project Zero Working Paper, 2016. http://www.pz.harvard.edu/sites/default/files/Towards%20a%20Pedagogy%20of%20Play.pdf.

Marvel, Kate. "We Need Courage, Not Hope, to Face Climate Change." *The On Being Project,* July 19, 2019. https://onbeing.org/blog/kate-marvel-we-need-courage-not-hope-to-face-climate-change.

Perry, Imani. "Racism Is Terrible. Blackness Is Not." *The Atlantic,* June 15, 2020. https://www.theatlantic.com/ideas/archive/2020/06/racism-terrible-blackness-not/613039.

Rosen, Michael. *Michael Rosen's Book of Play! Why Play Really Matters, and 101 Ways to Get More of It in Your Life.* London: Wellcome Collection, 2019.

TWENTY-SEVEN

Finding Hope in the Influence and Efficacy of Native/Indigenous Rights

KATE REAVEY

> This week-by-week assignment centers on the efficacy of Native/Indigenous presence in climate stewardship and mitigation planning. The reflective assignment emphasizes hope and positive change through a study of treaty rights, existing legislation, collaborative resistance, and planning.

This assignment invites students to research treaty rights and the power, efficacy, and agency that tribal nations are asserting to mitigate climate change. It was shaped and revised with bell hooks's concept of "engaged pedagogy" and Shawn Wilson's work on Indigenous methodologies in mind, and it reflects Robin Wall Kimmerer's emphasis on relationality and reciprocity. Because our campus serves six tribes directly, some of the content is specific to the Pacific Northwest. The framework of the course is adaptable to many other geographies.

Notes: The assignment has been taught in IS (Integrated Studies) 109: Introduction to Indigenous Humanities and is adaptable to English 101 and other courses in various disciplines.

This assignment requires continual engagement with legislation, both state and federal. Therefore media resources, news journals, and expert lectures (sometimes accessible online as TED talks or YouTube videos) can be vital resources along with updates to websites administered by Tribal Nations and other councils led by Native/Indigenous stewardship.

All the required readings and videos are accessible through open-sourced, no-cost access.

OUTCOMES AND GOALS

1. To learn about the positive ways that treaty rights, trust lands, and/or co-management of lands and waters can (and have) mitigate(d) impacts from climate change and how Native/Indigenous leadership continues in these areas
2. To engage place-based learning and situate ourselves as participants in positive change and to establish a sense of self-efficacy within the strength and resilience of groups, tribal communities, neighbors, advocates, and activists
3. To build understanding of Native nations, treaty rights, and tribal sovereignty

The basic structure of the assignment is to research, reflect in writing and discussions, create a final presentation, and contribute to the learning through continual engagement with the class community throughout the quarter.

Students choose a *Focus for Study*:

- An individual tribe's Climate Action Plan or mitigation efforts/practices
- A large-scale mitigation project that required tribal efficacy/agency (some examples include the following: Elwha Dam Removal, Undevelopment of Jimmycomelately Creek, Quileute "Move to Higher Ground")
- A group that serves to promote awareness of climate change and/or mitigation that includes at least one tribal nation (local example: Dungeness River Management Team)

The resources listed below provide a basic framework for engaging the following ideas: Shawn Wilson's methodology presents a theoretical shape for Robin Wall Kimmerer's presentation of the practice(s) of relationality, and Zoltan Grossman's TED Talk presents working examples of collaborative relationality. The article by Anna Smith in *Indian Country Today* is an example of a useful essay in a popular, accessible journal; updated articles may be chosen as they become available. The website for the Affiliated Tribes of Northwest Indians (ATNI) is a frequently updated source with a page dedicated to climate change.

REQUIRED READINGS AND VIEWINGS

- Affiliated Tribes of Northwest Indians. Climate Change Page. https://atnitribes.org/climatechange/.
- Grossman, Zoltan. Native Resilience, TED Talk, 2012. https://www.youtube.com/watch?v=U0CRa9mT4Qs.

- Kimmerer, Robin Wall. "The Honorable Harvest: Lessons from an Indigenous Tradition of Giving Thanks." *Yes,* November 26, 2015. https://www.yesmagazine.org/issue/good-health/2015/11/26/the-honorable-harvest-lessons-from-an-indigenous-tradition-of-giving-thanks. Optional TED Talk: Kimmerer, Robin Wall. "Reclaiming the Honorable Harvest." https://www.youtube.com/watch?v=Lz1vgfZ3etE.
- Smith, Anna V. "How Do Tribal Nations' Treaties Figure into Climate Change?" *High Country News,* May 14, 2019. https://www.hcn.org/articles/tribal-affairs-how-do-tribal-nations-treaties-figure-into-climate-change.
- Wilson, Shawn. "What Is an Indigenous Research Methodology?" *Canadian Journal of Native Education* 25, no. 2 (2001). https://www.researchgate.net/publication/234754037_What_Is_an_Indigenous_Research_Methodology.

RATIONALE

Researching treaty rights and the power, efficacy, and agency that tribal nations are asserting to mitigate climate change is the heart of this assignment. Challenging the mistaken claim of "special rights" for Native/Indigenous peoples, this assignment brings students into the requirement of learning specific examples of resistance and resilience. This assignment requires individual research completed by students, then a collaborative opportunity for learning through sharing knowledge and reflecting on common themes during the final week of the quarter. Indigenous peoples have often faced the impacts of climate change earlier than other people. They also began organizing, mitigating impacts, and working together to influence a positive future. Adaptation and resilience have been crucial. Hope is evident in the studies that individual students complete through this assignment. Engagement as a community—learning from each other—has the potential to strengthen our resolve to remain inspired by what we have come to know.

ELEMENTS OF THE ASSIGNMENT

- weekly readings and/or videos
- written reflections
- synchronous and asynchronous discussions
- an annotated bibliography
- a final presentation to the class (in the form of an oral presentation, video, story)
- a one-page-per-presentation reflection in response to two other students' final presentation of their learning
- weekly reflections: 350–500 words each. The assignment flows through the entire scope of the course.

WEEK ONE: Overview of the assignment. Discussion and opportunity to ask questions and share ideas and prior knowledge. Create shared agreements if your course does not already have a covenant or set of ground rules, because this assignment will include social and emotional learning with dialogues for building trust and confidence.

WEEK TWO: Read article by Robin Wall Kimmerer.
Some questions posed by the author that can guide the discussions for the week: What is our responsibility in return for our gifts from the earth? When we imagine social, political, economic decision-making based on the "Honorable Harvest," what do we envision as the positive outcomes and what do we envision as the challenges? Kimmerer reminds us of the shared agreements that humans have made with regard to how we act and how we take without asking, and she suggests we can change these ways. Questions and dialogue reflect upon this change and prepare us for Shawn Wilson's article later in the quarter.
Deadline: Students choose their *Focus of Study* and post a one-sentence intention for what they will pursue.

WEEK THREE: Read Anna Smith's "How Do Tribal Nations' Treaties Figure into Climate Change?" and consider this quotation from the article: "U.S. courts rarely favor environmental protections as a right—except when it comes to tribes expressing their treaty rights" as part of the discussion.
Reflection on *Focus of Study* including further discussion: What led you to choose this plan or legislation or group? If the group has a mission or vision statement, include that. If there is a link to the primary source for the legislation, include that. If you are studying a tribe's Climate Mitigation Plan or Climate Action Plan, note when it was established.
What connections can we make with Zoltan Grossman's TED Talk?

WEEK FOUR: Return to the online discussion from last week and respond to two other students' posts. Complete a brief reflection (350-500 words) that focuses on the concept of relationality and returns to the ideas presented by Kimmerer in the first week.

WEEK FIVE: Personal reflections may include the following: Has your research changed your focus or impressions of climate change and resilience? Where do hope and inspiration emerge? How are your experiences different or similar to what you were thinking and feeling before you began this quarter? How much did you know about Native/Indigenous treaty rights, collaborative resistance, legislation that supports climate mitigation efforts?
What further connections are arising from the readings and individual research?

WEEK SIX: Read Shawn Wilson's "What Is an Indigenous Research Methodology?" and post two questions to the discussion board (or bring them to the synchronous discussion). Engage some of the complexities of Wilson's ideas, which will continue to inform the learning during the rest of the quarter. Reflect on the meaning of "epistemology" and the questions, How do we come to know what we know? and How have our laws, our practices, and our social structures been shaped by our ways of knowing?

Reflect on how different these approaches would be (could be) if Kimmerer's emphasis on the personhood of plants and animals and the reciprocity of the "honorable harvest" were the epistemological foundations of most laws and practices.

WEEK SEVEN: Continue to reflect upon the theoretical and practical considerations engaged so far, including relationality, reciprocity, and Indigenous research methodologies. Write a 500-word-minimum reflection.

WEEKS EIGHT to TEN: Submit final draft of annotated bibliography, make presentation, and offer continued reflection on key takeaways from the research. Engage with other students' presentations through dialogue and written reflection.

WEEK TEN: Complete a brief self-assessment of the work completed for the entire term (500 word minimum). Submit written responses to at least two other students' final presentations with one-page reflections on the learning accomplished or co-created through the listening.

NOTE ON RESOURCES

For this assignment, encourage students to go directly to the websites posted and updated by Tribal Nations.

To adapt this assignment to other areas/geographies, encourage students to read and cite the books written by and for the local and regional tribes.

REFERENCES

"Affiliated Tribes of Northwest Indians (ATNI)." *Affiliated Tribes of Northwest Indians (ATNI) Tribal Climate Change Guide*, December 29, 2015. https://tribalclimateguide.uoregon.edu/climate-programs/affiliated-tribes-northwest-indians-atni.

Brown, Alex. "Once-Ignored Promises to Tribes Could Change the Environmental Landscape." *Stateline*, December 1, 2020. https://stateline.org/2020/12/01/once-ignored-promises-to-tribes-could-change-the-environmental-landscape/.

Clapperton, Jonathan, and Liza Piper. *Environmental Activism on the Ground: Small Green and Indigenous Organizing*. Calgary: University of Calgary Press, 2019.

Enduring Legacies: Native Case Studies. The Evergreen State College. Accessed May 13, 2023. https://nativecases.evergreen.edu.

Grossman, Zoltan. *Unlikely Alliances: Native Nations and White Communities Join to Defend Rural Lands*. Seattle: University of Washington Press, 2017.

hooks, bell. *Teaching to Transgress: Education as a Practice of Freedom*. New York: Routledge, 1994.

Northwest Indian Fisheries Commission (NWIFC). https://nwifc.org.

Olympic Peninsula Intertribal Cultural Advisory Committee. *Native Peoples of the Olympic Peninsula: Who We Are*. Norman: University of Oklahoma Press, 2003.

Parker, Alan, and Zoltan Grossman. *Asserting Native Resilience: Pacific Rim Indigenous Nations Face the Climate Change*. Corvallis: Oregon State University Press, 2012.

Smith, Anna V. "How Do Tribal Nations' Treaties Figure into Climate Change?" *High Country News*, May 14, 2019. https://www.hcn.org/articles/tribal-affairs-how-do-tribal-nations-treaties-figure-into-climate-change. Accessed 22 August 2023.

Wilson, Shawn. "What Is an Indigenous Research Methodology?" *Canadian Journal of Native Education* 25, no. 2 (2001).

Zuckerman, Jocelyn C., and Ted Williams. "How Native Tribes Are Taking the Lead on Planning for Climate Change." Yale E360. Accessed May 13, 2023. https://e360.yale.edu/features/how-native-tribes-are-taking-the-lead-on-planning-for-climate-change.

TWENTY-EIGHT

Teaching Climate Change Resilience through Play

JESSICA CREANE

> In order to help educators remain creative and energized in environmental work, this chapter explores the role of playfulness and games in environmental education and emotional resilience.

At first glance, grief and play go together like peanut butter and a record player, which is to say, not at all. Grief is the loss of what was once possible. Play is the embrace of what is now possible. To hold these two diametrically opposed ideas in one breath can be jarring, yet they are not so far apart as we might think.

My first foray into the relationship between grief and play followed the death of my father in 2016. I wasn't a game designer yet. In fact, the idea of becoming one had never occurred to me, nor could I imagine becoming anything at all that year. Uncertainty reigned over my life like a despot. But here's the thing about uncertainty—it's not wholly bad. It turns out to be a key ingredient in many generative emotions, in joy as well as grief. In our pursuit of joy we actively seek uncertainty in the form of jokes we don't know the punch line to, books with plots we know nothing about, and recipes we haven't tried before. When we open a box to play a board game, we are filled with anticipation of how the game will go. If we already knew, we probably wouldn't play. In fact, we would be incapable of it. If we knew in advance how the game would unfold, any actions taken would be fulfilling a prophecy rather than playing a game. Play is only possible in the face of uncertainty (for more on play, see Meehan's chapter in this volume).

This does not mean that play is inevitable or easy in the face of the unknown, but merely that it can be worth the effort. I was as surprised as anyone that from the depths of my grief in 2016 I began to study game design, and what I came to learn was that our responses to the things we take most seriously, be it grief, love, fear, justice, climate disruption, or just regular old change, are the things that benefit most from our playfulness. In fact, there is a whole branch of game design dedicated to the playful exploration of serious subject matter. The products of this exploration are often referred to as "serious games" or "impact games," such as The Oregon Trail, Math Blaster!, Spirit Island, or That Dragon, Cancer, a game where the player takes on the role of a parent caring for their dying child.

It is often assumed that games are *fun,* but games are a genre, just like books, movies, or paintings, and we would never dream of limiting books or movies to providing just one emotional experience to the audience. The emotional and intellectual outcomes of games are as diverse as any other art form. Games *can* be fun, but more precisely they are *engaging* and *interactive*. In his book *The Grasshopper: Games, Life and Utopia,* Bernard Suits wrote that games are "the voluntary attempt to overcome unnecessary obstacles," meaning that we enter into the challenges games set for us voluntarily.[1] What differentiates games from other mediums is not their commitment to fun but their commitment to human choice and agency.

It is this commitment to agency that sets games and play at odds with grief. In fact, conventional wisdom suggests that we curtail our agency following a major loss and avoid making any big life choices. Don't sell your house, change jobs, or move to a new city. Wait.

Yet when it comes to climate change-induced grief, waiting is a luxury we can't afford. Time does not heal all wounds if those wounds are open and bleeding. We must mourn the losses we endure year after year while simultaneously using our agency to staunch the bleeding. To survive, we must learn to embrace our peanut butter and record player combo and engage in the possibility of a bright future with an open mind even while we grieve the loss of the possibility of the bright future we imagined. To do these things together is a necessary paradox to be embraced. Fortunately, games are useful tools in reconciling such paradoxes.

Games themselves may be voluntary, or, as James P. Carse put it in *Finite and Infinite Games,* "whoever *must* play cannot *play,*" but in overcoming in-game obstacles, players are in a literal playground and a metaphorical training ground.[2] The rules and constraints of a game provide a safe space for players to take risks and see the effects of their choices in real time, a feedback loop we often lack in the real world.

Game systems are microcosms of real-world systems. Monopoly, for instance, is built on the principles of capitalism. (If you haven't looked up the backstory for Monopoly, I highly recommend you do!) Game systems are simpler than those of the real world, which is why they make for excellent training grounds, especially with a facilitator present to debrief with students about how to tie in-game learning to their real-world actions.

The centering of humans as players, first and foremost, was put forth by Johan Huizinga in *Homo Ludens* (first published in 1938), in which he argues that humans are built for play and rely on clear rituals in order to behave playfully.[3] These rituals usually come in the form of game constraints (i.e., rules and roles) that enhance player agency by establishing clear guidelines for players' actions. This clarity is empowering. As Jane McGonigal lays out in

Reality Is Broken: Why Games Make Us Better and How They Can Change the World, "Gamers just don't give up. We aren't used to feeling so optimistic in the face of things that are extremely difficult to us. That's why so many gamers relish wickedly hard game content."[4]

When it comes to climate change, wickedly hard problems are abundant and clear guidelines are a precious few, which makes climate change hard, but potentially profoundly satisfying, to engage with. Even the brightest and most knowledgeable among us have different ideas of what constitutes a win, whether a win is even possible, how long we'll be at this, what actions are off limits and what actions are absolutely necessary if we are to survive, or even figure out whose turn it is to act. Without agency training, we are easily confused and apathetic. *Is it my turn to do something about climate change? Is it the older generation's turn? The younger generation? People who know more than me? The government? The UN? Who is, uhh, up next . . . ?*

This confusion can be immobilizing, and while there are many lessons to draw from the clarity of game rules and constraints, where games really shine is in fostering human agency not only by setting forth clear constraints for us to play within, but by inviting us to play with the constraints themselves, bringing our own creativity, values, identity, and even sense of mischief into the mix.

What follows are three rounds of an open-ended, cooperative, climate-focused game that stems from my work with The Marsh Billings Rockefeller National Park in Woodstock, Vermont. Each round was designed to be played by park visitors on a hike, while waiting for a tour to start, or as a part of in-park workshops, classes, and events. I offer these rounds as templates that can be adapted to a variety of contexts and environmentally focused content. Following each round of the game is a suggestion for how it might be adapted for a classroom or group setting. Educators are encouraged to debrief with students after each round, especially if playing them in succession. The nature of the games is to be constrained enough to create a shared, collective experience among the players while remaining open enough to invite students to formulate their own meaning from the games and begin the process of personal, creative problem-solving that games encourage.

Feel free to play each round multiple times to encourage students to choose a different path and compare the results. If it is helpful, ask your students directly how they brought their creativity, values, identity, and mischief to each round of the game!

ENVIRONMENTALISMS: ROUND ONE

(See Ladino's essay in this volume for an example of how this game can be used in a class setting.)

Round One of the game* invites players, in groups of 3–4, to complete each of the following steps in order. For each step, players may speak *only* in questions. (Example questions for Step 1: *How much rainforest will be left in 20 years? Will my family lose their home to climate change? Will The Great Coral Reef live on?*)

Step 1
What are your environmental fears? (3 minutes)
Step 2
What are your environmental hopes? (3 minutes)
Step 3
What are you willing to do for the environment? (3 minutes)
Step 4
What are you willing to do for each other? (3 minutes)
It's helpful to note that silence is welcome during this round of the game, even to an uncomfortable degree. We do some of our best thinking in silence.

The interactivity in this round is simple and modular. It invites players to enter into a communal experience of voicing their fears and anxieties without any pressure, or indeed any incentive, to resolve them. Their only goal is to be present with one another and with their own thoughts and feelings. Time limits can be easily adjusted for varying group sizes and questions can be added, subtracted, and swapped out to specifically address content that is particularly relevant to the group.

ENVIRONMENTALISMS: ROUND TWO

Complete each of the following steps.*
Step 1
Individually, research a plant that is native to the land you are on. Keep your plant's name secret. (10 minutes)
Step 2
Go for a stroll and silently reflect on what you have learned about your plant. What surprised you about the plant you chose? What are the qualities of this plant that help it to thrive? (3 minutes)
Step 3
Think and move like your plant. Imagine how it grows, responds to light, responds to air, responds to predators, sets down roots, etc. Imagine your body is your plant's body. (3 minutes)
Step 4
As a group, have a conversation with the other players as if you were your plants. What would your plant say in a conversation with other plants? Do not name your plant species to the group. (5 minutes)
Step 5
Guess the plant species chosen by the other players. (1 minute)
Bonus round: Repeat Steps 1–5 choosing, instead, a non-native plant competing for resources with the native plant you originally chose.

Round Two invites a higher degree of playfulness, interactivity, and empathy from the group. For some groups, this round would be fruitful on day one of class, for instance, if the silliness of it wins out. For other classes, trust might need to be built up over time before students are comfortable pretending to be a plant. The ultimate goal of this round is to begin empathizing with non-human species in an embodied way and, in the bonus round, to experience unintentionally harming an ecosystem and attempting to survive in a world one didn't choose, but was dropped into and must survive in. This primes the players for the third round, in which they must adopt this mindset from a more complex and politically contentious vantage point of other humans.

ENVIRONMENTALISMS: ROUND THREE

Played in groups of three, one player at a time.

Step 1

State something you believe to be true about the environment.

Step 2

Beginning with "On the other hand . . . " convince yourself otherwise. (3 minutes)

Step 3

Defend the stance you have talked yourself into against the rest of the group's questions. (3 minutes)

Step 4

Once everyone has gone through steps 1–3,* have a polyphonic conversation (a conversation in which everyone is both listening to each other *and* speaking, all at once) in which players simultaneously defend their new stance and challenge each other's stances. (3 minutes)

Note: This last step is extremely challenging. It is best for players to keep talking without stopping as they attempt to navigate the conversation. It takes practice to get good.

Round Three of the game explores empathy and the role of personal narrative in shaping our understanding of climate change. Some players might discover that it is easier than they anticipated to slip into the mindset of someone with a diametrically opposed worldview. Having put themselves in the mindset of someone with different priorities, reasons for that prioritization spring to mind, thereby diminishing their ability to "other" folks who think differently. Players may also find that it is easy to prioritize "winning" a conversation rather than engaging in a conversation with an open mind. Others might recognize that it is nearly impossible to listen and speak at the same time and consider the ramifications of folks talking over each other rather than sitting down to truly listen to one another. In running this game, I have observed each of these outcomes, and more.

In my personal and professional journey to blend the creativity and community that emerge from games with the grief and anxiety that accompany loss, I come back time and again to the role of agency and our need to practice it in low-stakes, friendly environments in order to feel comfortable exercising it in high-stakes environments, like our time on Earth. One way to incentivize such practice, be it in regard to external progress on climate change or internal growth around eco-anxiety, is for that practice to be enjoyable. Counteracting climate change requires our deepest creativity and connectivity, which is unlocked in our students, and ourselves, when we feel that change is possible and we have the clarity and confidence to achieve it: feelings that games are masterfully designed to evoke.

NOTES

1. Suits, *Grasshopper*, 54–55.
2. Carse, *Finite and Infinite Games*, 4.
3. Huizinga, *Homo Ludens*, 1.
4. McGonigal, *Reality Is Broken*, 67.

REFERENCES

Carse, James P. *Finite and Infinite Games*. New York: Ballantine Books, 1987.
Davidson and Associates. *Math Blaster!* Apple II, Atari 8-bit, C64, IBM PC. 1983.
Huizinga, Johan. *Homo Ludens: A Study of the Play-Element of Culture*. London: Routledge and Kegan Paul, 1949.
McGonigal, Jane. *Reality Is Broken: Why Games Make Us Better and How They Can Change the World*. New York: Penguin Press, 2011.
Minnesota Education Computing Consortium. *The Oregon Trail*. 1971, 1974.
Numinous Games. *That Dragon, Cancer*. Microsoft Windows, OS X, Ouya, iOS, Android. 2016.
Reuss, Eric. *Spirit Island*. Greater Than Games, 2017.
Schell, Jesse. *The Art of Game Design: A Book of Lenses*. Amsterdam: Elsevier/Morgan Kaufmann, 2008.
Suits, Bernard. *The Grasshopper: Games, Life and Utopia*. Toronto: University of Toronto Press, 1978.

TWENTY-NINE

Building Capacity for Resilience in the Face of Environmental Shocks

ABOSEDE OMOWUMI BABATUNDE

Drawing on my teaching on the intersection of environmental security and peacebuilding at the University of Ilorin, Nigeria, this chapter explores how people are positively adapting to ecological crisis in the environmentally ravaged Niger Delta.

The climate crisis is a major existential problem that poses significant risk to planetary sustainability and all forms of life. The scale of the climate crisis is, of course, magnified by its interaction with social, economic, and political factors, thereby heightening or "multiplying" the risk of poverty, disease, food insecurity, political instability, and conflict.[1]

According to the report of the Global Spotlight of the Climate Scorecard Organization, oil-rich Nigeria is one of the most vulnerable countries in Africa—and indeed, the world—to climate change. Most of Nigeria's coastal areas are also exposed to perennial and chronic flooding due to rising sea levels—likewise linked to pollution from human activities.[2] The Nigeria's National Emergency Management Agency[3] report showed that from 2019–2020, 2.2 million people were directly affected by flooding, including 158 fatalities.

This article explores how people are positively coping with and adapting to ecological crisis, drawing on my research in the environmentally ravaged Niger Delta and on my teaching on the intersection of environmental security and peacebuilding at the University of Ilorin, Nigeria. Studies on the intersection between ecology and human health have shown that climate change impacts on mental health range from altruism, a sense of personal growth, and strong sense of community to post-traumatic stress disorder, panic, and anxiety.[4] Researchers have also identified the social impacts of climate changes, noting that they can heighten existing social inequities in which Indigenous people, the poor, elders, children, and people of color are the most affected by climate change impacts.[5]

Yet there is good news from research that has indicated that about 80 percent of people have the capacity for resilience, can overcome hardships, do not develop long-term psychological consequences, and can build their ability to develop new skills in dealing with painful and distressing experiences.[6]

CLIMATE CHANGE AND ENVIRONMENTAL RESILIENCE EDUCATION

The emergent nature of climate change has led to uncertainty among many educators about how best to enable students to navigate the complex emotions that they experience in response to their proximity to, and responsibility for, a myriad of injustices and environmental catastrophes.[7] The recent attention given to the educational aspect of emotional responses to climate change, and how learners navigate, negotiate, contest, and interpret the climate crisis, is a necessary starting point for effective climate education.[8]

Thus, scholarship in the field of climate change has increasingly focused on the shift from climate actions aimed at "preventing and mitigating responses in most climate affected environments, towards building the skills, capacities and knowledge to respond, cope with or transform the societies confronting climate change."[9] In relation to climate change, resilience is built around "a set of attributes or capacities that enable a society or community to endure, respond or bounce back from environmental shocks," particularly the "endogenous strengths in systems, structures and peoples within climate change impacted societies."[10]

> Climate change education provides a vital opportunity for students to express and reflect critically on their emotional responses to global warming.

Without this capacity, students are likely to remain locked in states of emotional paralysis or denial, thereby preventing engagement, action, and response-ability.[11] This position is further elaborated by the emphasis placed on the need to foreground emotion in any pedagogical response to the climate crisis, taking cognizance of the fact that emotions have been identified as "the missing link" in effective communication about climate change. It is important that climate education provides the opportunity for students to freely express their emotional despair so as to be able to develop the mindset to cope with grief. Emotion, it is argued, is embedded in "difficult" learning encounters and in various social injustices and inequalities.[12]

Climate change education is also embedded in the notion of learners' positioning as "implicated subjects" in the climate crisis, rather than merely victims of, bystanders to, or the actual perpetrators of, the harms associated with global warming.[13] Thus, a complex set of affective, socio-cultural, economic, and perceptual factors interact to shape people's engagement (or lack thereof) with the climate crisis[14] (see also Bryan's essay on implicated subjects in this volume). Such positionality is perceived as an approach that can enable students to look critically and reflexively at themselves in terms of their proximity to, and responsibility for, climate-related harms and injustices. Using a positionality approach to my teaching on climate and environmental resilience-building entails creating awareness about those human activities that adversely impact the environment, drawing student attention to their roles as implicated actors—that is, as bystanders who fail to act in tackling the climate crisis even when they have the capacity to do so. The positionality approach can engender a sense of collective actions in students and serve as motivators for them to see themselves as vital "agents of change."

In another study, Poland et al.[15] consider climate change educational training as an important avenue for reckoning for many who sympathize with people on the front lines of climate change, but who may not recognize the role that their privilege plays in shielding them from many climate impacts. Climate education can provide the missing links in the sense that it can stimulate "awareness" and "reawakening" for young people to understand the importance of collective actions in tackling the climate crisis (see chapters by Holmes and Frame et al. on community-building as resilience in this volume).

INSIGHTS FROM MY FORUM ON ENVIRONMENTAL AND EMOTIONAL RESILIENCE

The discussion forum on environmental and emotional resilience and self-care for my postgraduate students at the Centre for Peace and Strategic studies, University of Ilorin, Nigeria, is designed as a problem-solving strategy to provide emotional support along with resources, tools, and skills to help students address eco-anxiety, grief, and the range of feelings that accompany climate change impacts. Emotional resilience and self-care are important concepts that are linked to the educational training activities to build the skills and capacity of students to cope with eco-grief and anxiety. In my practical class on environmental resilience, I have therefore worked to enhance student well-being and mental health by creating a forum for discussion for them to express their emotions, generate ideas, and build their capacity to respond positively, creatively, and healthily to climate impacts.

Storytelling is considered one of the main ways to develop emotional resilience and self-care: it activates empathy, agency, and collective action and skills necessary for responding well to climate change. Through storytelling, educators can build ideas and creativity and develop more climate-conscious behavior. The learning process also ultimately helps build the capacity of students to be responsible global citizens who take actions to create a sustainable future for all.

As the Climate Therapy Alliance[16] explains, sharing eco-anxiety stories and related everyday experiences with climate change can be both healing and overwhelming. Such collective efforts can help to develop individual histories of grief and trauma, as well as build individual strengths. The main aim of the environmental resilience-building training for my postgraduate students is to provide a forum for the students to express and discuss their feelings of fears, doubt, guilt, despair, and concerns about climate change, support each other in building their agency and resilience, and stimulate ideas and plans of action to cope, adapt, and transform their communities. It can also help students confront their emotional despair and transform their hopelessness to hopefulness through attitudinal and behavioral changes in lifestyles.

In my classroom, I created a forum for group discussions, where my postgraduate students (some of whom are from localities where people deal with challenges of chronic flooding, environmental and health hazards linked to poor drainage, exposure to industrial waste and oil pollution), brainstorm among themselves, offering practical suggestions and applying these to their specific realities. This involves story-telling sessions of relatively successful cases of communities that have coped and developed their capacity to mitigate environmental degradation. There is also scenario building of the worse- and best-case outcomes, along with strategies to mitigate the worst cases. These sessions helped students to better cope with their emotions, raising their hope and capacity to develop mitigation measures. The students have been able to build their capacity as advocates within their communities, creating awareness about human actions that can worsen the challenges and about how to navigate the environmental problems.

The environmental resilience-building training for students also helps to address what Petersen et al.[17] refer to as ideological denialism relating to the inability to envision alternative social futures or an alternative to our current economic system (see chapters by Gray, Anson, and Stabinksy and Oesterby in this volume for more on envisioning alternative futures). Thus, the training can help to prevent and forestall such harmful threats by building students' capacity to take positive actions and be creative in developing ideals about an alternative economic system that can mitigate climate-related changes. It also helps them to develop the capacity to take effective climate actions and responses through an appropriate diagnosis of the human-induced drivers of climate changes and its consequences.

The discussion forum also entails what Poland et al.[18] refer to as meditative and interactive practices, which involve the use of the imagination to stimulate creativity and cultivate empathy. The students are able to develop a joint problem-solving approach that enables them not only to build their capacity to cope with their emotional anxiety and despair, but also to develop their imaginative capacity to take action. This approach resonates with similar learning practices that connect students emotionally to one another and the larger world. Some of the students have offered insights into impacts of the forum on their everyday realities. As one student explained, "the forum has raised my understanding of our collective responsibility to be a change agent, stimulating a sense of responsibility and generating ideas and creativity about how to bring about the much-needed change to tackle the climate crisis."

My experiences and extensive research in the Niger Delta region offer further examples of how some communities have been able to develop their capacities to cope and adapt to climate-related changes. In the Niger Delta, oil pollution resulting from the activities of multinational oil companies and the resultant massive adverse impacts on local people's main source of livelihoods, farming and fishing, and on their health and cultural life have provoked grievance. The people's grievances have mostly been expressed through violent reactions that include destruction of oil facilities, pipeline vandalism, and artisanal oil bunkering, which not only directly disrupt oil production, but also worsen the ecological crisis.

I shared these examples with my students during our story-telling sessions. Insights from local people and Environmental Right Actions, a major NGO working with those communities afflicted by the operations of oil companies, show ways that communities have managed to cope with incessant oil spills by devising strategies to curtail the spread of the spills. This is done by digging holes so that much of the oil sinks into the ground, limiting the spread to farmland and fishing waters. In some communities, the people have been able to mitigate adverse effects of oil spills on farmland by relocating farms to hilly places. In other communities where attacks on pipelines are rampant, vigilante groups are set up to guard oil pipelines and hand over vandals to the security agents. These cases demonstrate how communities are becoming more adept at coping with and managing environmental shocks. It also offers a ray of hope to the students that can help them confront and cope positively with their environmental crisis.

CONCLUSION

The environmental resilience training for my postgraduate students has positively impacted their ability to cope with eco-grief and anxiety. At the same time, it is important to know that there is no one-size-fits-all approach to the many complex emotional responses to climate crisis and resilience-building. Nevertheless, Bryan[19] emphasizes that a psychosocial-inflected climate education that engages productively with emotions and directly confronts socially sanctioned forms of denial and ignorance is vital to the broader public response to the climate crisis.

The insights from the case of the oil communities in the Niger Delta that I shared with my students also demonstrate that people can build capacity for resilience in the face of daunting environmental dislocation, rather than focusing attention on the doom and gloom of environmental problems—a focus that tends to lead to reactive outbursts such as oil pipeline sabotage that worsens ecological challenges.

NOTES

1. Peters and Vivekananda, *Topic Guide: Conflict, Climate and Environment*.
2. Princewill, "Africa's Most Populous City."
3. National Emergency Management Agency, "Flood Forecasting."
4. Hayes et al., "Climate Change and Mental Health"; Poland et al., "5 Ways Communities Are Coping."
5. McMichael, *Climate Change and the Health of Nations*.

6. Collier, "Growth after Trauma."
7. Bryan, "Affective Pedagogies."
8. Adams, *Ecological Crisis*.
9. Pihkala, "Climate Anxiety"; Bryan, "Affective Pedagogies."
10. Obi and Babatunde, "Challenge of Building Resilience," 3.
11. Hamilton, "Emotions, Reflexivity."
12. Bryan, "Affective Pedagogies."
13. Rothberg, *Implicated Subject*.
14. Bryan, "Affective Pedagogies."
15. Poland et al., "5 Ways Communities Are Coping."
16. Climate Therapy Alliance, *Emotional Resilience Toolkit*.
17. Petersen et al., "Reconceptualizing Climate Change Denial."
18. Poland et al., "5 Ways Communities Are Coping."
19. Bryan, "Affective Pedagogies."

REFERENCES

Adams, M. *Ecological Crisis, Sustainability and the Psychosocial Subject*. New York: Palgrave Macmillan, 2016.

Bryan, Audrey. "Affective Pedagogies: Foregrounding Emotion in Climate Change Education." *Policy & Practice: A Development Education Review* 30 (2020): 8–30.

Climate Therapy Alliance. *Emotional Resilience Toolkit for Climate Work*. Pacific Northwest Chapter. Climate Therapy Alliance, 2019. https://retooling.ca/resources/emotional-resilience-toolkit-for-climate-work.

Collier, Lorna. "Growth after Trauma: Why Are Some People More Resilient Than Others—And Can It Be Taught?" *American Psychological Association* 47, no. 10 (2016): 1–48.

Hamilton, J. "Emotions, Reflexivity and the Long Haul: What We Do about How We Feel about Climate Change." In *Climate Psychology: On Indifference to Disaster*, edited by P. Hoggett. Basingstoke: Palgrave, 2019.

Hayes, Katie, G. Blashki, J. Wiseman, S. Burke, and L. Reifels. "Climate Change and Mental Health: Risks, Impacts and Priority Actions." *International Journal of Mental Health Systems* 12, no. 28 (2018): 1–12.

McMichael, Anthony. *Climate Change and the Health of Nations: Famines, Fevers, and the Fate of Populations*. New York: Oxford University Press, 2017.

National Emergency Management Agency. "Flood Forecasting, Management: NEMA Provides Automatic Weather Observatories to NIHSA." Abuja, Nigeria: NEMA, 2021.

Obi, Cyril, and Abosede O. Babatunde. "The Challenge of Building Resilience in Post-Conflict African States: What Role for Local Institutions?" *African Conflict & Peacebuilding Review* 9, no. 2 (2019): 1–8.

Peters, K., and J. Vivekananda. *Topic Guide: Conflict, Climate and Environment: Evidence on Demand*. London: Foreign, Commonwealth and Development Office, 2014.

Petersen, B., D. Stuart, and R. Gunderson. "Reconceptualizing Climate Change Denial: Ideological Denialism Misdiagnoses Climate Change and Limits Effective Action." *Human Ecology Review* 25, no. 2 (2019): 117–41.

Pihkala, P. "Climate Anxiety." Helsinki: MIELI (Mental Health Finland), 2019.

Poland, B., K. Hayes, and M. Hathaway. "5 Ways Communities Are Coping with Climate Anxiety." *Yes Magazine*, August 22, 2018. https://www.yesmagazine.org/environment/2018/08/22/5-ways-communities-are-coping-with-climate-anxiety.

Princewill, N. "Africa's Most Populous City Is Battling Floods and Rising Seas." *CNN News*, August 1, 2021. https://edition.cnn.com/2021/08/01/africa/lagos-sinking-floods-climate-change-intl-cmd/index.html?utm_term=link&utm_source=twCNN&utm_content=2021-08-01T15%3A32%3A28&utm_medium=social.

Rothberg, M. *The Implicated Subject: Beyond Victims and Perpetrators*. Palo Alto: Stanford University Press, 2019.

THIRTY

Releasing Growth

TERRY HARPOLD

In this chapter readers learn to embrace the power of tree planting to release transformative truths about human-plant conviviality and world-making co-conspiracies.

TREES!

My environmental humanities courses at the University of Florida are closely linked with a faculty and student initiative I founded in 2019, "Trees!"[1] In collaboration with the Alachua County Department of Land Conservation and Management, Trees! organizes volunteer tree plantings in and around the city of Gainesville, Florida, where the university's main campus is located. The plantings emphasize restoring forest canopy and increasing biodiversity in economically challenged locations: Section 8 housing developments, public school grounds, and city-owned parks and trails for which canopy restoration funding is limited.[2] We welcome participation by members of the community; many of the plantings have included elementary, middle, and high school student groups. Provisions are made for the welfare of trees after planting—regular watering, biologically appropriate pruning, site maintenance, etc.—until they will thrive without further human intervention. Our most recent census (2021) showed that 98 percent of the trees planted in the initiative are in excellent health.[3]

The COVID-19 pandemic interrupted our work for much of 2020; pandemic-related safety precautions slowed us during Spring 2021. In Fall 2021, we resumed our normal tempo of planting 100–150 saplings per semester and have maintained this through to the present (Spring 2023).[4] We could plant more if we were not also committed to keeping the scale of each planting small, in the range of a dozen to two dozen human helpers and as many trees. These numbers allow us to train new participants in best practices for specific tree species and locations. A cohort of returning planters mentors the newcomers, while the arborist, her assistants, and I circulate between the teams (usually groups of two to four participants) giving advice and encouragement. Keeping team sizes and overall participation limited helps to foster a sense of collective purpose and ensures that everyone has a significant role to play. With repetition these modest numbers scale over time: despite the pandemic slowdowns, during the four years of the initiative to date participants have shared more than 2500 human-and-tree contact hours.

The plantings are, very simply, a lot of fun: a festive morning spent in the company of enthusiastic, like-minded humans, dirtying one's hands and knees in the service of a common good. My students often observe that they find these contributions to our region's ecological resilience satisfying and emotionally restorative, a sort of living laboratory of the methods and ideas that we discuss in the classroom. At the height of the pandemic, when my courses met only online, these opportunities to gather safely out of doors and in a natural setting—to *do* something of lasting benefit together in one place—eased our anxieties a little and boosted our morale. These things would be enough in themselves to justify the Trees! initiative. But something else happens in the plantings that I believe to be of greater and lasting importance.

VEGETALIZING OUR SENSORIUM

For many students, the plantings are both their first experiences of putting anything into the soil and occasions for mindful reflection, perhaps also for the first time, on the more-than-human lives of trees. Digging a hole, working the earth through one's fingers, untangling the knotted root ball, seating and propping the sapling upright as the hole is refilled, mulching, and the first careful watering . . . these activities bring students into intimate contact with a living being of a kind that is all around them, but which they may not have apprehended in this way before. The work of planting presents opportunities for discussion of the characteristics of different tree species, soil ecology, the role of subsurface mycorrhizal networks, our growing awareness of plant perception, learning, and sociality, and the importance of trees as agents of resistance to biodiversity collapse and anthropogenic climate change.[5] Often our conversations turn to the topic of the extreme longevity, in human terms, of the trees we're planting. A live oak, for example, can live for twenty human generations, a bald cypress for twice as long as that, all the while sequestering atmospheric carbon, exhaling oxygen, storing and releasing water as vapor, seeding its offspring and collaborating with its neighbors, and providing shelter and food to numberless other living beings.

Our morning's labors, I remind the planters, nudge these wondrous legacies forward, but they are not of our making. While still in the nursery pot the sapling is *already* alive, *already* photosynthesizing and respiring, *already* aware of and responsive to its environment in ways that elude our immediate perception and understanding.[6] What we do at most is (re)introduce the sapling to an expansive, vital order of the already and the ongoing that is specific to its way of life.

Such an encounter with a tree, and an awareness of its already and ongoing roles in its ecosystem, can be a first step toward what critical plant studies scholar Natasha Myers calls the "vegetalization of our sensorium": more than witnessing the omnipresence of plants in our lives—"they literally breathe us into being!" Myers reminds us—this means opening our sensibilities to their more-than-human collectivity:

> Once you have set in motion efforts to decolonize your common sense, it is time to vegetalize your sensorium so that you too can learn with and alongside the plants. As a co-conspirator supporting their world-making projects, you will need to apprentice with them. What do plants want? What do plants know? What can a plant do? We do not yet know. But you could reach toward them with the openness of not knowing, and forgetting what you thought counts as knowledge. . . . Let their planty sensitivities inflect your own. . . . Pay attention to the ways they defy all-too-human notions of individuality, bodily integrity, subjectivity, and agency. Let the plants redefine what you mean by the terms sensing, sensitivity, and sentience.[7]

I have many times observed the reactions of a newcomer when this unknowing and forgetting takes root, when their "planty sensitivities" are inflected and awakened. It looks like this. They rise, brush the soil and mulch from their hands and knees. They step away from the planted sapling, anticipating a warm flush of satisfaction at a job well done. More often than not, there is a slight deflation, a feeling of loss borne of the realization that the tree is shifting away from their attention. And then, if they're tuned in, there comes a burst of joy in recognizing the sapling's vitality and uncanny sovereignty: a living being that is *already* active within its ecosystem and may continue to be active for decades and centuries to come. In that moment the tree appears to them as fully *itself,* standing apart from their solicitous care, perhaps unaware or unconcerned with their care: doing its own thing in its own way and already building alliances with others more like itself than them. Though they cannot know if the sapling senses liberty in the moment of release—they can only *not* know, Myers might say—something *in them* is liberated. Their ecological fatalism and learned passivity, products of the endless drumbeat of terrible news about the planet's future, fall away. Newcomers are shocked by this experience the first time; the returning planters tell me that they keep coming back for its euphoric jolt.

GREENING SELVES

There are more knots to untangle here. A sapling's "uncanny sovereignty" is not uncanny to it; the forms of sovereignty specific to plants are their first nature. (I use the term "sovereignty" here not in its typical sense, meaning a unidirectional relation of power, really an imbalance of power, but to describe an unstable relation that is in the process of balancing: contingent and apprehended precisely where control over or responsibility for another being is transitioning to reciprocity.) Our awakening to the sapling's sovereignty opens us to an important truth: it would be the height of hubris to assume that the imbalances of power that have historically shaped human relations with our photosynthetic kin must, or will always, remain in place, or that our narrow experiences of rootless selfhood describe the only selves that matter.[8] There are more extended and diverse expressions of selfhood in the world than we humans are accustomed to. As I have written elsewhere, a revolution in botanical thought is forcing redefinitions of concepts like self and community:

> [Trees] exchange airborne biochemical signals with other trees, collectively monitoring resource availability and the presence of predators and parasites. Their notionally individual bodies extend outward to, tangle with, and merge with other plant bodies, and provide shelter and sustenance, above and below ground, to numberless organisms. By way of vast, mutualistic mycorrhizal (fungus) networks beneath the forest floor—comprising half of a forest's biomass—they share water, carbon, and other nutrients with other trees and plants. In contrast to our impoverished isolation and melancholy autonomy . . . theirs are lives of exuberant *heteronomy*, an ever-vital interdependence with the lives of many selves.[9]

In the moment of embracing and then releasing our solicitous care for a newly planted sapling, we acknowledge its plural self-realization within unthinkably complex ecosystems, *within which we also may abide and connect.* What happens next is a question of tuning in and apprenticing. On the model of vegetal others, we may (re)imagine the terms of our self-realization, not as melancholy autonomy, but as equally contingent, exuberant heteronomy.

Beyond practices of stewardship ("I plant this tree because it is the responsible thing to do"), restoration ("I plant this tree to help repair a world in crisis"), or self-care ("I plant this tree because its thriving will ease my anxiety"), the release of the sovereign sapling to grow in its own way is an act I would describe as *more-than-self* care. It acknowledges that the planter and the tree are linked by their brief contact and their world-making co-conspiracy, even if it is unclear that they are aware of this in comparable ways. The act of more-than-self care cultivates the expression of our wider ecological selfhood.

I'm drawing here on terminology introduced by the Norwegian philosopher Arne Næss, who also coined the term "deep ecology." *Ecological* selfhood, according to Næss, is the self's ongoing, self-organizing process of identification with its environmental situation, with which other selves, perhaps many other selves, may also identify.[10] An ecological self is not necessarily an environmentally aware self; it may have no discernible consciousness of its situation. Nor does ecological selfhood require feelings of responsibility for the welfare of others, though Næss asserts that the pursuit of ecological self-realization ("the fulfillment of potentials that each of us has, but that are never the same for any two living beings") widens and deepens the human self and tends to increase its compassionate concern for others.

Environmental activist Joanna Macy associates this ecological entanglement of self and other with the Buddhist concept of "dependent origination" or "dependent co-arising" (Pali:

paṭiccasamuppāda, Sanskrit: *pratītyasamutpāda*): all phenomena (*dharmas*) arise and cease to arise in relation to other phenomena; nothing truly stands apart or holds fast; there may be no discernible end to the mutual dependencies of any ostensibly single phenomenon on other phenomena. Seen in this light, selfhood as such is only a convention of thought or language ("I plant *this tree* . . . "); in the deeper reality of dependent co-arising, the appearance of an unconditioned, autonomous self (*this* I, *this* tree) is an illusion, and sovereignty is always transitioning to something else. It's a strange and unsettling form of conviviality, co-conspiracy, and interimplication, and it seems to me a powerful model for deepening Myers's call to co-develop our planty sensitivities.

Macy proposes that acknowledging the extensibility of co-arising human selfhood moves us toward the "greening" of the self, by which she means a recognition of something very much like the heteronomy of plant being, only more so. Greening the self expands our grasp of our "profound interconnectedness with all life," human and more-than-human, and our compassion for others increases, not out of feelings of moral responsibility or dread but in our heightened sense of existential solidarity with all beings.[11] Næss is more circumspect in that he attributes self-realization only to animals, plants, and possibly other biological entities, whereas Buddhist thinkers such as Macy, Stephanie Kaza, and Jason Wirth, influenced by the teachings of Eihei Dōgen, the thirteenth-century founder of Sōtō Zen Buddhism, expand this widening and deepening of the self to include nonsentient and abiological entities.[12] But Næss is sympathetic to what I take to be the fundamental insight of Macy's greening self. (Not a *green* self, but instead a green-*ing* self: a process that is always-already underway.) Self-realization doesn't inhere in acts of more-than-self care, such as planting a sapling and releasing it to its own way of world-making, but in our awakening to the ecological interimplications that such acts affirm: to the array of dependent relations whose compass is greater than the selves that arise and cease to arise therein.[13]

In the context of a morning spent doing something as unpretentious, pleasant, and efficacious as planting trees, the planters' surrender of their feelings of responsibility for the sapling, and their recognition of its qualities of the *already* and *ongoing*, exemplify the greening of the human self. We're drawn away, in joyous wonder, from the evidence of the plant's self-realization, then back again, more deeply into an embrace of a general conviviality that unites us. We're more than tuning in; moving away from the lonely-making illusion of human autonomy, we liberate the transformative truth of co-arising human-plant heteronomy.

My phrasing of these concepts renders them more speculative than I would like. I find it difficult to describe the joys of releasing care and greening the self without also invoking their existential and ethical significance. This is important, no doubt, for teaching and writing about what happens in the already and the ongoing. But all of this can be, if you wish it, more directly and simply felt. If you create the practical space for others to embrace the exuberance of this moment, if you introduce them to the living laboratory of it, the jolt will come to them. There will be time after to untangle the knotted changes that grow from that.

ACKNOWLEDGMENT

Trees! would not be possible without the expert support of Lacy Holtzworth, County Arborist, Alachua County Department of Parks and Conservation Lands.

NOTES

1. "Trees!," Imagining Climate Change, University of Florida, accessed May 23, 2023, https://imagining-climate.clas.ufl.edu/trees.

2. Named a "Tree City USA" in 1983 by the Arbor Day Foundation, and a "Tree City of the World" by the Food and Agriculture Organization of the United Nations in 2020, Gainesville is an unusually forested American city for its size, about 64 square miles in territory, with a human population of about 141,000. The most recent analysis (Andreu et al., *Urban Forest Ecological Analysis*) estimates that the city is home to seven million trees, with 47 percent overall canopy cover. The canopy has substantially declined during the last two decades, a period of accelerating urban development in the region.

3. Working with a qualified arborist and relevant local authorities is essential to the safety and success of a project such as this. Determining appropriate tree species and planting sites, identifying utility locations, and securing legal permissions is complicated and time-consuming. Your municipality probably already has a tree restoration program in need of volunteer labor. They will welcome your offer to help.

4. We typically don't plant during the summer because temperatures in North Central Florida are hard on juvenile trees.

5. Chamovitz, *What a Plant Knows*; Ryan et al., *Mind of Plants*; Simard, *Finding the Mother Tree*.

6. Hall, *Plants as Persons*.

7. Myers, "How to Grow Livable Worlds."

8. Marder, *Plant-Thinking*.

9. Harpold, "Kenneth A. Kerslake, *Tree Dreams*."

10. Næss, "Self-Realization."

11. Macy, "Greening of the Self."

12. Kaza, *Green Buddhism*; Okumura et al., *Mountains and Waters Sūtra*; Wirth, *Mountains, Rivers, and the Great Earth*.

13. Næss, "Gestalt Thinking and Buddhism."

REFERENCES

Andreu, Michael G., David A. Fox, Shawn M. Landry, Robert J. Northrop, and Caroline A. Hament. *Urban Forest Ecological Analysis. Report to the City of Gainesville*. Gainesville, FL: City of Gainesville, 2017.

Chamovitz, Daniel. *What a Plant Knows: A Field Guide to the Senses*. New York: Farrar, Straus and Giroux, 2017.

Hall, Matthew. *Plants as Persons: A Philosophical Botany*. Albany: State University of New York Press, 2011.

Harpold, Terry. "Kenneth A. Kerslake, *Tree Dreams*." In *Plant Life: Exploring Vegetal Worlds in the Harn Museum Collection*. Gainesville, FL: Samuel P. Harn Museum of Art, 2021.

Kaza, Stephanie. *Green Buddhism: Practice and Compassionate Action in Uncertain Times*. Boulder, CO: Shambhala, 2019.

Macy, Joanna. "The Greening of the Self." In *Dharma Gaia: A Harvest of Essays in Buddhism and Ecology*, edited by Allan Hunt Badiner, 53–63. Berkeley: Parallax Press, 1990.

Marder, Michael. *Plant-Thinking: A Philosophy of Vegetal Life*. New York: Columbia University Press, 2013.

Myers, Natasha. "How to Grow Livable Worlds: Ten Not-So-Easy Steps." In *The World to Come: Art in the Age of the Anthropocene*, edited by Kerry Oliver-Smith, 52–64. Gainesville, FL: Samuel P. Harn Museum of Art, 2018.

Næss, Arne. "Gestalt Thinking and Buddhism." In *The Ecology of Wisdom: Writings by Arne Næss*, edited by Alan Drengson and Bill Devall, 195–203, 321. Berkeley: Counterpoint, 2008.

———. "Self-Realization: An Ecological Approach to Being in the World." In Næss, *The Ecology of Wisdom*, 81–96, 318.

Okumura, Shohaku, Carl Bielefeldt, Gary Snyder, and Issho Fujita. *The Mountains and Waters Sūtra. A Practitioner's Guide to Dōgen's "Sansuikyo."* Somerville, MA: Wisdom Publications, 2018.

Ryan, John C., Patricia Vieira, and Monica Gagliano, eds. *The Mind of Plants: Narratives of Vegetal Intelligence*. Santa Fe, NM: Synergetic Press, 2021.

Simard, Suzanne. *Finding the Mother Tree: Discovering the Wisdom of the Forest*. New York: Knopf, 2021.

Wirth, Jason M. *Mountains, Rivers, and the Great Earth: Reading Gary Snyder and Dōgen in an Age of Ecological Crisis*. Albany: State University of New York Press, 2017.

THIRTY-ONE

Ecotopia versus Zombie Apocalypse

Collaborative Writing Games for Existential Regeneration

MARNA HAUK

> Playing collaborative writing games, such as "Ecotopia versus Zombie Apocalypse," can generate messy co-adventures while surfacing creative imaginaries that undermine binary (disaster/utopian) thinking. This gamified approach supports the release and evolution of climate grief and denial toward grace, connection, and regenerative creativity.

This chapter discusses the "complementary and recursive use of artistic, experiential, embodied, symbolic, spiritual, and relational learning" that Fumiyo Kagawa and David Selby mandate for meaningful climate change education.[1] Collaborative, experiential, and creative curricula nurture learners in the context of climate change and climate justice. Well-designed climate crisis and climate justice games fit the bill: they regenerate the imagination while building resilience and collaborative possibility-thinking.

CHALLENGING THE "PERFECT MORAL STORM"

Stephen Gardiner identifies three critical incapacities that generate the "perfect moral storm" of current Western failures to engage with the climate crisis in an effective way.[2] This perfect storm is caused by the incapacity to understand interconnections across time as well as space/place, along with a preference for maintaining personal comfort. Staying in denial places us in ethical corruption, preventing effective action on behalf of creatures, systems, ourselves, and larger wholes. I think of this moral corruption Gardiner presages as creating a kind of gravity well of inertia to act.

FIGURE 31.1 The Spiral of the Work That Reconnects. Image by Dori Midnight used with permission of the artist.

A powerful alternative to this cycle of minimization, denial, and inaction is Joanna Macy's Spiral of the Work That Reconnects.[3] Macy is a systems-informed ecopsychologist who has spent decades moving those encountering existential dread into their powerful feelings and then through into action. Her generative work has a homeplace as an activist way of perceiving and helps those encountering the existential threat of nuclear threat and annihilation to move from immobilization to action. Her experiential research and tool development has come to further flowering in the ages of ecological crises and now in climate crisis.[4]

CLIMATE ECOPHOBIA

Macy noticed that those who were aware of systems-level difficulties, and even those wanting to be solutionary in response to them, were still encountering incapacitating despair and grief. Young learners can be particularly sensitive to and incapacitated by perceiving world-scale challenges.[5] Macy furthermore noticed that this despair and grief were roadblocks for effective action, in two ways. First, the emotional and existential load of the grief was itself incapacitating. Also, many were actually blocked from effective action and collaborative organizing by deep aversion to encountering the scale of their grief and despair. Folks will do a lot to avoid this pain and grief. There was ecophobic avoidance of working on these issues because of the recoiling effects of the catastrophic existential angst and despair at species death, at the ravages of industrial extraction and mining, at the threat of nuclear annihilation, and now at the threat of human and multispecies annihilation from global climate change. David Sobel has described this as "climate change meets ecophobia."[6] How do we design curricula to avoid this climate-ecophobic aversion?

OUR PAIN IS OUR CONNECTION

Macy's work hinges on a central insight: as we engage with our feelings of despair in supported contexts, we come to experience this despair and overwhelm as evidence of deep interconnection and empathic relationship with larger systems and wholes. Together, grappling with this grief, rather than avoiding it, births perceptual breakthroughs. For learners and teachers who need to, as we regain our capacity to feel fully and deeply, and encounter our existential feelings, we are reconnected with the same source that lends the wide quality to world-scale challenges: the living world system as evolving emergence in which we are embedded and of which we are a part. We gain expanded perceptions and are supported in going forth in creative action. Our creative and emotional capacities are freed up.

MESSY, IMAGINATIVE EDUCATIONAL EXPERIENCES OPEN UP FEELINGS AND POSSIBILITY-THINKING

This is in fact the kind of educational experience Selby and Kagawa emphasize is needed, requiring "lively and messy . . . emotional, imaginative, and creative entanglement with the world" to generate critical, transformative sustainability education.[7] Two key dimensions of this shift are working collaboratively and exercising our imagination. Working with arts-based, writing-based, or other modes of accessing the river of imagination opens this generative fount. It sparks a contagion, or amplifying feedback loop, of possibility-thinking. The ability to imagine other possible futures kickstarts and sustains an "active hope," a reorientation from pushing against grim noticings and feelings towards harvesting larger emergent connectedness, freeing up generative motion and energy towards long-term actions of sustainment and care.[8] Imaginative and ecological approaches to teaching do foster holistic, interdisciplinary experiential learning with a "pedagogy that is learner-centered, collaborative, discovery-based, and authentic."[9] In particular, the ecological imagination is a gift of our embeddedness that can support us "catching an inspiring vision."[10]

Collaborative creativity and possibility-thinking foster strength that girds learners for facing what has been avoided and reclaiming an underlying sense of connection. Macy developed a series of experiential approaches and a framework for individuals and groups to grapple with and metabolize systems-scale despair, turning despair into affirmative clarity, with its root truth of connectedness.[11] Designing curriculum that tends to despair without overwhelming learners beyond their capacity is a developmental challenge. Sobel advocates for developmentally appropriate curriculum to avoid "the overwhelmingness of environmental problems [that] can breed a sense of ennui and helplessness . . . [because] too much knowledge about environmental tragedies actually discourages environmental behavior."[12] Developmentally appropriate interventions foster connection, and range from species kinship through experiences such as school gardening and outdoor immersion on through to more direct community and project-based work as developmental capacities deepen.[13] Imaginative play and writing games can create meaningful and developmentally appropriate ways of supporting students engaging with the existential dimensions of climate change and climate justice to access connectedness.

I designed a collaborative writing game, "Ecotopia versus Zombie Apocalypse" to be exactly this kind of multi-dimensioned, developmentally connective, creative exploration. It generates messy co-adventures while surfacing complex and creative imaginaries that undermine binary (disaster/utopian) thinking. The creative space opened up by a gamified approach supports imaginal encounter and the evolution of grief to grace, possibility, and co-emergent regenerative creativity (see chapters in this volume by Creane and Meehan for more on play, games, and creativity).

THE GAME

Ecotopia versus Zombie Apocalypse helps groups explore the fertile intersection of the mythic, the monstrous, and the possible at the convergence of the past, the present, and the future. This game heralds the future and serves as a clarion call for transformers. As a romp through our cultural materials, it awakens the assistance of the mythic, while helping us encounter rather than deny the monstrous in our midst. Ecotopia versus Zombie Apocalypse plumbs powers within the deep imagination and calls out to future beings, lending us courage and affirmation in a time of challenge, contention, danger, and—yes—possibility.

The intent of the game is to regenerate culture and our sense of the possible. Game play invokes what Edward deBono calls "lateral thinking," juxtaposing highly unusual, dissimilar ideas to generate fresh thinking.[14] Writers pull an "ecofractal regeneration pattern card" to inspire their writing in each round of play. Each participant creates three characters: an Ecotopian (near-future communitarian, sustainability-culture dweller), a zombie, and a mythic figure. The group writes a linked set of stories together within the same universe through dynamic interactions of the characters and emergent landscapes. During each round, or cycle, participants add to the story and also write meta-reflections on their emerging experience.

Within each cycle of narrative play, each player pulls one or more Earthflow cards from one of three decks and over 50 cards inspired by biomimicry, nature and bioculture fractals, and regeneration patterns.[15] Using a shared document, each student on each day adds to the emerging story at least 300–500 words of narrative, using at least two of their characters and interacting, after the first cycle, with at least one of the other players' characters. Taking into consideration the cycle's theme, and emergently responding to the unfolding text, players take turns writing. Game participants are encouraged to hook their account and activity into what has come before. When composing their portion of the story for each cycle, they add in details and narrative that bring the situation to life. They are encouraged to start with a transition from the previous text and then provide continuity in addition to setting, detail, sensory details, and action. Players can also add soundtracks and visuals to enhance the depth of the narrative. Each cycle of play has a theme, one of the layers of the systems-thinking iceberg:

- Cycle 1. Sounding the Alarm: Habits, Habitats, and Encounters (Events layer)
- Cycle 2. Inside the Encounters, Rivalries, and Unlikely Partnerships (Patterns and Trends layer)
- Cycle 3. Eruption and Reconnection: Liberating Structures—with Hidden Strengths Emerging (Structural layer)
- Cycle 4. Homecoming: Regenerating Cultures and Mindsets (Paradigm layer)

Participants also respond to the cycle's theme. Detailed instructions for play appear at https://earthregenerative.org/ecotopiavzombie/game.[16]

TRANSTEMPORAL EMBRACE: STRENGTHS-BASED APPROACHES ACROSS TIME AND WITH LIVELY ALLIES

The Ecotopia versus Zombie Apocalypse game creates a synergistic convergence accessing multiple strengths-based and creative resources. The game features several strengths-based dimensions, including solutionary, mythic, and regenerative-thinking resources. The game consciously includes solutionary elements (Ecotopians) yet does not deny their foibles and humanity (or, as Donna Haraway might say, their "humusity").[17] We are immediately invited into imagining future beings within a fantasy of a more ecologically healthy future. We allow ourselves the fantastic possibility that humanity gets it together and realigns for the further flourishing of all beings and species. Being able to inhabit this possibility, even if only imaginally, starts to fill the storehouse and provides a sanctuary experience for the exhaustion and anxiety of bathing in the grim statistics of climate crisis.

Alexis Pauline Gumbs's creative scholarship is important here. Engaging with the future and future beings helps strengthen and reconnect us in deep time from the ancestral (imagined, familial, intellectual ancestors) and the future, resituating us in a transtemporal embrace. Gumbs's creative future fictions, including *M Archive: After the End of the World*[18] and her story "Evidence" from *Octavia's Brood: Science Fiction Stories from Social Justice Movements*,[19] describe trans-generational conversations between past and future beings. Gumbs conceives, in an archive of cross-generational letters, a justice-framed engagement: "You affirm our collective agreement that in the time of accountability, the time past law and order, the story is the storehouse of justice"—a justice that is not about punishment but about "brilliance on the scale of the intergalactic."[20] In her account, to "consciously exist in an ancestral context" is an act of justice and remembrance.[21] Writing stories is conceived as a balm and repository of relational accountability: "the story is storehouse."[22] This addresses one of Gardiner's main identified deficits of the perfect storm of climate crisis: a lack of capacity for trans-temporal thinking.[23] Haraway also endorses fictional work about the future to leverage imaginings and relatings with future beings as kinds of acts of healing and annealing the timeline.[24] This Harawayian act of decomposing and composting invents fresh compositions in collaboration with the creatures with whom we are kin. It conveys the heart of Ecotopia versus Zombie Apocalypse: an act of re-emergence.

CREATING FREE SPACE FOR THE EMERGENT SYNERGIES

Importantly, no characters are perfect. We are falling from projective perfections to dice-rolled complex characters with strengths and issues. Along with the imperfect future Ecotopians and the powerful mythic figures, we include rather than avoid the nightmarish figures in the current collapse, embodied in the Zombies. Of course, this must be done in a developmentally appropriate way, as there is a danger of compounding the effects of what we discussed as climate-ecophobia aversion. So we must dance a subtle line here, encouraging our own projections, fears, despair, and emotions about our collective situation without flattening or collapsing into doomsday scenarios or the resultant avoidance or annihilative responses to overwhelm.

Zombies, as a repository for fears about collapse, are a wonderful, ready-made imaginary for engaging with and processing submerged feelings and finding refreshed perspectives. We come to find that zombies in particular not only represent the doomsday scenarios, but also the doomsday mindset itself. How can the insufferable, incessant, half-alive, and seemingly mindless help us encounter our feelings of dread? Literally, and also, projectively, through their desire for consciousness—"Brains!"—how might zombies' gnawing also represent our persistent existential overwhelm and angst at our current predicament? Perhaps we, too, in climate denial have become zombies, metaphorically.

> Zombies, as a repository for fears about collapse, are a wonderful, ready-made imaginary for engaging with and processing submerged feelings and finding refreshed perspectives.

And, further, how might the presence of zombies suggest and indicate the "way through": mindful, metacognitive, caring, co-presencing with attentional awareness, slowness, and a breathing, fully lived embodiment? In the permaculture tradition, we say the solution is embedded in the constraints. In this game, some of the characteristics we explore for both Ecotopians and Zombies (the qualities rolled for) are generated by imagining common traits of Ecotopians and Zombies: Perseverance. Regeneration. Resilience. Creativity. Along with strength, dexterity, and charisma, old standbys in multiplayer role-playing games.

The third role, the mythic characters, reinforce and affirm our capacities for alliance with greater forces and resources. Some students select friendly or benefic mythos, and others invoke retributive powers. The inclusion of the mythic affirms expanded power with and agency for learners while modeling accessing larger forces and resources for healing.

REFLECTIONS

The game has taken interesting turns, whether with four or fourteen co-authors across a five-day period of playing. I have facilitated this game in sustainability leadership and altruism graduate courses, though I see that it could be played in other community contexts as well. The storylines often break down binaries and generate innovative solutionaries. Some of the surprising turns of events include a whole narrative about zombies becoming healthy after a shift to veganism, as well as conversations about vaccine anxiety through characters wrestling with the complex ethics and unknowns of the anti-zombie vaccine. Some students come to wonder, What is the deeper learning below the binary of the zombie zeitgeist? Following are some key themes that emerged from my experience facilitating the game.

COLLECTIVE EMERGENCE

Many climate ecopsychologists affirm the value of gaining skills in becoming comfortable with the unknown, in fact, "to become strengthened by uncertainty."[25] Games, if engaged in deeply, can nurture our sense of mystery and ongoing adventure. We are drawn out of the past and the future, and into the present, reconnecting with a sense of vitalizing engagement with an ongoing, unfolding emergence. Creativity and presence practices can nurture intrinsic hope.[26] Coming more fully alive and present, in an increasingly connective and creative way, sustains us and helps us befriend uncertainty, "grounding in our conscious connection with all life."[27]

> As we write our potential futures, entanglements, messy struggles, and emergent becomings together, we regenerate our stamina for the world the future is calling us to.

The collaborative dimension of the game is very important, as we nurture "a richer experience of community."[28] We are undertaking this uncertain adventure together. We can learn to create together. This collaborative approach affirms that there are larger strands of innovation, creativity, and endeavor with which we are bound. This draws in Macy's notion of a perceptual shift of "widening circles of connection" within the structure of the game. By experimenting with improvisational creativity in groups, we leverage the gifts of what R. Keith Sawyer has researched as phenomena within improvisational creativity: collaborative emergence, group flow, or group genius.[29] By extension, the game is an act of what I call Earthflow, coming to collaborate with the living energies of other humans, mythos, ecosocial patterns, and co-creative living systems from the personal to the multispecies and planetary (see also Harpold's chapter on our photosynthetic kin in this volume). In fact, in this way, the game supports us learning to collaborate. These collaborative, creative approaches are a way of practicing embodying what Macy and Johnstone describe as the Great Turning toward a life-giving society[30] (for more activities using the Great Turning, see Gray's chapter in this volume). As we write our potential futures, entanglements, messy struggles, and emergent becomings together, we regenerate our stamina for the world the future is calling us to. One participant summarized their experience along these lines in this way:

> I am beginning to understand the necessity of presence and participation in the reality of other people. Writing my characters into a weaved tapestry of narrative seems much more difficult at times, though results in greater storytelling. Without the recognition of another's reality, the story may go stale. There is great excitement to get involved in the storylines of others. It is bringing more life to my own characters and allowing for deeper development. And in a sense I am beginning to see how "my" characters are not really mine. We are all helping each other develop all these beings. Though they have sprung from our own individual imaginations, we all now have collective narration through all of them.

SYNTHESIS

In this time of peril and possibility, the future calls us to reconnect, to take our places in the flows of time and space so we can learn to be good ancestors while gaining strength from those who have come before us. Engaging with deep imagination in collaborative climate change writing games can shift learners from aversion to presence, from grief to co-creative action. As much as we might want to avoid it, engaging with seemingly intolerable possibilities and re-perceiving our pain as evidence of interconnectedness re-sources us. As we more fully resource, we can sustain considering the monstrous and diving into messy entanglements with the lively dimensions of life, conceiving of fresh transformations and pathways of thriving. Our children's lives and the fates of many species are not written yet or stonehewn. We carry these vision seeds, cast into our mutual garden of reciprocity, justice, and becoming.

NOTES

1. Kagawa and Selby, "Climate Change Education," 243.
2. Gardiner, *Perfect Moral Storm*.
3. Macy and Brown, *Coming Back to Life*.
4. Macy and Johnstone, *Active Hope*.
5. Ambrose, "Utopian Visions."
6. Sobel, "Climate Change Meets Ecophobia."
7. Selby and Kagawa, "Drawing Threads Together," 278.
8. Macy and Johnstone, *Active Hope*.
9. Judson, "Re-Imagining Sustainability Education," 210.
10. Macy and Johnstone, *Active Hope*, 159.
11. Macy and Brown, *Coming Back to Life*.
12. Sobel, "Climate Change Meets Ecophobia," 17.
13. Bigelow, "Teaching the Climate Crisis," 87–89.
14. deBono, *Lateral Thinking*.
15. Hauk, "Earthflow Cards."
16. Hauk, "Ecotopia versus Zombie Apocalypse"
17. Haraway, "Symbiogenesis, Sympoiesis, and Art Science Activisms," M45.
18. Gumbs, *M Archive*.
19. Gumbs, "Evidence."
20. Ibid., 33.
21. Ibid.
22. Ibid.
23. Gardiner, *Perfect Moral Storm*, 32.
24. Haraway, "Symbiogenesis, Sympoiesis, and Art Science Activisms."
25. Macy and Johnstone, 231.
26. Davies, *Intrinsic Hope*, 73–88, 160.
27. Macy and Johnstone, *Active Hope*, 231–32.
28. Ibid., 121.
29. Sawyer, "Group Flow and Group Genius."
30. Macy and Johnstone, *Active Hope*, 26–33.

REFERENCES

Ambrose, Don. "Utopian Visions: Promise and Pitfalls in the Global Awareness of the Gifted." *Roeper Review* 30, no. 1 (2008): 52–60.

Bigelow, Bill. "Teaching the Climate Crisis." In *A People's Curriculum for the Earth: Teaching Climate Change and the Environmental Crisis*, edited by Bill Bigelow and Tim Swinehart, 79–91. Milwaukee: Rethinking Schools, 2014.

Davies, Kate. *Intrinsic Hope: Living Courageously in Troubled Times*. Gabriola Island, BC, Canada: New Society, 2018.

deBono, Edward. *Lateral Thinking: Creativity Step by Step*. New York: HarperCollins, 2015.

Gardiner, Stephen Mark. *A Perfect Moral Storm: The Ethical Tragedy of Climate Change.* New York: Oxford University Press, 2013.

Gumbs, Alexis Pauline. "Evidence." In *Octavia's Brood: Science Fiction Stories from Social Justice Movements,* edited by Walidah Marisha and adrienne maree brown, 34–41. Oakland: AK Press, 2015.

———. *M Archive: After the End of the World.* Durham, NC: Duke University Press, 2018.

Haraway, Donna. "Symbiogenesis, Sympoiesis, and Art Science Activisms for Staying with the Trouble." In *Arts of Living on a Damaged Planet: Monsters of the Anthropocene,* edited by Anna Tsing, Heather Swanson, Elaine Gan, and Nils Bubandt, M25–M50. Minneapolis: University of Minnesota Press, 2017.

Hauk, Marna. "Earthflow Cards, Decks I–III." Hood River: Institute for Earth Regenerative Studies, 2014/2023. Retrieved from https://earthregenerative.org.

———. "Ecotopia versus Zombie Apocalypse: A Collaborative Writing Game." Hood River, OR: Institute for Earth Regenerative Studies, 2018/2023. Retrieved from https://earthregenerative.org/ecotopiavzombie/game.

Judson, G. "Re-imagining Sustainability Education: Emotional and Imaginative Engagement in Learning." In *Sustainability Frontiers: Critical and Transformative Voices from the Borderlands of Sustainability,* edited by David Selby and Fumiyo Kagawa, 205–20. Opladen, Germany: Barbara Budrich, 2015.

Kagawa, Fumiyo, and David Selby. "Climate Change Education: A Critical Agenda for Interesting Times." In *Education and Climate Change: Living and Learning in Interesting Times,* edited by David Selby and Fumiyo Kagawa, 241–43. New York: Routledge, 2010.

Macy, Joanna, and Molly Young Brown. *Coming Back to Life: The Updated Guide to the Work that Reconnects.* Rev. ed. Gabriola Island, BC, Canada: New Society, 2014.

Macy, Joanna, and Christopher Johnstone. *Active Hope: How to Face the Mess We're in with Unexpected Resilience and Creative Power.* Rev. ed. Novato, CA: New World Library, 2022.

Sawyer, R. Keith. "Group Flow and Group Genius." *NAMTA* 40, no. 3 (2015): 49–52.

Selby, David, and Fumiyo Kagawa. "Drawing Threads Together: Transformative Agenda for Sustainability Education." In *Sustainability Frontiers: Critical and Transformative Voices from the Borderlands of Sustainability,* edited by the article authors, 277–280. Toronto: Barbara Budrich, 2015.

Sobel, David. "Climate Change Meets Ecophobia." *Connect* 21, no. 2 (2007): 14–21.

PART SEVEN

COMMUNITY, COLLABORATION, AND KINSHIP

THIRTY-TWO

Facilitating "R&R"

Student-Led Climate Resilience and Resistance

JESSICA HOLMES

This classroom assignment provides an opportunity for students to directly lead resilience and/or resistance practices. It integrates affective exploration of environmental issues and mental health support into the existing framework of any given course.

Serves:

Any class size (or community group) that is small enough to comfortably foster student-to-student interaction but large enough to split into smaller groups.

Pairs well with:

This activity series can be carried out in conjunction with a course text or as a free-standing assignment prompt. In the example prompt below, the assignment is integrated with required course reading: *Emotional Resiliency in the Era of Climate Change,* by Leslie Davenport.[1] This text, initially written for clinicians, first responders, and mental health practitioners, provides easy-to-understand research, theory, and hands-on methods for developing emotional resilience in the context of the climate crisis. Because the book is written for practitioners, it is well suited to integrate into this assignment, which places students in the role of classroom and community facilitators of care. If the assignment is to be carried out without an accompanying text or specific research component, I recommend providing direct in-class guidance and groundwork around climate emotions and affective responses to trauma prior to asking students to lead an activity. (This volume contains many helpful resources in this regard, particularly in Part One: Getting Started with Emotions in the Climate Classroom.)

Preparation time required:

The activity itself takes at least 20 minutes per small group. I recommend spreading the assignment series out on a weekly or biweekly basis (depending on the length of the course), enabling one group to lead an activity for each respective week. However, this format can be adapted to fit the length, frequency, and structure of the course. Preparation time for instructor and students will depend on the nature of the content and whether the activity series is integrated with a course text or research component.

Ingredients:

Depending on the nature of the activities students develop, some materials may be helpful—such as a whiteboard, a projector, paper and pens, an open space on the floor or outside, comfortable clothing, etc. Teachers should let students know about available resources and students should let the teacher know in advance if such materials will be needed for their activity.

Oven temperature:

Very hot . . . the world is on fire!

Notes before you start:

This assignment prompt is situated at the intersection of academic pedagogy, mental health support, and direct action. It provides an adaptable model and curated opportunity for students to directly lead resilience and/or resistance practices in the context of studying, processing, and living through the climate crisis. The immediate purpose of the activity series is to integrate affective exploration of environmental issues and mental health support into the existing framework of any given course. Students will create and lead their own resilience and resistance exercises, usually constructing some brief framing or presentation of research, philosophy and/or objective to precede and anchor the activity. Most diverse student populations, whether or not they realize it, enter the classroom well-equipped with manifold, sometimes even transformative, resilience and resistance tools; however, in my experience, students have rarely been asked to directly harness those tools within the context of their scholarship. This assignment asks them to do just that.

> Most diverse student populations, whether or not they realize it, enter the classroom well-equipped with manifold, sometimes even transformative, resilience and resistance tools.

Why resilience and resistance? Because the current existential environmental crisis demands both an ability to withstand or rebound from challenges *and* to fight back—in other words, to resist patterns of injustice and harm. If we focus solely on resilience, we risk furthering the physical and mental burdens that disproportionately affect the working class, women, and people of color. The pedagogical goal of fostering student resilience alone fails to address the conditions that necessitate increasingly chronic resilience in the first place.[2] On the other hand, the singular goal of resistance through direct action sometimes fails to sufficiently attend to affective experience. Activists of all kinds can be prone to burnout, compassion fatigue, and pre-/post-traumatic stress.[3] Only through a combination of resilience and resistance practices can we successfully effect radical change with our mental and physical well-being intact.

> Only through a combination of resilience and resistance practices can we successfully effect radical change with our mental and physical well-being intact.

This assignment asks students to consider affective intersections between resilience and resistance. How might practices of direct action support *both* climate justice and mental health? And how might we productively rethink definitions of "action" to include practices of self-care and community-building? Climate justice and ecological health can only be achieved through diverse tactics—some reflective, meditative, or artistic in nature (examples of student-led activities include exercises in communal storytelling, visualization practices, mindfulness techniques, and nature walks), others more conventionally "active" (examples include writing postcards to elected officials, creating educational materials to share, and working in a community garden or natural space). By encouraging students to explore and define the parameters of climate resilience and resistance themselves, we avoid making assumptions about which forms of affective experience, expression, and engagement are worthy; and we seek to honor diverse student experiences, positionalities, knowledge, family histories, and ways of processing emotion.

> How might practices of direct action support *both* climate justice and mental health?

While the assignment is content-based in some respects (ideally students will encounter new ideas and learn about specific practices of care), perhaps more important is the spirit of collaboration and community-building that the series fosters over time. Teaching climate change in the humanities is different than teaching students to solve an algebra equation, learn a language, or craft an academic thesis statement. As the teacher, I can guide students through various texts, discourses, and approaches to environmental topics, but I don't have the answers to the climate crisis. Today and in the years to come, young people will need to lean on each other, drawing on the variety of skills, attributes, and approaches they collectively embody. If this assignment achieves its goal, students will walk away armed not only with new information and a wider array of available care practices, but also with a mutual appreciation for and validation of the unique strengths each individual brings to this fight.

Directions for students:

In groups of four and over the course of the academic term, you will be responsible for facilitating peer learning and other activities linked to the six central chapters of Davenport's *Emotional Resiliency in the Era of Climate Change*.

Begin by reading and discussing (as a group) your assigned chapter. Think about the key words and central ideas in that chapter. What do you find most compelling? Can you connect Davenport's ideas to class concepts or topics? Can you think of examples in your own lives? How do (and don't) Davenport's strategies help build resilience and resistance on a personal, local and/or global level? How do (and don't) they serve the goals of affective well-being and/or climate justice?

You are responsible for engaging with this chapter's material through a group-led activity. In order to situate other students in the class, I recommend briefly providing some framework, vocabulary, examples and/or analysis; but you can (and should!) use the textual material as a jumping-off point to engage the class directly in a discussion, activity, or training. In other words, *bring the material to life*!

You are welcome to use the end-of-chapter worksheets in Davenport's book as a starting point (or draw on the exercises in Part Two of her book). You can have your peers complete one as part of your activity. You might want to adapt one, or more likely, come up with your own exercise entirely. Be creative and think about how other students could best engage with the material—*especially affectively*. This might look like a traditional academic discussion or debate, or it might look totally different. Maybe you ask the class to breathe and meditate. Maybe you ask the class to write to someone or draw something creative. Maybe you make the students jog backwards around the quad singing. Just be sure you explain why and how the exercise relates to the course and material at hand. Ask your peers to reflect on the activity afterwards, especially with regard to the concepts of "resilience" and "resistance."

Please note that you don't have to agree with all of Davenport's points. Feel free to critique them or allow the class to critique them (respectfully, of course). You are welcome to situate your activity in opposition to a particular concept or argument as best suits your group goal(s).

Student-led "R&R" sessions should total 20 minutes (and no more than 30). At least half of this time should be used to facilitate some type of activity, discussion, or training.

Due dates: Presentations will take place throughout the term. A sign-up sheet will circulate during the first week of classes.

NOTES

1. Davenport, *Emotional Resiliency in the Era of Climate Change.*
2. Holmes, "The Trouble with Resilience."
3. jones, *Aftershock.*

REFERENCES

Davenport, Leslie. *Emotional Resiliency in the Era of Climate Change: A Clinician's Guide.* Philadelphia: Jessica Kingsley, 2017.
Holmes, Jessica. "The Trouble with Resilience." In *Communicating in the Anthropocene: Intimate Relations,* edited by Alexa M. Dare and C. Vail Fletcher, 53–68. Lanham: Lexington Books, 2021.
jones, pattrice. *Aftershock: Confronting Trauma in a Violent World: A Guide for Activists and Their Allies.* New York: Lantern Books, 2007.

THIRTY-THREE

Climate Justice and Civic Engagement Across the Curriculum

Empowering Action and Fostering Well-Being

SONYA REMINGTON DOUCETTE and HEATHER U. PRICE

This chapter describes the Climate Justice Across the Curriculum program in Washington State. Faculty from over twenty colleges and universities have participated in this curriculum development program to create lessons that incorporate climate justice and civic engagements into their STEM, humanities, arts, business, allied health, and other courses.

When our students learn about the climate crisis, many feel emotions such as anxiety, grief, fear, anger, motivation, hope, hopelessness, and resignation.[1] Taking community and civic action to address the climate crisis can help students hold and process these emotions. Faculty can incorporate climate justice alongside civic and community engagement (C&CE) into our courses to foster these values in students and ourselves. Weaving engagement in civic and community activities into our teaching can cultivate empowerment and hope, while developing authentic connections between students and community.[2]

The Climate Justice and Civic Engagement Across the Curriculum project is focused on faculty professional development that provides intellectual and financial support, space, and time for faculty to create and implement climate justice lessons, with a civic engagement component, into their courses. Faculty participants come from a wide range of disciplines. The faculty experience involves an introduction to climate justice and C&CE, interdisciplinary group brainstorming to develop lesson drafts, "experiencing" a climate justice lesson as an example to facilitate further collaborative lesson development and refinement, and finally drafting lesson presentations with supportive feedback from colleagues (Table 33.1). Faculty go on to implement their lessons and submit them to a Climate Justice Curriculum Repository, which now collectively contain over 70 lessons. Some of these lessons are

TABLE 33.1 Climate Justice and Civic Engagement Across the Curriculum Project: Components and Workshop Activities

Components	Description of Workshop Activities
	Activities can occur in a 1-day (2-hour) workshop, or over a 3- to 4-day (1.5- to 2-hour/day) institute or learning community
1. Introduction to Climate Justice, and Community and Civic Engagement (C&CE)	Getting participants into the "mindset" to create lessons that integrate climate justice and C&CE requires baseline knowledge of climate justice and C&CE. Therefore, the first step is for faculty to learn about and discuss climate justice and C&CE. Through readings and group discussions they learn how climate justice and C&CE are different from, but related to, climate science and service learning, respectively. Participants also learn and discuss tips for facilitating difficult conversations that can come up in the classroom.
2. Importance of a Solutions Focus and Climate Emotions	Participants also learn the importance of holding emotions and acknowledging climate grief, anxiety, and fostering hope through a solutions focus. Through readings and discussion participants explore ways to move away from avoidance and a doom-and-gloom mindset and toward actively talking about climate change and fostering hope by seeking out positive stories about solutions to climate change that are working and also by taking action, which inspires hope and in turn inspires more action in a positive feedback.
3. Experience a Climate Justice Lesson and Collaborative Reflection	The facilitator walks through or shares an example lesson that includes a climate justice issue and C&CE, and illustrates a variety of pedagogical elements (lectures, activities, videos, discussion). Participants continue to brainstorm and work on their lesson drafts in small interdisciplinary groups as they reflect on the pedagogical elements they may adopt, determine resources and information still required, and explore how to situate the lesson in their course based on students' prior knowledge.
4. Interdisciplinary Faculty Group Brainstorm to Create a Draft Module	The next step is for faculty to work together in an interdisciplinary group brainstorm to come up with an outline for a module (lesson, assignment, lab, and/or activity) for their chosen class. Faculty first identify basic disciplinary concepts and skills specific to their discipline and chosen class. Next, they connect these concepts and skills to climate justice and C&CE. Faculty create and share their draft lessons using a visual that triangulates their disciplinary concepts and skills with the climate justice issue(s) and the C&CE they will weave into their courses. Throughout the brainstorm exercise, faculty give and receive feedback as they share out their lesson ideas and drafts.
5. Faculty Present Draft Module with Feedback and Reflection	Each faculty presents a short overview of their lesson to the interdisciplinary faculty group, including an overview of all lesson materials (i.e., learning outcomes, PowerPoint presentations, videos, grading rubrics, activity sheets, assignments). After each presentation, the group offers feedback. Faculty are also asked to offer their reflections and insights on how different disciplines connect climate justice and C&CE to their disciplinary content.
6. Faculty Submit their Module to a Curriculum Repository	Faculty implement their climate justice lessons. One other faculty person from the interdisciplinary group observes their lesson. After lesson implementation, faculty and students complete surveys. Faculty make final lesson revisions and create a lesson plan. Finally, lessons are uploaded to a public Climate Justice Curriculum Repository.

TABLE 33.2 Types of Civic Engagement for Climate Justice

Policy	Outreach
• Vote • Encourage or register others to vote • Contact government officials • Canvassing • Volunteer • Raise funds for a campaign • Attend lobby days at state or federal capital • Testify or public comment (written or in-person) • Run for political office • Join or advise government council or task force	• Talk with friends and family • Create PSA video, article, or comments for social media or publication (Medium, TikTok, YouTube, Mastodon, LinkedIn, Instagram, etc.) • Write an op-ed or letter to the editor • Community problem solving • Organize or educate your community • Contact the media • Wear information that sparks conversation: button, jewelry, clothes
Volunteer	Protest
• Join or advise government council or task force • Join or advise government council or task force • Participate in an event/run/walk to raise funds for a community organization • Collect, create, and/or donate food or goods • Advise or join the board of a climate organization	• Organize or participate in a protest • Sign or organize a petition • Take part or organize a boycott • Participate in nonviolent direct action • Create art, fashion, music, or film for protest • Create/display buttons, logos, signs, stickers

published on the Science Education Resource Center (SERC) site (https://serc.carleton.edu/bioregion/activities.html?q1=sercvocabs__191%3A25), an internationally recognized source for higher education curriculum materials for all disciplines.

WHAT IS CIVIC ENGAGEMENT AND CLIMATE JUSTICE?

Civic and community engagement (C&CE) involves collaboration with others to improve your community and is done through political or nonpolitical means and by anyone no matter age or citizenship status. It can involve a wide range of actions, such as posting on social media, talking with a friend about climate, volunteering with a community organization, testifying before city council, or organizing a protest (Table 33.2). In our classrooms, C&CE can also mean instilling in our students the knowledge, skills, responsibility, efficacy, commitment, and values important for the engaged citizen.[3]

There is no one definition for climate justice. A working definition is taking action for a just transition to a sustainable future by recognizing the disproportionate effects of climate

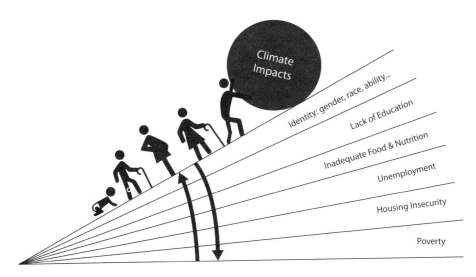

FIGURE 33.1 Intra-Generational Climate Justice Wedges. Wedges represent factors affecting an individual's or community's capacity to mitigate or adapt to climate impacts, represented by the boulder. Adapted from Making Partners: Intersectoral Action for Health 1988 proceedings and outcomes of a WHO Joint Working Group on Intersectoral Action for Health, The Netherlands.

disruption on marginalized groups and future generations. It is about asking questions such as: Who has power and access to resources, and who doesn't? As a result, who feels the disproportionate effects of things like climate change, and who will weather climate changes better? It also means taking action to address these inequities and disrupt systems of oppression through C&CE.

Climate justice includes inter-generational and intra-generational aspects.[4] Inter-generational climate justice is the impact of today's generations on future ones. Intra-generational climate justice refers to impacts within a generation. Everyone living today plays a role in intra-generational climate justice. Factors such as education, gender, race, ability, employment, food and nutrition, healthcare, and housing all impact a person or a community's ability to contend with climate impacts that are happening right now (Figure 33.1). Those most affected by intra-generational climate injustice are people in the Global South, communities of color, and those of lower socioeconomic status.

An example of an intra-generational climate justice issue is particulate matter (PM 2.5) pollution that comes from climate-exacerbated wildfire smoke and from the burning and extraction of fossil fuels like oil, coal, and gas. Burning fossil fuels is responsible for an estimated 1 in 5 deaths worldwide, and over 950 deaths each day in the US.[5] A person or community's ability to contend with PM 2.5 pollution depends on many factors. This is also where the benefits of C&CE are apparent. In Seattle, a BIPOC-led community organization, Got Green, set up a trading space for people in the community to donate or pick up air filters and box fans during the 2020 smoke season. The organization used social media to advertise to the community. In this case, C&CE, in the form of mutual aid, helped individuals and communities address climate-driven inequities.

DESCRIPTION OF CLIMATE JUSTICE FACULTY CURRICULUM

The Climate Justice with Civic Engagement Across the Curriculum project for faculty development was adapted from a local sustainability curriculum infusion model developed by the Curriculum for the Bioregion Initiative under the leadership of Jean McGregor.[6] The program originated at Bellevue College and was adopted by North Seattle College and the University of Washington. It arose from grassroots efforts of faculty and staff presenting proposals to administrators for funding, and included support from the offices of Instruction, Sustainability, Equity, Diversity, and Inclusion, and Student Leadership. At North Seattle College, a teaching assistant was also employed to provide a STEM student perspective and voice. Most recently, this work is being disseminated to the entire Washington State community and technical college system of 34 institutions. The statewide effort is being funded by the Washington State legislature; STEM focus work is funded by a National Science Foundation Improving Undergraduate STEM Education grant.

The curriculum offers in-person and online learning communities, summer institutes, workshops, and other modalities that provide interdisciplinary faculty cohorts with the financial support and time to co-create and weave climate justice and civic engagement into their courses. We developed what we believe have been the key elements making this a successful faculty learning community:

- Financial support, paying faculty for their time in lesson development and implementation, is important especially for community college faculty, who are tethered to very high teaching loads with little time "to ponder, rethink, relearn, and redesign courses."[7]
- It's also important to protect meeting times for collaborative work and discussion so that most lesson creation happens in-community. Three to four online workshop meeting days, 1.5 to 2 hours each, take place with one to two weeks between each meeting day. This spacing allows time for faculty to reflect on their readings and discussions, and to design their curriculum.
- Faculty institutes and workshops should be kept relatively small; organizers might also allow for online or in-person breakout spaces where faculty feel more comfortable discussing the challenges of incorporating issues of justice, equity, and community action into their courses. This is especially important for STEM faculty for whom incorporating such issues is often a new skill; pairing them with humanities and arts faculty (who may be more familiar with this way of teaching) is also helpful. Faculty benefit from working with small groups of 8–16 colleagues, learning from each other, and building bridges across disciplines.

- An online learning system, in our case Canvas, is used to organize faculty into quarterly learning community groups, where they can find and share resources on climate justice case studies and civic and community engagement.
- A public-facing website is also available to point passersby to the curriculum repositories and upcoming opportunities to participate.

SOME ADDITIONAL RESOURCES

More details and information about the Climate Justice and Civic Engagement Across the Curriculum projects can be found at https://northseattle.edu/climate-justice.

Teaching-and-learning activity ideas can be found in the Curriculum for the Bioregion's Curriculum Collection, which is housed at the Science Education Resource Center (SERC) at Carleton College (https://serc.carleton.edu/bioregion/activities.html?q1=sercvocabs__191%3A25). Visit the Curriculum for the Bioregion website (http://bioregion.evergreen.edu) and click on Curriculum Collection.

The Curriculum for the Bioregion's Sustainability Courses Faculty Learning Community spent several meetings in 2012 brainstorming big ideas in sustainability that would be relevant to college and university learners. Robert Turner, a faculty member at University of Washington Bothell, created this website to collect our work; there are some excellent resources here: http://faculty.washington.edu/rturner1/Sustainability/Big_Ideas01.htm.

NOTES

1. Atkinson, *Facing It*; Clayton et al., "Mental Health and Our Changing Climate."
2. Doucette et al., "Teaching STEM through Climate Justice."
3. Wang and Jackson, "Forms and Dimensions of Civic Involvement."
4. Kanbur, "Education for Climate Justice"; Stapleton, "Case for Climate Justice."
5. Vohra et al., "Global Mortality."
6. Washington Center, "Curriculum for the Bioregion Initiative."
7. Hiser, "2020 Edition Introduction: Community Colleges Responding," 1.

REFERENCES

Atkinson, Jennifer. *Facing It: A Podcast about Love, Loss, and the Natural World*. 2021. https://www.drjenniferatkinson.com/facing-it.

Clayton, Susan, Christie Manning, Kirra Krygsman, and Meighen Speiser. "Mental Health and Our Changing Climate: Impacts, Implications, and Guidance." American Psychological Association, Climate for Health, and ecoAmerica, March 2017. https://www.apa.org/news/press/releases/2017/03/mental-health-climate.pdf.

Doucette, Sonya Remington, Heather U. Price, Deb L. Morrison, and Irene Shaver. "Teaching STEM through Climate Justice and Civic Engagement." *Science Education for Civic Engagement* 15, no. 1 (2023). http://seceij.net/wp-content/uploads/2023/04/Doucette2.pdf.

Hiser, Krista. "2020 Edition Introduction: Community Colleges Responding to the Climate Crisis." Global Council for Science and the Environment, 2020. https://www.gcseglobal.org/sites/default/files/inline-files/2020%20Edition%20Introduction-Community%20Colleges%20Responding%20to%20the%20Climate%20Crisis.pdf.

Kanbur, Ravi. "Education for Climate Justice." Working Papers 250015, Cornell University, Department of Applied Economics and Management, 2015. https://ideas.repec.org/p/ags/cudawp/250015.html.

Stapleton, Sarah Riggs. "A Case for Climate Justice Education: American Youth Connecting to Intragenerational Climate Injustice in Bangladesh." *Environmental Education Research* 5, no. 25 (2018): 732–50. https://doi.org/10.1080/13504622.2018.1472220.

Vohra, Karn, Alina Vodonos, Joel Schwartz, Eloise A. Marais, Melissa P. Sulprizio, and Loretta J. Mickley. "Global Mortality from Outdoor Fine Particle Pollution Generated by Fossil Fuel Combustion: Results from GEOS-Chem." *Environmental Research* 195 (2021). https://doi.org/10.1016/j.envres.2021.110754.

Wang, Yan, and Golden Jackson. "Forms and Dimensions of Civic Involvement." *Michigan Journal of Community Service Learning* 11, no. 39 (2005): 39–48. https://eric.ed.gov/?id=EJ848471.

Washington Center for Improving the Quality of Undergraduate Education. "Curriculum for the Bioregion Initiative: Building Concepts of Sustainability into Undergraduate Curriculum." Curriculum for the Bioregion Initiative, May 27, 2020.

THIRTY-FOUR

Come for Climate, Stay for Community

Acting, Emoting, and Staying Together through the Climate Crisis

ALISSA FRAME, CHARLOTTE GRAF, LYDIA O'CONNOR,
JILLIAN SCANNELL, AMY SEIDL, and EMMA WARDELL

> This chapter illustrates how and why students formed a community to deliberately cultivate the emotional resources needed to live with climate change.

The summer we wrote our testimony for this volume, the Sixth Assessment Report from the Intergovernmental Panel on Climate Change (IPCC) was published along with an ActNow press release from the United Nations. The ActNow message, picked up by hundreds of news outlets, was this: there's still time for individual action, but the urgency is undeniable.

Campaigns and mobile apps linked to the ActNow press release asked people from affluent countries to engage in sustainable lifestyles in order to be less harmful to the planet. It asked that individuals reduce consumption of fossil fuels through changes in diet and travel, and "repair and reuse clothes and electronics."

This press release, and the intimation that climate change be addressed at the individual level, is, in part, why in 2018 students at the University of Vermont formed CCALL (Climate Communication, Advocacy and Literacy Lab). While the Sixth Assessment Report from the IPCC details the planetary nature of the climate crisis and the fact that reliance on fossil fuels will exacerbate its impacts, it, somewhat ironically, urges action not by international governments and industry, but among individuals. In sum: eat plants and fly less.

There are two dominant difficulties with this: (1) ActNow lacks complex systems thinking in its message, leaving the individual to conclude that she is responsible for effecting change; and (2) ActNow does not anticipate the affective skills individuals need to act on the subject of climate change. Not only are the actions non-normative in a culture that depends systemically on fossil fuels for nearly every purchase and act of consumption, but the effect of individual action is essentially negligible when compared to the scale of corporate or government impacts.

CCALL members, most of whom are Environmental Studies majors or minors, are well aware of the urgency, and emergency, of the climate crisis. We have studied the IPCC's previous assessment reports and know about the social and ecological outcomes that will accompany a warming planet, chaotic weather, and the health effects of both. Furthermore, we understand the physical limitations of Earth's atmosphere and what will unfold as it becomes more carbonated. Indeed, CCALL members have understood these dynamics since learning about the physical science of climate change in high school. Therefore, the findings of the Sixth Assessment Report are not new to us, nor is the admonition to "act now"—a phrase as familiar as "scroll down."

Clearly, the impetus behind the UN's framing is to galvanize people to build an international caucus for action from the bottom up. This plea is not unlike others that have been made over the entirety of our lives. Bill McKibben made the plea to an older generation in his book *The End of Nature*. Al Gore popularized the message with his film *An Inconvenient Truth*. Yet it was Greta Thunberg, a person who leveraged her youth, vision, and persistence to publicly and unabashedly demand that government leaders, not individuals, change the rules by which the energy industry, and other major carbon polluters, functioned.

We know we need to mitigate and adapt to the climate crisis. Yet we also know that action requires motivation and a feeling of agency. Furthermore, acting relies on knowing that the effort one expends will make a positive change—in social arenas like climate justice, and in ecological arenas like hardwood forests. Moreover, acting takes place with the full knowledge that if we don't act, humanity will suffer, and marginalized, frontline communities will suffer the most. In other words, acting on the climate crisis is not scrolling and liking. It is not a to-do item on a list nor is it accomplished through consumption. Action requires that individuals grasp the complexity of the challenging reality that industrialization relies on cheap, plentiful, carbon-emitting fossil fuels and has caused an unprecedented change to the Earth's climate. Human and non-human life is threatened by continuing with business as usual. As important, action requires that individuals' affective natures be developed, and tended to, while acting during a state of crisis.

The climate crisis is an existential one. Yet climate dread is deepened when people believe the crisis can be addressed mainly through individual action because it feels so "drop-in-the-bucket." This is especially true for the climate generation—college-aged people who, having grown up with this "long emergency," as Rob Nixon calls it, see it shaping their future.[1] For many it feels like self-preservation to develop an affective self, able to withstand the crisis and maintain motivation to act. Like other moments in history where the systems that predominate must radically change, the climate crisis requires individuals to cultivate emotional resources—empathetic listening, safe spaces, and trusted communities—to simultaneously live with and proactively mitigate the climate crisis. It is in this spirit that CCALL originated.

ORIGINATION OF CCALL

CCALL began as an outgrowth of a large introductory course at University of Vermont, Introduction to Environmental Studies. This course is taken by all environmental studies majors and minors and aims to introduce first-year students (about 200 each year) to the grand challenges in the field from an interdisciplinary and systems-thinking perspective using lecture (for content and context) and field instruction (for experiential and discussion opportunities). Amy Seidl has been the course's instructor for a decade. While Seidl is the lecturer of the course, she teaches it with 15 undergraduate teaching assistants who facilitate three-hour long field/lab exercises each week, forming a team approach.

CCALL was envisioned as a way to provide a place for students to continue discussions that emerged in the field and as a community in which students applied what they learned to campus-scale changes. In this way, individual efficacy as well as collaborative skills could be developed. As one of the founders put it: "It was our hope [that CCALL would] spearhead various initiatives with a goal of communicating mitigation and adaptation strategies to climate change. We saw a need for people to come together for campus climate projects, discuss science and policy, and advocate for climate and energy issues on campus and beyond" (Scannell).

The planned U.S. withdrawal from the Paris Accord in 2016 was a catalyst for the incipient group. Here was evidence that one of the world's largest emitters refused to employ its economic and political power and mediate the unjust impacts that frontline communities would bear. The terror that accompanied the withdrawal from the accord was palpable, for students and instructor alike. Founding members asked: How could the U.S. government not support international action on climate? What did withdrawal signal to other nations? Could cities and states maintain the nation's climate commitment on their own? CCALL became a place to talk through the implications of this historic moment.

The political moment catalyzed CCALL and elevated a campus need for students: a communal space where members responded to "the call" of the climate crisis. The group aimed to listen to one another, raise climate justice and seek out unrepresented voices, develop a nonhierarchical approach among members, and form a supportive community that valued emotional expression along with pragmatic action, mutually and symbiotically.

COME FOR CLIMATE, STAY FOR COMMUNITY

Six members of CCALL co-wrote this essay. Our method was to solicit willing members to reflect on the following prompts: What affective development have you experienced in CCALL? What examples can you provide? How does your experience in CCALL encourage you to address the affective and emotional dimension of climate justice and the climate crisis elsewhere? How has being a member of CCALL shaped you?

Once the testimonies were written, Seidl collected the responses and identified two themes among members' affective relationship to CCALL. These are: (1) CCALL provides a safe space to express emotions as well as to innovate collaborative responses to the crisis at the campus level; and (2) CCALL has become a support network and a trusted community wherein members feel a sense of belonging and shared efficacy.

Safe Space for Expression and Collaboration

"Our intimate club meets once per week to discuss our feelings on the latest climate or political news, to share ideas for future projects, to collaborate on proposals and applications for research funding, to read literature on climate psychology, but mostly to take the time to listen to one another in a safe space" (Graf).

"Oftentimes, emotions are the last factor to be given attention, despite the power that they can hold. I feel that CCALL is a resource of expression" (Wardell).

These quotes represent what members report after coming to CCALL—it is a safe space. Security is accomplished, in part, by the conditions members set for our meetings. For instance, the group meets weekly in the early evening (5–6 pm), a time that rarely conflicts with classes. We drink tea and bring snacks that everyone can eat. We gather around a single table and make room for all who show up. We are nonhierarchical and start with a check-in, often inquiring about members' self-care. The first 15 minutes are social and relaxed. Sometimes we do a contemplative practice focused on the breath. Only after social ease is established, do we co-create an agenda and encourage everyone to speak.

In this way, CCALL sets itself up to be a safe, social, and interactive space, one where members talk "through their feelings" (Frame), "be unburdened" (Frame), and "find comfort talking to others" (O'Connor). In effect, the space that CCALL creates allows for emotional growth. While nearly all of the CCALL members have taken the introductory environmental studies course (ENVS 001), and have been introduced to, if not become expert in, the social, scientific, historical, and ethical dimensions of the climate crisis and its justice implications, having a space to respond affectively to what is usually offered as information, is "cathartic" (Frame). Furthermore, bringing in voices that are not found around the CCALL table provides the opportunity to listen to experiences members have not had and to encourage us to see the intersection of the climate crisis with a myriad of contemporary political and social issues.

"There seems to be a collective appreciation for developing and progressing by listening to diverse perspectives. We have so much to learn from others in this movement and I can see the ways in which my own perspective has expanded and become more open" (Frame).

"In designing courses, clubs, or programs based on a group like CCALL, it is wise to incorporate breakout groups or safe spaces dependent on racial identity as race informs so much of how we cope with and perceive the dangers of the climate crisis" (Graf).

For instance, in the spring of 2021, CCALL invited Wanjiku "Wawa" Gatheru to speak in her role as a climate justice advocate and recent college graduate. Gatheru emphasized, as a young person of color in the climate movement, that the climate crisis is an "emotional conversation that demands emotional transparency." Safe spaces like CCALL were imperative to students seeking an entryway to action. "Climate change is disorienting," Gatheru said, stressing that CCALL members find climate stories and tell them using personal narratives. "We need a movement that tells the whole (emotional) truth," Gatheru said.

Several of the testimonies cited Gatheru's visit as seminal to their affective development. "Despite being an emotional and empathetic person, listening to Wawa was the first time I noticed the plethora of emotion that exists within the climate crisis and greater environmental field" (Wardell).

"Through listening to others, I am reminded of the reality that some groups of people have faced major crises in their history. For many, the climate crisis is not the first serious threat to their lives" (Frame).

Support Network through Trusted Community

CCALL intended to be a support network for its members, a plank articulated in its mission statement: an interdisciplinary "collaboratory" of thinkers and learners seeking to create and influence effective change by workshopping ideas, connecting with others, and conducting research in an attempt to navigate the grand environmental challenge of climate change. We hope to foster and encourage conversation about climate change in order to find purpose and strength in the midst of crisis. This intent is born out in the following testimonies:

"Becoming a part of CCALL felt like joining a family of people who are passionate, supportive, dedicated, and purposeful about the climate crisis, and who are determined to address it. This realization gives me the energy to continue with what I'd like, and feel I need, to pursue" (Frame).

"As much as our current climate crisis requires scientific attention, emotional attention is just as necessary. Personally, I've faced a great deal of anxiety in regards to the climate crisis and the acceleration of climate change. However, these emotions of disconnection are sparsely discussed, both in my own life and in our society as a whole, thus paving the way for feelings of solitude in the face of eco-anxiety. . . . The opportunity to feel and act in the company of a supportive community breeds motivation and persistence against even the toughest of obstacles" (Wardell).

Over the years, CCALL's emphases on action and connection, research and conversation, efficacy and expression, have emerged. Its actions include advocating for fossil fuel divestment at the University of Vermont and doubling the institution's investment in Green Bonds, both of which were manifested while one of our members was president of the Student Government Association (Scannell). Others advocated through protest, social media, and letters to university board members. In addition, CCALL's research has included a comparative analysis of carbon offset programs at other universities and a proposal that the university institute voluntary offsets, an idea that has been embraced by the Travel Study Office. CCALL is currently researching the viability of local, peer-verified offsets to address the concern that offsets be traceable and accomplish multiple social and ecological goals.

CCALL also advocated that the university expand its Clean Energy Fund and re-name it the Sustainable Campus Fund. Now the annual $250,000 allocation (that comes from a $10 fee per student per semester) supports projects that accelerate the campus towards a sustainable future, with explicit inclusion of environmental and social justice dimensions.

While CCALL members have been successful in securing funds for its actions and research, member efficacy has also developed through our podcast series, "Ripple Effect," where students interview recent graduates, faculty, and staff who identify as agents of change vis-à-vis the climate crisis. Collecting these narratives differs from action-based work as it allows reflection on how other people feel about the climate crisis, where their hope lies, and how they have become agents in the systemic change being called for.

"Navigating college, and all of the stresses it can bring, while simultaneously bearing the constant mental weight of the climate crisis, is exhausting. CCALL provided . . . an outlet for release, for discussion of emotion without pressure to create a tangible result or 'action.' Creating CCALL, and discussing our affective responses to climate change was an expression of our collective environmental grief" (Graf).

CONCLUSION

The United Nations press release notes that "harnessing the collective power of individual choices, the ActNow campaign aims to move leaders in government and business to follow suit and take bold steps to radically cut greenhouse emissions and reach net zero by 2050."[2]

The reliance on a bottom-up social movement, based in individual climate action, to catalyze top-down action from governments and industry, is one that many in the climate community have long waited for, and been disappointed with. Not because the grassroots movement hasn't made good—it hosted the largest citizen demonstration in history in 2019—but because systemic change has yet to be embraced by the top-down entities.

Therefore, remaining involved in the climate movement demands emotional resilience to the crisis itself as well as to the inertia of inactivity in the dominant political and economic systems that largely create the crisis individuals live in. In effect, CCALL members expect to be overwhelmed by assessment reports, like the one released in 2021, and to be further frustrated by the lack of consonance between climate science and action at the largest, and most contributing, scales.

CCALL members belong to a safe space in which members trust one another with their emotional and affective response to intensely burdensome information, yet remarkably find efficacy at the same time. Spaces like this are critical to living in the Age of Warming, especially for people of the climate generation, because the overwhelm of the climate crisis is not going away nor is its existential dimension. Facing the climate crisis in communities like CCALL provides the opportunity to bond with others and become agents of change who collectively emote, act and, ultimately, stay together.

NOTES

1. Rob Nixon, *Slow Violence and the Environmentalism of the Poor* (Cambridge, MA: Harvard University Press, 2013).

2. "The UN Campaign for Individual Action," United Nations ActNow, accessed May 23, 2023, https://www.un.org/en/actnow.

THIRTY-FIVE

The Climate Imaginary

Reading Fiction to Make Sense of the Climate Crisis

BENJAMIN BOWMAN, CHLOÉ GERMAINE, POOJA KISHINANI, and CHARLIE BALCHIN

This chapter explains how to produce and run climate justice reading groups, which can help students and educators learn from each other's climate experiences and support each other to imagine what a better, sustainable world could look like. The essay is co-authored by two academic researchers and two young people who worked together to produce and run their own climate justice reading group.

In this chapter, we introduce our project of climate fiction reading groups, which we shared with young people in the UK in 2021. Reading together helped participants in the groups to explore, share, and craft transformative climate imaginaries, while learning about and developing concepts of climate justice. For us, a climate imaginary is a way of imagining a future world significantly transformed by climate change. That future world will be "sculpted by anthropogenic changes to the environment, and perhaps restructured by social, political, economic and cultural reform" and we agree with others that our new world, "already characterised by interdependence, must be characterised by togetherness."[1]

The reading groups were partnerships, based on togetherness, between academics and young people. Young people were more than just readers and research participants. They also led the groups, facilitated discussions, and conducted research. Two of the young people from the groups are also authors of this chapter. This chapter is co-authored by two academics, Dr. Benjamin Bowman and Dr. Chloé Germaine, and two young co-researchers. Pooja Kishinani is a student at the University of Manchester and a member of Climate Emergency Manchester, and Charlie Balchin is a student at the University of Birmingham, both in the UK. By writing together, we share our reflections on the impact of these reading groups.

THE CLIMATE IMAGINARY: WORKING TOWARDS CLIMATE JUSTICE

A climate imaginary based on climate justice is a vision of the future that can be different from the present but that does not ignore the injustices and difficulties of the present. It may consider ways of protecting the environment, or even rewilding degraded habitats, or imagine different social, economic, and political systems. More specifically, climate imaginaries are what feminist scholars call "figurations," ways of imagining and living that are grounded in material realities and actually existing political conditions.[2] That is, they are neither naively utopian nor apocalyptic, but concerned with ways of what the philosopher Donna Haraway calls "staying with the trouble" of climate change.[3]

> A climate imaginary based on climate justice must also recognize that young people experience the "trouble" of climate change in their own way.

Among the many ways young people are hurt by climate change, one is that young people around the world report distress, anxiety, and other serious psychological effects.[4] The experience of young people intersects with other structural inequalities such as environmental racism, sexism, and global economic unfairness. Climate change is one outcome of an unfair, unequal, and unjust world. For this reason, climate justice is a systems-level response to unfairness, inequality, and injustice, and it identifies climate change as the legacy of human regimes of inequality. Therefore, to imagine climate justice means to imagine transformative change to the human world, and vitally, the liberation of most of the world after 500 years of colonization and oppression.[5]

In our work, we are challenged by the fact that young people experience older adults, like politicians, telling them "the youth are our future" when they, as young people, share a far different vision of the future. We are motivated by the experience of so many young people who are asked to participate and celebrated for having a voice, but who then find their voice is ignored when it comes to taking action on climate change.[6] We hope to amplify the climate imaginaries being developed by communities of young people and uphold the ways they can not only learn about climate change, but share in experiences of climate change, cooperate in envisioning climate justice, and join together as they imagine what a better world could look like.

This chapter explains how we organized and facilitated the reading groups. Two young people who were members of the reading group team provide testimonials to explain how the reading groups turned out. After our conclusions, we include a selected reading list of works that future reading groups might consider. We hope that our writing, our reading groups, and our reflections on climate justice can support other educators around the world, as we work together for positive and transformative change.

BACKGROUND TO THE READING GROUPS: FICTION AND CLIMATE JUSTICE

Elsewhere, in climate fiction where the imagined future is less bleak, children and young people remain central and are constructed either as objects of care or as carrying the burden of bringing about a better (and, greener) future.[7] Children and young people are, then, all too often the object of representation in climate fiction, but very rarely consulted about such fiction. At the same time, fiction and writing on the climate is being held as a tool through which young people might be engaged with and educated on the climate crisis, a mode that tends to construct young people as having a deficit of knowledge or capacity to act. Our project identified climate fiction as an imaginative resource through which young people might engage with and explore their own climate imaginaries, rather than as a tool to educate them. We also wanted to provide a space for young people to critique the kinds of representations that persist in climate fiction. We want to support them as they build on these challenges and as they construct alternative climate imaginaries.

> Fiction is the principal way in which the future of climate change is imagined. Much of these imaginings are ecophobic or apocalyptic, and not conducive to fostering ethical relationships between humans and the planet.

RATIONALE

Our reading groups build on work already trialed by critics working in literature who are keen to bring participatory research methods into their practice. Justyna Deszcz-Tryhubczak has taken the first steps to connect reading with the context of the climate strikes, suggesting that we need to "catch up with Thunberg and her peers by developing new practices of thinking and doing our research that both reflect children and adult's joint vulnerability and affirm the possibility of shared agency in the face of earth's finitude."[8] Her proposal for intergenerational collaboration is a participatory research project—ChildAct—which includes children as active contributors to all elements of the process. Deszcz-Tryhubczak introduced the children's book *Un Lun Dun* by China Miéville to focus groups of primary-school-age children (up to age 11) in the UK. The project was process-oriented, and Deszcz-Tryhubczak worked with the book and the children, generating creative outcomes on the theme of the climate crisis which were different in each group. Like Deszcz-Tryhubczak, we centered our reading groups on the process and not the outcome. We did not aim to use the text in a particular way or derive from its reading a particular skill or understanding, which is the norm in classrooms. Instead, we used the process-oriented approach to make space for young people to talk about the books while exploring, sharing, and challenging how the literature represented young people and climate change. Guided questions encouraged readers to use their book discussions as a springboard to develop alternative climate imaginaries.

SETTING UP

Practically, the reading groups began with a call to young people interested in the environment and reading. We did not address our offer solely to "activists" because many young people do not define themselves as such or might face barriers to participating in direct climate action such as the Fridays for Future climate strikes. Our open offer was shared across various stakeholders with whom we work, including schools and youth organizations. After recruitment, young people were offered the chance to choose the books they wanted to read in the groups, though we provided some options and guidance to support this process. The project included secondary school pupils, university undergraduates. and, on occasion, some postgraduates. Our final reading list is included at the end of this chapter; each book could be identified as either speculative fiction or climate fiction, with some aimed at a youth readership, but not all. These books were provided for free. We didn't pose any specific questions but merely asked them to read as much of the book as they wished; therefore, completing the book wasn't a requirement for taking part in discussions. We held four climate reading groups over the course of 2020–2021, and at time of writing our reading groups continue in several secondary schools and youth groups in the North of England.

BOOK DISCUSSIONS

In our group, discussions were supported by young research assistants, who used open, supportive questions to help participants interrogate the books and express their ideas about them. We alternated between reading portions of the text that we liked, discussing incidents that occurred in the stories, the characters, and the broad themes and ideas the texts suggested to us. Discussion often wandered far from the text itself, as participants wanted to talk about real-world racism, sexism, and other political issues related to climate justice. Sometimes participants discussed speculative ideas about what the future might actually hold in contrast to the futures imagined in the books. Following the reading groups, we held a focus group discussion in which participants and research assistants identified and reflected on key points of discussion and areas for future research with young people. These included exploring further the purpose of climate art, reflecting on the intersectional nature of climate justice (i.e., how it interacts with other forms of oppression), and the challenge of imagining a future that is different from the present.

> Testimonial: Charlie Balchin, undergraduate student at the University of Birmingham

Climate fiction seemed to hold a space for us as readers and researchers to explore emotions and ideas we might not usually explore. The fictional narratives we experienced presented the group with a "neutral ground" that was somewhat removed from reality, which enabled participants to question their preconceived ideas of justice and to engage in genuine debate. This provided a space for changing one's mind or being wrong about things, which is critical in learning yet can be difficult to find when discussing issues of justice.

This distance from reality helped participants explore alternative concepts of climate justice to the ones they had first considered, by being able to emotionally connect to a fictional character and to read from perspectives that did not mirror their own. In particular, when discussing issues of inequality and struggle, it is often difficult to have an experiential understanding of these issues. Even if there is an objective understanding of the intersections between inequality and climate change, it is very difficult to convey an embodied understanding of that experience.

Reading climate fiction seems to allow a level of exploration of different lived experiences on an unusually emotional and affective level, as readers are immersed into a character's experiences and development, often in the first-person. Reading perspectives that are different from your own cannot be thought to provide the same experiential knowledge as having lived experience of climate justice issues, but may offer a more diverse and unique understanding of these experiences and curate greater willingness to listen to others' voices.

This was something I did not expect to discover from our reading group, and yet I think it makes reading groups like ours a really powerful tool in teaching about climate justice and encouraging receptiveness to others' struggles.

Testimonial: Pooja Kishinani, undergraduate student at the University of Manchester

As a research assistant for the YoCli project, I was thrilled to co-create a space that uses fiction as a tool to explore questions about climate emotions and justice. The purpose of these reading groups was to listen and learn from each other since every participant's story is shaped by their positionality. With each discussion, I realized that our reading groups were more than just a space to envision alternate worlds; they also served as a space to sit with rage, frustration, and the helplessness of witnessing the multiple injustices in the world. Wealth, racial, and gender inequalities were common themes across every novel we discussed, emphasizing the fact that no conversation about climate justice can avoid power and privilege in our current world.

The most exciting part of the reading group discussions was mapping the intersections of different injustices across time and space to reveal how struggles for power are always connected. In most educational environments, the pursuit of "correct" answers is the prime focus of discussing books. However, we spent more time probing deeper into the complex emotions that each novel evoked in us, a key aspect of reading fiction that is often overlooked in formal education settings. Climate fiction offers a space for readers to reflect on inequalities and power in a rapidly changing world. Reading fiction without the lens of justice leaves us with an impoverished view of how to navigate and advocate for systemic change in our deeply unequal society.

Climate fiction reading groups are often (mis)perceived as a space for climate doomers and worriers to discuss the anxieties of living through an age of ecological destruction. Contrary to this, what I observed—and my most important takeaway from this project—was the process of community-building in our reading groups. As writer and botanist Robin Wall Kimmerer put it, "All flourishing is mutual."[9] Our discussions demonstrated that any movement for climate justice requires us to cultivate relationships based on reciprocity, kindness, and compassion towards each other and towards the non-human world. We can embark on this journey by listening and learning from each other's stories.

HOW TO RUN YOUR OWN CLIMATE FICTION READING GROUP

Above all, we recommend that those who want to run a climate fiction reading group design their group around the three fundamental principles of equality, support, and creativity.

First, equity: Try to address hierarchies of power in your group at the very beginning. For instance, in a research or educational setting there can be an assumption that readers should respond to the questions of the researcher or educator. On the contrary, let members know in advance that they are encouraged to ask questions of each other.

Remind readers to note things that occur in the discussion that they find interesting, and then give them space and support to ask about them. In our groups we provided examples for how group members could raise a topic and direct the conversation, such as, "I was really interested in the way the group talked about topic X" and "Earlier on I noted down topic Y, which made me think and I wanted to share." Facilitators can promote equity among group members by designing ways to bring down barriers to access; for instance, the group can co-author an agreement on how participants want to conduct sessions.

Second, support: Ask your group at the outset what they want to get out of the process. We conducted our reading group online during the COVID-19 pandemic and many participants reflected that they gained an opportunity to talk and share with others during a time of social isolation. Others received professional support from members in the form of job references and help with applications. Creating community almost always offers opportunity for mutual support, so we recommend starting your groups by allowing participants to voice what that would look like and how the community might best serve their needs while enabling them to serve others.

Finally, creativity: We hope that our account encourages readers to start their own creative discussions around art. We emphasized that our sessions were above all as a space for creating ideas, without the requirement to comment directly on the books (although participants often did!) or complete comprehension tests or quizzes. If educators must evaluate participation using tools like testing and grading, this should be dissociated from the discussion itself as much as possible so as not to dampen free creative expression.

SUGGESTED READINGS

In the groups that we wrote about above, we read three books:

Butler, Octavia E. *The Parable of the Sower*. London: Headline Books, 1993.

Dimaline, Cherie. *The Marrow Thieves*. London: Jacaranda, 2019.

James, Lauren. *The Quiet at the End of the World*. London: Walker Books, 2019.

In other groups, we read and recommend:

Kimmerer, Robin Wall. *Democracy of Species*. London: Penguin, 2021.

Okorafor, Nnedi. *Binti*. New York: Tor Books, 2015.

VanderMeer, Jeff. *Hummingbird Salamander*. London: 4th Estate Books, 2021.

We also considered:

Brahmachari, Sita. *Where the River Runs Gold*. London: Hachette, 2019.

Gayton, Sam. *The Last Zoo*. London: Anderson Press, 2019.

Itäranta, Emmi. *Memory of Water*. London: HarperVoyager, 2015.

Le Guin, Ursula K. *The Dispossessed*. London: Gollancz, 1974.

Qiufan, Chen. *Waste Tide*. London: Head of Zeus, 2013.

Singer, Nicky. *The Survival Game*. London: Hachette, 2018.

Vachharajari, Bijal. *A Cloud Called Bhura: Climate Champions to the Rescue*. New Delhi: Speaking Tiger, 2019.

NOTES

1. Bowman, "Imagining Future Worlds," 296; Hayes et al., "In it together!," 13.
2. Braidotti, *Nomadic Subjects*, 14.
3. Haraway, *Staying with the Trouble*.
4. Marks et al., *Young People's Voices*.
5. Delegates to the First National People of Color Environmental Leadership Summit, "The Principles of Environmental Justice."
6. Thew et al., "Youth Is Not a Political Position"; O'Brien et al., "Exploring Youth Activism."
7. Johns-Putra, *Climate Change and the Contemporary Novel*; Curry, *Environmental Crisis in YA Fiction*.
8. Deszcz-Tryhubczak, "Thinking with Deconstruction," 186.
9. Kimmerer, *Braiding Sweetgrass*, 20.

REFERENCES

Bowman, Benjamin. "Imagining Future Worlds alongside Young Climate Activists: A New Framework for Research." *Fennia—International Journal of Geography* 197, no. 2 (2019): 295–305. https://doi.org/10.11143/fennia.85151.

Braidotti, Rosi. *Nomadic Subjects*. New York: Columbia University Press, 2011.

Curry, Alice. *Environmental Crisis in YA Fiction: A Poetics of Earth*. London: Palgrave Macmillan, 2013.

Delegates to the First National People of Color Environmental Leadership Summit. "The Principles of Environmental Justice." Recorded October 24–27, 1991, in Washington, DC. https://www.ejnet.org/ej/principles.pdf.

Deszcz-Tryhubczak, Justyna. "Thinking with Deconstruction: Book-Adult-Child Events in Children's Literature Research." *Oxford Literary Review* 41, no. 2 (2019): 185–201. https://doi.org/10.3366/olr.2019.0278.

Estok, Simon C. *The Ecophobia Hypothesis*. London: Routledge, 2018.

Haraway, Donna J. *Staying with the Trouble: Making Kin in the Chthulhucene*. Durham, NC: Duke University Press, 2016.

Hayes, Tracy, Catherine Walker, Katie Parsons, Dena Arya, Benjamin Bowman, Chloé Germaine, Raichael Lock, Stephen Langford, Sean Peacock, and Harriet Thew. "In It Together! Cultivating Space for Intergenerational Dialogue, Empathy and Hope in a Climate of Uncertainty." *Children's Geographies* (2022). https://doi.org/10.1080/14733285.2022.2121915.

Johns-Putra, Adeline. *Climate Change and the Contemporary Novel*. Cambridge: Cambridge University Press, 2019.

Kimmerer, Robin Wall. *Braiding Sweetgrass: Indigenous Wisdom, Scientific Knowledge, and the Teachings of Plants*. Minneapolis: Milkweed, 2013.

Marks, Elizabeth, Caroline Hickman, Panu Pihkala, Susan Clayton, Eric R. Lewandowski, Elouise E. Mayall, Britt Wray, Catriona Mellor, and Lise van Susteren. *Young People's Voices on Climate Anxiety, Government Betrayal and Moral Injury: A Global Phenomenon*. SSRN [Social Science Research Network] (2021). Accessed October 1, 2022. http://dx.doi.org/10.2139/ssrn.3918955.

Miéville, China. *Un Lun Dun*. London: Del Rey Books, 2007.

O'Brien, Karen, Elin Selboe, and Bronwyn M. Hayward. "Exploring Youth Activism on Climate Change: Dutiful, Disruptive, and Dangerous Dissent." *Ecology and Society* 23, no. 3 (2018): 42. https://doi.org/10.5751/ES-10287-230342.

Thew, Harriet, Lucie Middlemiss, and Jouni Paavola. "'Youth Is Not a Political Position': Exploring Justice Claims-Making in the UN Climate Change Negotiations." *Global Environmental Change* 61 (2020). https://doi.org/10.1016/j.gloenvcha.2020.102036.

PART EIGHT

THESE SKILLS ARE NEEDED IN THE WORLD

Career Planning for the Climate Generation

THIRTY-SIX

How Will Climate Change Affect My Career?

DEBRA J. ROSENTHAL, JEFFREY JOHANSEN, and RUTH JACOB

Focusing on career discernment, this assignment asks students to reflect on ways climate change may impact their future vocation, and helps them manage resulting sober and uneasy emotions.

We created an assignment we call "How Will Climate Change Impact My Career?" in order to encourage our students to reflect on the impact of climate change on their future careers, and then to manage their resulting sober emotions as they realize that climate change will likely have a dramatic, and mostly negative, impact on their careers. We teach two courses: Dr. Jeffrey Johansen and Dr. Ruth Jacob teach an introductory-level biology course on the science of climate change, and Dr. Debra J. Rosenthal teaches an introductory literature course on climate-change fiction (cli-fi). These biology and English courses are linked together with students jointly enrolling in both. Through our linked courses, students engage in a six-hour-a-week interdisciplinary immersion in the mechanisms of global climate change in the United States and how American poets and fiction writers portray climate catastrophe. Through this assignment, students gain a deeper understanding of fossil fuel capitalism's direct impact on their future occupations. Since students presumably pursue an undergraduate degree to qualify them for the job market, we feel this assignment deliberately channels and challenges their discernment about the future, and how they might consequently direct their career preparation towards climate remediation or amelioration.

The assignment below first asks students to define their career area. If students are not yet sure about their professional aspirations, we ask that they identify some aspect of a vocation that they gravitate towards. If students express interest in an emerging field, we ask them to describe briefly what they know about what a professional in that career might do. For more career discernment, we refer them to our Career Placement Center or to the online tool O*NET (https://www.onetonline.org).

Once students define or describe their career area, we ask them to cast ahead to life after graduation. Using the science of climate change they have studied all semester, and the ways the literary imagination has helped us think of a threatened human existence, students must then predict how global heating will affect their chosen careers. This assignment comes at the end of the semester in order for students to synthesize their knowledge about climate change and realize how it will impact their future livelihood.

For example, for students who intend to pursue a management career, we ask them to think about the impact of carbon taxes, degraded environment, or productivity on supply chains. We direct wealth management majors to imagine how future clients' portfolios might be affected by investment in fossil fuel multinationals or in green energy solutions. Athletes write about how climate change will affect their sport. Education majors must decide how climate change will negatively impact their future pupils. One particularly strong paper came from a student aiming for a career in human resources—she realized part of her job will entail dealing with employees left homeless due to drought or flooding.

Over the years we have received numerous lengthy, thoughtful, deeply personal responses as students express anxiety and concern about the rapidly deteriorating world they will step into. Since the cli-fi we read in class tends to emphasize socioeconomic inequality exacerbated by climate change, students' career papers often reflect an awareness of increasing social injustices. Students often struggle with anger at their predecessors' wasteful and culpable generation, and with despair at the environmental misery that awaits them. For us as professors, these papers make us feel validated that our pedagogical goals are successful, but we also feel concern for our students as they wrestle with worry and distress about their futures and careers. Thus, the second part of the assignment unfolds in class to help students reflect on their emotional responses to ecological degradation, social injustice, and their vocational choice. We discuss students' feelings about their career-choice papers in relation to the emotional registers that fiction writers and poets elicit. We try to see how literature can imbue students' existential doubt with narrative purpose.

Drawing on recent research, we name and define the terms "eco-grief" and "eco-anxiety," and lead students in a discussion of their personal emotions elicited by their career assignment paper. Since feeling alone can be demoralizing, we aim to model for students that talking about climate change is communal and creates group solidarity. We divide students into small groups and have them share with each other one particularly salient fear or anxiety about what Michael Richardson terms the "affect of the apocalypse." We hope that even these few minutes of conversation decrease feelings of isolation and show that we can glean a lot by considering talk as action.

We then briefly discuss Lauren Berlant's idea of "cruel optimism"—that the profligate "American dream" lifestyle that students have been conditioned to want will inhibit their long-term flourishing. Students experience despair when they realize that the wasteful fossil-fueled aspirational American Dream that their parents want for them actually has been responsible for global temperature rise and planetary devastation. We hope to turn students' feelings of hopelessness into empowered feelings of obligation. Drawing upon our university's Jesuit mission of reflection and discernment, we discuss how scientists alone cannot be tasked with solutions; we need students of all majors and passions to generate and sustain space for conversation and change.

After interpersonal sharing of climate grief, we ask students to consider the emotions of grief and hope and untangle how they can be mutually constitutive. Following Elin Kelsey's argument that focusing solely on the negative ramifications of climate change can instill a sense of fear and paralysis with harmful psychological implications, we end the course with forward-looking solarpunk fiction that imagines a hopeful world that has already transitioned to solar energy, and build on Ashlee Cunsolo and Neville Ellis's work on acknowledging grief as a legitimate response to ecological loss. We also discuss Jennifer Atkinson's point that students' eco-anxiety indicates that they want to live in a just world; the anxiety suggests that students do not align with climate denialists. Tempering students' sense of betrayal and "moral injury" from inheriting such a damaged planet, this assignment helps students realize that their generation has fantastic leverage: they are the only generation with both knowledge about climate change and the agency to do something about it.

ASSIGNMENT

Instructor prep time: 30 minutes

Materials: Typed-out assignment, uploaded to your course's online learning platform

Implementation time: Students complete assignment in advance (1.5 hours), classroom discussion time, variable grading assessment time

Class size and level: Any size, introductory level

How will climate change impact your career? This assignment asks you to think about how climate change in the United States and around the world will affect your chosen profession.

Write a paragraph describing the career you plan to pursue. If you aren't sure yet about a career, select a group of closely defined vocations that you find interesting. If you choose an emerging field, please define what professionals in this career currently do, but keep it to a single paragraph. Then address the question of possible impacts on your career choice based on your understanding of anthropogenic climate change.

This paper does not necessarily have correct or incorrect answers. Aim for a thoughtful response and a realistic understanding of likely projected futures. A prediction of a far worse future than is realistically projected will hurt your grade because you must situate your paper within the current IPCC estimates of climate change. A prediction of almost no change from the present will also hurt your grade. This assignment asks you to reflect on your understanding of the future, and on how this could impact your professional life.

This assignment is meant to be reflective, in the spirit of Jesuit reflection, as well as realistic and engaging to read. Minimum length: 500 words.

REFERENCES

Atkinson, Jennifer. *Facing It: A Podcast about Love, Loss, and the Natural World.* 2021. https://www.drjenniferatkinson.com/facing-it.

Berlant, Lauren. *Cruel Optimism.* Durham, NC: Duke University Press, 2011.

Cunsolo, Ashlee, and Neville R. Ellis. "Ecological Grief as a Mental Health Response to Climate Change-Related Loss." *Nature Climate Change* 8 (April 2018): 275–81.

Kelsey, Elin. *Hope Matters: Why Changing the Way We Think Is Critical to Solving the Environmental Crisis.* Vancouver, BC, Canada: Greystone Books, 2020.

Richardson, Michael, "Climate Trauma, or the Affects of the Catastrophe to Come." *Environmental Humanities* 10, no. 1 (May 2018): 1–19.

THIRTY-SEVEN

Fostering Student Agency for Climate Justice through Vocational Exploration

RACHEL F. BRUMMEL

Students' climate anxieties are often intertwined with anxieties about who and how they want to be in the world. This chapter presents a series of vocational and career exploration assignments to help students develop a stronger sense of self and envision possible futures in the context of climate justice.

FRAMING THE ASSIGNMENTS

For the climate generation, the weight of typical developmental questions such as Who am I? and Who can I be? is made heavier by climate disruption. Our students grapple with additional existential questions— What can be done?, Can I do enough?, and Does it matter, anyway?—which may affect their sense of self and their capacity to develop visions for their future.[1] Students of the environment may have *particularly* murky future-sense about their vocations because many are drawn to environmental studies for ethical reasons, with their careers not front-of-mind. Further, environmental studies is an interdiscipline with many possible vocational paths, many of which are unknown to students. Taken together, our students face radical uncertainty that can lead to feelings of anxiety and powerlessness.[2]

Over the past several years, I have started to recognize my own responsibility to help students attend to the dual uncertainties of Who can I be? (vocational anxiety) and What can be done? (climate anxiety). I also have come to understand that students' vocational and climate anxieties are often interdependent; if students don't have a few working drafts of who they might want to be in the world, it is difficult to envision a just future at all.

> Students' vocational and climate anxieties are often interdependent; if students don't have a few working drafts of who they might want to be in the world, it is difficult to envision a just future at all.

So, if there is a "thesis" to this set of assignments, it is this: building vocational exploration into our environmental studies curricula helps students envision possible futures for themselves, which ultimately fosters their agency to support climate justice. Indeed, psychologists argue that identifying pathways to address a problem and developing agency—or a vision of how to participate in that solution—are central to developing hope[3] and counteracting climate despair.[4]

In making vocational exploration part of my work as a professor, I've become more certain about its benefits. While educators may worry that giving class time to affective and personal exploration could lead to cutting "material," I've found it gives salience to course content. Further, I've come to believe vocational exploration needs to occur *within* our curricula and not be seen only as the work of career centers. Integrating discussion of vocation can have particular benefits for first-in-family college students, Black, Indigenous, and People of Color (BIPOC) students, and others that have been historically underserved by higher education institutions. Ultimately, I've found that all my students have more agency to pursue opportunities—from activism to internships—which enhances their sense of vocation and furthers climate justice.

ACTIVITIES

Below, I present four activities to support vocational exploration within climate justice education. Across this chapter, I intentionally use the word "vocation" rather than "career" to capture the wholeness of students' current and future lives, including but not limited to career, service, activism, and volunteerism. I organize activities into two areas: (1) developing a *sense of self* in the context of climate justice work, and (2) exploring *possible futures* to promote climate justice.

Activities for Developing Sense of Self

These first two activities ask students to reflect on their values, strengths, and skills to develop a stronger and more specific sense of themselves as agents of change. I use these activities in workshop settings, where students work before class and then reflect in small groups. However, both activities could be modified as individual written reflections to accommodate larger class sizes.

Activity 1—Skills Audit

I sometimes hear students say, "I feel like I know less than when I started!" Asking students to conduct skills audits encourages awareness that they are, in fact, learning things that can make them effective agents of change for climate justice.

Individual preparation and reflection: Have students look over their academic transcript—or alternatively a class syllabus—and ask them to respond to the following:

- How would you communicate to someone unfamiliar with your education the skills you've developed along the way? Start brainstorming, and as you do, link each skill to an educational experience (e.g., policy research and analysis, learned in environmental policy class).

If you find students are struggling to articulate skills, present a few categories such as communication, research, and analysis. Grounding this activity in your program learning goals is another fruitful approach.

Discussion: Put students in small groups to discuss:

- Share with your group your list of skills that you brainstormed individually. Do you notice any commonalities? Any differences? Any surprises? Anything missing?
- What are some of the most important skills that emerge? Were you able to add new skills to your list?

Additional reflection: To encourage deeper reflection and help students carry this thinking forward, end the activity by asking:

- Thinking back, what are the courses, topics, experiences, or skills that you've enjoyed or been energized by the most? What about these experiences energizes you? Is there something that unites them? If so, what might that be?
- What might your reflections mean for you and your decisions about your future?
- What additional skills do you want to build or explore? What might you *do* to build or explore them?

Follow-up homework: Have students develop résumé entries based upon their skills audits. Providing feedback on these entries will give students confidence to actually use them.

Activity 2—Reflecting on Strengths and Values Inventories

Many institutions subscribe to assessment services that synthesize a personalized set of strengths for students. If your institution doesn't, your career services office will have tools that also work well, such as "values sorts" which ask students to consider and rank their core values. While I acknowledge these inventories can be limiting if viewed as deterministic, I have found them great jumping-off points for vocational reflection.

Individual preparation and reflection: Make sure all students have taken the same assessment of strengths and values before class and start with some basic reflection. Ask students:

- What are your first reactions to your assessment? What resonates with you? What doesn't?
- How are these strengths or values already present in your life?

Connecting to climate justice: Next, ask students:

- In what ways might your particular strengths or values be helpful in promoting positive environmental change and climate justice?
- If students are struggling to make connections, model your own strengths and explain how they've been present in your trajectory. Alternatively, discuss the strengths of climate justice leaders to make connections concrete.

Sharing out: Small group discussion at this stage allows students to see how their strengths or values are unique, but also how others are making connections. Ask students:

- Share at least two of your own insights about your assessment, as well as the connections you've made between your strengths and climate justice.
- Work with your group to help them make new connections—your strengths may help you see things they initially didn't.

Assignments for Exploring Possible Futures

The next two assignments help students envision possible futures for themselves with the goal of fostering agency and connecting with personal theories of change. While there is ample opportunity to tailor these assignments, do ask students to explore *multiple* futures; doing so alleviates the anxiety of discerning a single "right" path.

Activity 3—Developing a Possible Futures Portfolio

This assignment provides opportunity for structured vocational exploration within a class context. I often add a final "connection phase" where students conduct informational interviews with people working in the fields they are interested in exploring. You may also choose to link Activity 3 and Activity 4 and engage students in a mock interview for the job description they identify in their possible futures portfolio. I include the example assignment text for students below.

Phase 1—Exploration: The goal of your exploration is to identify three different careers that you are interested in. Think about the sectors you might want to work in (e.g., nonprofit, government, private), consider the substantive areas you might want to work on (e.g., climate justice, biodiversity, water, urban areas/cities), but think also more broadly about your "vocation" or your "fit" in the world—weigh places you might want to live, your strengths and values, and your relationship to community.

Here are a few more resources and strategies to get you started:

- Visit websites of organizations you are interested in and look for "employment opportunities" or the "staff" page.
- Explore the Occupational Outlook Handbook: https://www.bls.gov/ooh
- Browse the O*NET: https://www.onetonline.org
- Search for jobs at federal environmental agencies: https://www.usajobs.gov
- Choose a city or region where you are interested in living and search for environmental and climate justice organizations
- Have a conversation with me!

Written product for the exploration phase: Present the three careers that you want to explore further and reflect on the following in writing:

- Why is this a career you are drawn to? What excites you about this type of position?
- How does this career fit with your sense of "vocation"? How does it draw upon your strengths, skills, and values?
- How might this kind of career promote climate justice? How does it connect with your own theory of change?

Phase 2—Planning: In this next step of your possible futures portfolio, you'll work to make the exploration you did in the previous phase more concrete.

- Find an actual job description you could see yourself in five years from now. This will be the focal point of the "planning" phase of this assignment. (Optional link to Activity 4: Note that I will ultimately "interview" you for this position!)
- After settling on one possible "5 years from now" future, identify a "stepping stone" to help you explore that career. Your stepping stone should be something that you can do in the next year. This might be grad school, a service year, an internship, a job shadow, work study, research experience, or something else entirely. Choose something you would actually do.

Written product for the planning phase: Share your "five years from now" job description and your "stepping stone" with me. Additionally, reflect on the following in writing:

- Why did you choose this position? Why is this an appealing "possible future"?
- Why did you select your stepping stone? What might it provide in terms of experience, insight, or skills?
- How might this career intersect with your sense of vocation? How might this career be a part of a meaningful life that includes service, community, and leisure?
- How does this "five years from now" future fit into your own theory of change related to climate justice?

Activity 4—Mock Interview Assignment

Working with students on interview preparation will enhance the value of this assignment and reduce anxiety (students get nervous about this). I partner with my institution's career center on class workshops and to provide additional interviewing resources. After the mock interview, I immediately debrief with the student—both on how they think it went, but also on how I think it went—which has great value and builds student confidence. Perhaps most importantly to my students, I make the interview a small part of their course evaluation. The primary impact of the assignment is the experience itself and the reflections that follow. The example assignment text begins below.

Mock interview: In this final assignment, you will interview with me for the "five years from now" job description you selected in your possible futures portfolio assignment.

To prepare for your interview:

- Share your job/program description and tailored résumé with me.
- Know your job description—what are the skills, experiences, and responsibilities the employer is looking for?
- Review your notes on visits to the career center and take advantage of the opportunities to practice.
- Reflect upon your résumé—be prepared to explain your skills, experiences, values, strengths and to tell stories that demonstrate them.
- Review common interview questions and practice your responses.

What I'll be looking for:

- How well are you able to communicate the ways your values, strengths, skills, and experiences align with the position's description? Do you give concrete examples of skills, experiences, values, and strengths in context?
- Can you demonstrate curiosity about the position and explain why you want to pursue it?
- Can you think on your feet and respond to follow-up questions?

NOTES

1. Ray, *Field Guide to Climate Anxiety*, 2.
2. Clayton, "Climate Anxiety."
3. Rand and Cheavens, "Hope Theory."
4. Stevenson and Peterson, "Motivating Action."

REFERENCES

Clayton, Susan. "Climate Anxiety: Psychological Responses to Climate Change." *Journal of Anxiety Disorders* 74, no. 102263 (2020).

Rand, Kevin L., and Jennifer S. Cheavens. "Hope Theory." *Oxford Handbook of Positive Psychology* 2 (2009): 323–33.

Ray, Sarah Jaquette. *A Field Guide to Climate Anxiety: How to Keep your Cool on a Warming Planet*. Oakland: University of California Press, 2020.

Stevenson, Kathryn, and Nils Peterson. "Motivating Action through Fostering Climate Change Hope and Concern and Avoiding Despair among Adolescents." *Sustainability* 8, no. 1 (2016): 6.

APPENDIX

Chapters Sorted by Themes

BY TYPE OF SUBMISSION

Theoretical Essays

"A Pedagogy for Emotional Climate Justice," Blanche Verlie (1)
"Balancing Feelings and Action: Four Steps for Working with Climate-Related Emotions and Helping Each Student Find Their Calling," Andrew Bryant (2)
"Joyful Climate Work: The Power of Play in a Time of Worry and Fear," Casey Meehan (26)
"Overcoming the Tragic," Peter Friederici (19)
"Why Worry? The Utility of Fear for Climate Justice," Jennifer Ladino (23)
"Beyond the Accountability Paradox: Climate Guilt and the Systemic Drivers of Climate Change," Marek Oziewicz (25)
"Releasing Growth," Terry Harpold (30)
"Come for Climate, Stay for Community: Acting, Emoting, and Staying Together through the Climate Crisis," Alissa Frame, Charlotte Graf, Lydia O'Connor, Jillian Scannell, Amy Seidl, and Emma Wardell (34)

Assignments

"Using Poetry to Resist Alienation in the Climate Change Classroom," Magdalena Mączyńska (16)
"Facilitating 'R&R': Student-Led Climate Resilience and Resistance," Jessica Holmes (32)
"From Existential Crisis to Action Planning: Building Individual and Community Resilience," Jessica D. Pratt (4)
"Empathy and Care: Activities for Feeling Climate Change," Sara Karn (5)
"The Emotional Impact Statement," Christie M. Manning (6)
"From Principles to Praxis: Exploring the Roots and Ramifications of the Environmental Justice Movement," Shane D. Hall (13)
"Practicing Speculative Futures," April Anson (20)
"Cultivating Radical Imagination through Storytelling," Summer Gray (21)

"Building Somatic Awareness to Respond to Climate-Related Trauma," Emily (Em) Wright (15)
"Prompts for Feeling-Thinking-Doing: Somatic Speculation for Climate Justice," Sarah Kanouse (17)
"Infrastructure Affects: Registering Impressions of Mega-Dams," Richard Watts (12)

Hybrid of Theory and Assignment

"Leveraging Affect for Climate Justice," Michelle Garvey (11)
"Transformative Psychological Approaches to Climate Education," Leslie Davenport (3)
"Preparing Students to Navigate a Harrowing Educational Landscape: Accessibility and Inclusion for the Climate Justice Classroom," Ashley E. Reis (9)
"Photovoice for the Climate Justice Classroom: Inviting Students' Affective and Sociopolitical Engagement," Carlie D. Trott (10)
"The Social Ecology of Responsibility: Navigating the Epistemic and Affective Dimensions of the Climate Crisis," Audrey Bryan (24)
"Building Capacity for Resilience in the Face of Environmental Shocks," Abosede Omowumi Babatunde (29)
"The Tool of Imagination," Doreen Stabinsky and Katrine Oesterby (18)
"Working with Ecological Emotions: Mind Map and Spectrum Line," Panu Pihkala (14)
"Climate Justice and Civic Engagement Across the Curriculum: Empowering Action and Fostering Well-Being," Sonya Remington Doucette and Heather U. Price (33)
"The Politics of Hope," Daniel Chiu Suarez, Sophie Chalfin-Jacobs, Hannah Gokaslan, Sidra Pierson, and Annaliese Terlesky (7)
"Unfucking the World," Leif Taranta (8)
"Critical Journalism, Creative Activism, and a Pedagogy of Discomfort," Kimberly Skye Richards (22)
"Finding Hope in the Influence and Efficacy of Native/Indigenous Rights," Kate Reavey (27)
"Teaching Climate Change Resilience through Play," Jessica Creane (28)
"The Climate Imaginary: Reading Fiction to Make Sense of the Climate Crisis," Benjamin Bowman, Chloé Germaine, Pooja Kishinani, and Charlie Balchin (35)
"How Will Climate Change Affect My Career?" Debra J. Rosenthal, Jeffrey Johansen, and Ruth Jacob (36)
"Fostering Student Agency for Climate Justice through Vocational Exploration," Rachel F. Brummel (37)
"Ecotopia versus Zombie Apocalypse: Collaborative Writing Games for Existential Regeneration," Marna Hauk (31)

BY DISCIPLINE/PROFESSION/METHOD

Composition/Literature/Poetry

"Using Poetry to Resist Alienation in the Climate Change Classroom," Magdalena Mączyńska (16)
"The Climate Imaginary: Reading Fiction to Make Sense of the Climate Crisis," Benjamin Bowman, Chloé Germaine, Pooja Kishinani, and Charlie Balchin (35)
"Cultivating Radical Imagination through Storytelling," Summer Gray (21)
"Practicing Speculative Futures," April Anson (20)
"Overcoming the Tragic," Peter Friederici (19)

STEM

"From Existential Crisis to Action Planning: Building Individual and Community Resilience," Jessica D. Pratt (4)

"Climate Justice and Civic Engagement Across the Curriculum: Empowering Action and Fostering Well-Being," Sonya Remington Doucette and Heather U. Price (33)

History

"Finding Hope in the Influence and Efficacy of Native/Indigenous Rights," Kate Reavey (27)
"Empathy and Care: Activities for Feeling Climate Change," Sara Karn (5)
"Critical Journalism, Creative Activism, and a Pedagogy of Discomfort," Kimberly Skye Richards (22)
"From Principles to Praxis: Exploring the Roots and Ramifications of the Environmental Justice Movement," Shane D. Hall (13)

Environmental Justice/Decolonization

"A Pedagogy for Emotional Climate Justice," Blanche Verlie (1)
"Climate Justice and Civic Engagement Across the Curriculum: Empowering Action and Fostering Well-Being," Sonya Remington Doucette and Heather U. Price (33)
"Photovoice for the Climate Justice Classroom: Inviting Students' Affective and Sociopolitical Engagement," Carlie D. Trott (10)
"Leveraging Affect for Climate Justice," Michelle Garvey (11)
"Building Capacity for Resilience in the Face of Environmental Shocks," Abosede Omowumi Babatunde (29)
"The Tool of Imagination," Doreen Stabinsky and Katrine Oesterby (18)
"Practicing Speculative Futures," April Anson (20)
"Cultivating Radical Imagination through Storytelling," Summer Gray (21)
"From Principles to Praxis: Exploring the Roots and Ramifications of the Environmental Justice Movement," Shane D. Hall (13)

Participatory Action Research

"Leveraging Affect for Climate Justice," Michelle Garvey (11)
"Facilitating 'R&R': Student-Led Climate Resilience and Resistance," Jessica Holmes (32)
"Climate Justice and Civic Engagement Across the Curriculum: Empowering Action and Fostering Well-Being," Sonya Remington Doucette and Heather U. Price (33)
"Photovoice for the Climate Justice Classroom: Inviting Students' Affective and Sociopolitical Engagement," Carlie D. Trott (10)

Social Science

"A Pedagogy for Emotional Climate Justice," Blanche Verlie (1)
"Building Capacity for Resilience in the Face of Environmental Shocks," Abosede Omowumi Babatunde (29)
"Finding Hope in the Influence and Efficacy of Native/Indigenous Rights," Kate Reavey (27)
"The Emotional Impact Statement," Christie M. Manning (6)
"Beyond the Accountability Paradox: Climate Guilt and the Systemic Drivers of Climate Change," Marek Oziewicz (25)

Journalism

"Critical Journalism, Creative Activism, and a Pedagogy of Discomfort," Kimberly Skye Richards (22)

Visual Arts

"Photovoice for the Climate Justice Classroom: Inviting Students' Affective and Sociopolitical Engagement," Carlie D. Trott (10)

"Infrastructure Affects: Registering Impressions of Mega-Dams," Richard Watts (12)

"The Social Ecology of Responsibility: Navigating the Epistemic and Affective Dimensions of the Climate Crisis," Audrey Bryan (24)

"Prompts for Feeling-Thinking-Doing: Somatic Speculation for Climate Justice," Sarah Kanouse (17)

Clinical Perspectives

"Balancing Feelings and Action: Four Steps for Working with Climate-Related Emotions and Helping Each Student Find Their Calling," Andrew Bryant (2)

"Transformative Psychological Approaches to Climate Education," Leslie Davenport (3)

Student-Led

"Photovoice for the Climate Justice Classroom: Inviting Students' Affective and Sociopolitical Engagement," Carlie D. Trott (10)

"Facilitating 'R&R': Student-Led Climate Resilience and Resistance," Jessica Holmes (32)

"The Politics of Hope," Daniel Chiu Suarez, Sophie Chalfin-Jacobs, Hannah Gokaslan, Sidra Pierson, and Annaliese Terlesky (7)

"The Climate Imaginary: Reading Fiction to Make Sense of the Climate Crisis," Benjamin Bowman, Chloé Germaine, Pooja Kishinani, and Charlie Balchin (35)

"Come for Climate, Stay for Community: Acting, Emoting, and Staying Together through the Climate Crisis," Alissa Frame, Charlotte Graf, Lydia O'Connor, Jillian Scannell, Amy Seidl, and Emma Wardell (34)

Experiential

"Photovoice for the Climate Justice Classroom: Inviting Students' Affective and Sociopolitical Engagement," Carlie D. Trott (10)

"Releasing Growth," Terry Harpold (30)

"Climate Justice and Civic Engagement Across the Curriculum: Empowering Action and Fostering Well-Being," Sonya Remington Doucette and Heather U. Price (33)

"Working with Ecological Emotions: Mind Map and Spectrum Line," Panu Pihkala (14)

"Teaching Climate Change Resilience through Play," Jessica Creane (28)

"Prompts for Feeling-Thinking-Doing: Somatic Speculation for Climate Justice," Sarah Kanouse (17)

"Building Somatic Awareness to Respond to Climate-Related Trauma," Emily (Em) Wright (15)

"Critical Journalism, Creative Activism, and a Pedagogy of Discomfort," Kimberly Skye Richards (22)

BY TYPE OF AFFECT

Fear/Worry

"Why Worry? The Utility of Fear for Climate Justice," Jennifer Ladino (23)

Hope

"Building Capacity for Resilience in the Face of Environmental Shocks," Abosede Omowumi Babatunde (29)

"Finding Hope in the Influence and Efficacy of Native/Indigenous Rights," Kate Reavey (27)
"From Principles to Praxis: Exploring the Roots and Ramifications of the Environmental Justice Movement," Shane D. Hall (13)
"The Politics of Hope," Daniel Chiu Suarez, Sophie Chalfin-Jacobs, Hannah Gokaslan, Sidra Pierson, and Annaliese Terlesky (7)
"Unfucking the World," Leif Taranta (8)
"Facilitating 'R&R': Student-Led Climate Resilience and Resistance," Jessica Holmes (32)
"Transformative Psychological Approaches to Climate Education," Leslie Davenport (3)
"Ecotopia versus Zombie Apocalypse: Collaborative Writing Games for Existential Regeneration," Marna Hauk (31)

Empathy

"Empathy and Care: Activities for Feeling Climate Change," Sara Karn (5)
"Cultivating Radical Imagination through Storytelling," Summer Gray (21)
"Teaching Climate Change Resilience through Play," Jessica Creane (28)

Guilt/Accountability

"Beyond the Accountability Paradox: Climate Guilt and the Systemic Drivers of Climate Change," Marek Oziewicz (25)
"The Social Ecology of Responsibility: Navigating the Epistemic and Affective Dimensions of the Climate Crisis," Audrey Bryan (24)

Joy

"Joyful Climate Work: The Power of Play in a Time of Worry and Fear," Casey Meehan (26)
"Teaching Climate Change Resilience through Play," Jessica Creane (28)

Regeneration

"Ecotopia versus Zombie Apocalypse: Collaborative Writing Games for Existential Regeneration," Marna Hauk (31)

Empowerment/Efficacy

"Finding Hope in the Influence and Efficacy of Native/Indigenous Rights," Kate Reavey (27)
"How Will Climate Change Affect My Career?" Debra J. Rosenthal, Jeffrey Johansen, and Ruth Jacob (36)
"Fostering Student Agency for Climate Justice through Vocational Exploration," Rachel F. Brummel (37)
"Come for Climate, Stay for Community: Acting, Emoting, and Staying Together through the Climate Crisis," Alissa Frame, Charlotte Graf, Lydia O'Connor, Jillian Scannell, Amy Seidl, and Emma Wardell (34)
"From Existential Crisis to Action Planning: Building Individual and Community Resilience," Jessica D. Pratt (4)

Trauma

"Building Somatic Awareness to Respond to Climate-Related Trauma," Emily (Em) Wright (15)
"Transformative Psychological Approaches to Climate Education," Leslie Davenport (3)

BY TYPE OF TEXT USED IN THE INSTRUCTIONAL METHOD

Poetry

"Using Poetry to Resist Alienation in the Climate Change Classroom," Magdalena Mączyńska (16)

Climate Fiction

"Why Worry? The Utility of Fear for Climate Justice," Jennifer Ladino (23)
"Practicing Speculative Futures," April Anson (20)
"Cultivating Radical Imagination through Storytelling," Summer Gray (21)
"The Climate Imaginary: Reading Fiction to Make Sense of the Climate Crisis," Benjamin Bowman, Chloé Germaine, Pooja Kishinani, and Charlie Balchin (35)

Film

"Infrastructure Affects: Registering Impressions of Mega-Dams," Richard Watts (12)

Games

"Prompts for Feeling-Thinking-Doing: Somatic Speculation for Climate Justice," Sarah Kanouse (17)
"Teaching Climate Change Resilience through Play," Jessica Creane (28)
"Joyful Climate Work: The Power of Play in a Time of Worry and Fear," Casey Meehan (26)
"Ecotopia versus Zombie Apocalypse: Collaborative Writing Games for Existential Regeneration," Marna Hauk (31)

Newspapers/Media

"Critical Journalism, Creative Activism, and a Pedagogy of Discomfort," Kimberly Skye Richards (22)

Political Document

"From Principles to Praxis: Exploring the Roots and Ramifications of the Environmental Justice Movement," Shane D. Hall (13)

The Body

"Working with Ecological Emotions: Mind Map and Spectrum Line," Panu Pihkala (14)
"Building Somatic Awareness to Respond to Climate-Related Trauma," Emily (Em) Wright (15)
"Releasing Growth," Terry Harpold (30)

CONTRIBUTORS

ABOUT THE EDITORS

JENNIFER ATKINSON, PhD, is Associate Professor of Environmental Humanities at the University of Washington, Bothell. She is author of the book *Gardenland: Nature, Fantasy and Everyday Practice* and creator of the podcast *Facing It*, which provides tools to navigate the emotional toll of climate breakdown. She leads public seminars on climate and mental health in partnership with youth activists, psychologists, climate scientists, and policy makers, and her workshops on eco-grief have been widely featured in national media. Jennifer received a PhD in English Literature from the University of Chicago and has lived in Seattle since 2010.

SARAH JAQUETTE RAY, PhD, works at the intersection of social justice and climate emotions and serves as Department Chair of Environmental Studies at Cal Poly Humboldt. She is the author and editor of multiple books, including *The Ecological Other: Environmental Exclusion in American Culture* and *A Field Guide to Climate Anxiety: How to Keep Your Cool on a Warming Planet*. Sarah is also a certified mindfulness teacher through the UCLA Mindfulness Awareness Research Center. She received a PhD in Environmental Sciences, Studies, and Policy from the University of Oregon.

ABOUT THE CHAPTER AUTHORS

APRIL ANSON, PhD, is an Assistant Professor of English at the University of Connecticut. She works at the intersection of environmental humanities, Indigenous and American studies, and political theory.

ABOSEDE OMOWUMI BABATUNDE, PhD, works at the intersection of extractive politics, environmental conflict, and peacebuilding, focusing on Nigeria's Niger Delta. She teaches at the Centre for Peace and Strategic Studies, University of Ilorin, Nigeria.

CHARLIE BALCHIN is a policymaker and strategist in renewable energy, interested in local initiatives reconnecting citizens with nature. Their work on climate justice includes supporting communities to explore the climate crisis through art and culture.

BENJAMIN BOWMAN, PhD, studies young people's democratic participation during climate crisis, including young people's environmentalist activism. He teaches sociology, creative research methods, and youth studies at Manchester Metropolitan University, UK.

RACHEL F. BRUMMEL, PhD, is Director of Environmental Studies at Luther College in Decorah, Iowa, where she teaches courses in environmental politics, environmental justice, and community-based sustainability.

AUDREY BRYAN, PhD, is Associate Professor of Sociology at Dublin City University, Ireland. Her current work explores the complex intersection between individual-level and more structural aspects of climate-related harm as well as innovative pedagogical approaches to the climate crisis.

ANDREW BRYANT, MPH, MSW, is a social worker based in Seattle. He is the founder of Climate & Mind, a project dedicated to gathering and promoting resources about climate psychology for clinicians, the media, and the public.

JESSICA CREANE, founder IKantKoan Play/s, creates interactive games and events that guide us to remain playful and creative in the face of uncertainty. She is Professor of Game Design and a TEDx speaker.

LESLIE DAVENPORT is a climate psychology consultant. She is author of four books including *Emotional Resiliency in the Era of Climate Change* and co-leads the Climate Psychology Certification program at California Institute of Integral Studies.

SONYA REMINGTON DOUCETTE, PhD, threads climate justice, civic engagement, systems thinking, and positive stories of change into her STEM teaching. She helps faculty do the same using a faculty professional development curriculum she collaboratively created.

PETER FRIEDERICI writes, teaches, and works on community initiatives involving ecology, food, and storytelling on the southern Colorado Plateau. His most recent book is *Beyond Climate Breakdown: Envisioning New Stories of Radical Hope*.

MICHELLE GARVEY, PhD, is an environmental justice educator and organizer rooted in Minneapolis, Minnesota. Her teaching, which weaves experiential, experimental, and engaged learning methods with interdisciplinary content, is geared toward growing the environmental justice movement.

CHLOÉ GERMAINE, PhD, works in children's literature and game studies at the Manchester Metropolitan University, UK. In her youth-centered research she uses creative approaches to explore young people's imaginaries during the climate crisis.

SUMMER GRAY, PhD, works at the intersection of climate justice and critical resilience. She is author of *In the Shadow of the Seawall* and teaches Environmental Studies at the University of California, Santa Barbara.

SHANE D. HALL, PhD, is Associate Professor of Environmental Studies at Salisbury University. He teaches classes on environmental justice, race and environment, and the environmental humanities.

TERRY HARPOLD'S research and teaching are focused on the poetics and ethics of ecological crisis and climate change, with particular attention to intersectional and interspecies approaches to climate equity, justice, and resilience.

MARNA HAUK, PhD, innovates learning and scholarship in regenerative futures, arts-based methods, climate justice, and leadership and imagination as Associate Director of the Doctoral Program in Visionary Practice and Regenerative Leadership at Southwestern College.

JESSICA HOLMES, PhD, teaches university-level English in the Pacific Northwest. She writes at the intersection of literature, environmentalism, and feminism. She is also a poet and holistic health educator.

RUTH JACOB has an MS from The Ohio State University and PhD from Case Western Reserve University, both in Geological Sciences. She is an advocate for sustainability and climate change while developing equitable solutions.

JEFFREY JOHANSEN has taught climate change for science majors and nonmajors for over ten years. He is an active scholar who has held joint appointments at John Carroll University and University of South Bohemia, Czech Republic.

SARAH KANOUSE is an interdisciplinary artist and writer examining the politics of space, landscape, and ecology. She is Associate Professor of Media Arts in the Department of Art + Design at Northeastern University.

SARA KARN, PhD, is a Postdoctoral Fellow at McMaster University in Hamilton, Ontario. Her research explores historical empathy in social studies and history education, and she teaches environmental education to pre-service teachers.

POOJA KISHINANI is a campaigner and researcher interested in the intersection between climate, social, and economic justice. She is also the co-author of *The Student Guide to the Climate Crisis*.

JENNIFER LADINO, PhD, is Professor of English, co-founder/co-director of the Confluence Lab, and a longtime seasonal park ranger. Her expertise is in American climate change fiction, affect studies, public memory, and the environmental humanities.

MAGDALENA MĄCZYŃSKA, PhD, is a professor of contemporary Anglophone literature at Marymount Manhattan College in New York City. She is an expert in climate fiction and a lover of mountains, trees, and lakes.

CHRISTIE M. MANNING, PhD, teaches Environmental Studies and Psychology at Macalester College. Her research explores the factors that support collective action in response to the climate crisis.

CASEY MEEHAN, PhD, champions the power of playfulness and playful learning to help others (re) engage in the Serious Business of climate work. He is Director of Sustainability at Western Technical College.

KATRINE OESTERBY is a Danish student and climate activist. She has been involved in climate justice activism for the past ten years and has worked in Norway, the United States, and most recently in Denmark.

MAREK OZIEWICZ, PhD, is Professor of Literacy Education and Director of the Center for Climate Literacy at the University of Minnesota. He empowers teachers to build young people's climate literacy across all subject areas.

PANU PIHKALA, PhD, specializes in eco-anxiety research at University of Helsinki, Finland. He often collaborates with educators, artists, and psychologists in practical work with eco-emotions. Pihkala has written two books on the topic in Finnish.

JESSICA D. PRATT, PhD, is an ecologist and teaching professor at University of California, Irvine in conservation, climate change education, and sustainability. Her scholarship explores the role of educators in addressing the impacts of ecological grief and climate anxiety in students.

HEATHER U. PRICE, PhD, works with faculty integrating climate justice and civic engagement across curricula, including within her chemistry classes. She's also co-founder of Talk Climate, a community hub fostering climate communication and supporting climate emotions.

KATE REAVEY, PhD, teaches interdisciplinary courses at Peninsula College and has taught in the Native Pathways Program (NPP) through the Evergreen State College. She values and practices critical pedagogies, centering co-teaching, collaboration, and Indigenous methodologies.

ASHLEY E. REIS, PhD, is Clinical Associate Professor at The University of North Texas. A member of New College, she specializes in Project Based Learning, while her research focuses on environmental justice narratives.

KIMBERLY SKYE RICHARDS, PhD, engages performance to resist extractivism, inspire just transitions, and move through intellectual, imaginative, and affective impasses. She teaches at the University of British Columbia.

DEBRA ROSENTHAL, PhD, Professor of English at John Carroll University, has published numerous books and essays on American literature, particularly climate change literature.

AMY SEIDL, PhD, is the author of two nonfiction books on climate change: *Early Spring* and *Finding Higher Ground: Adaptation in the Age of Warming*. She is a Senior Lecturer at the University of Vermont. Her students, **Alissa Frame**, **Charlotte Graf**, **Lydia O'Connor**, **Jillian Scannell**, and **Emma Wardell** served as members of the Climate Communication, Advocacy and Literacy Lab (CCALL), and co-authored the essay in this volume.

DOREEN STABINSKY, PhD, is Professor of Global Environmental Politics at College of the Atlantic and a social and climate justice activist.

DANIEL CHIU SUAREZ, PhD, is Assistant Professor of Environmental Studies at Middlebury College where he studies the contemporary politics of global ecological change. **Hannah Gokaslan, Sidra Pierson, Annaliese Terlesky,** and **Sophie Chalfin-Jacobs** are each former students of his and they all participated in the course "The Politics of Hope," which Prof. Suarez designed and continues to teach at Middlebury.

LEIF TARANTA (they/them) is an organizer and direct-action trainer with the Climate Disobedience Center, where they work to support fossil fuel resistance and abolitionist community safety networks. They live in the Green Mountains.

CARLIE D. TROTT, PhD, is Associate Professor of Psychology at the University of Cincinnati. Her community-engaged, action-oriented research aims to bring visibility to and work against the inequitable impacts of climate change, socially and geographically.

BLANCHE VERLIE, PhD, is a multidisciplinary social scientist whose work focuses on how people understand, experience, and respond to climate change. She is the author of *Learning to Live with Climate Change: From Anxiety to Transformation*.

RICHARD WATTS, PhD, Associate Professor of French at University of Washington, researches and teaches environmental humanities and postcolonial studies. He is the author of *Packaging Post/Coloniality: The Manufacture of Literary Identity in the Francophone World*.

EMILY (EM) WRIGHT, MA, is a queer somatic and design practitioner who guides individuals and organizations in embodied transformation as a foundation for climate justice and collective liberation. They teach environmental studies at Seattle University.

INDEX

Abdulkarim, Zeena, 205
abolition ecology, 76
"academic-industrial complex," 147
accessibility: accommodation and, 86; inclusion and, 81–86; photovoice and, 93–94; somatic practices and, 139; undocumented students and healthcare access, 85–86
accommodation, 86
accountability paradox, 210–15
action and activism: action planning tool, 47; art activism activity, 50; community messages on environmental racism, 54; creative activism as inspiration, 189; Feel-Talk-Unite-Act process, 24–29, 25*fig.*; groups, forming or joining, 28; guilt as substitute for, 212; leverage point, direct action as, 103*table*; narratives discouraging, 162–63; photovoice and, 92, 93; research activity, 50; as salve for despair, 99; types of civic engagement, 274*table*; WTO protests, 1999, Seattle, 22. *See also* apathy; resistance
Active Hope (Macy and Johnstone), 182
activities and exercises: agricultural exploitation, 54; alien ultimatum questionnaire, 212–13; art activism, 50; body mapping, 136–37; climate circles, 33–34; drinking-water access, 52; environmental activists research, 50; environmental privilege checklist, 197; extinct and endangered species research, 53; garden or walking trail, 54; grounding exercise, 36; human-caused disasters research, 51; imagining place and imaginary person, 53; Indigenous knowledge, 50; messages on environmental racism, 54; methods and leverage points, 99–104, 102*table*, 103*table*; military sites research, 51–52; Mind Map of Climate Emotions, 125–28, 127*fig.*, 128*table*, 129–30; nervous system, befriending, 40; NIMBY and environmental racism, 52; oil industry climate denial timelines, 51; oral history interviews, 51; partnering with community groups, 54; photography and collage, 51; photovoice, 89–95, 94*fig.*; policies and feelings, 52; possible futures portfolio, 302–4; renewable energy and sustainable technologies research, 52; resource exploitation, 50; skills audit, 300–302; Spectrum Line of Climate Emotions, 128–29; Stockholm Declaration, 51; strengths and values inventory, 301–2; synergistic thinking practice, 38–39; values and beliefs, changed, 53; videos for change, 53. *See also* group discussion; lesson plans and class modules; small groups; writing assignments
affect: agency, affective, 110; arc, affective, 43; dissonance, affective, 110; engagement with empathy and care, 49; epistemic and affective dimensions of climate change, 201–2, 205–6; films responses and alignment, 112; knowing, affective ways of, 100; poetry and, 142. *See also* emotions
Affiliated Tribes of Northwest Indians (ATNI), 229

agency: affective, 110; art as invitation to human agency, 202, 205–7; games and, 235–36; vocational exploration and, 299–304
agricultural exploitation activity, 54
agroecology as leverage point, 103*table*
Akomolafe, Bayo, 157
Albrecht, Glenn, 164
alexithymia, 125
alienation, 141–42
alien ultimatum questionnaire, 212–13
all-or-nothing view, 32–33
All We Can Save project, 8
American Gods (HBO), 210, 213–14
analytic thinking, 37
animal liberation as leverage point, 103*table*
Anson, April, 170–76
anthropocene: Anthropocene Curriculum (Haus der Kulturen der Welt), 148, 151; changing meaning of, 76; "Environmental Justice in the Anthropocene" course, 74–77
anxiety: as climate stress response, 135*fig.*; eco-grief lesson plan, 42–47; emotional climate justice and, 17–18; existential, 130; expanding conceptions of, 195; in Finland, 123–30; guest speakers on, 27; "How will climate change affect my career?," 296–97; Mind Map and Process Model of Eco-Anxiety and Ecological Grief, 125–28, 127*fig.*, 128*table*, 129–30; political ecology of, 18; silenced or ostracized, 124
apathy: as climate stress response, 135*fig.*; doom and gloom model and, 1; hopelessness as, 69; psychological defenses and, 33
apocalyptic stories, 171–72
art: art activism activity, 50; art essays, 118–19; body mapping, 136–37; as invitation to human agency, 202, 205–7; as not neutral, 171; OLUP cards curriculum, 148–51; resistance beginning in, 214
assessment, 84–85, 191n11
assignment policies, inclusive, 86–87
assignments. *See* lesson plans and class modules; writing assignments
Atkinson, Jennifer, 1–14, 66, 68–69, 297
attendance policies, 85–86
attunement, 224

Babatunde, Abosede Omowumi, 240–46
backgrounding, 142

Barrington-Leigh, Christopher, 44
Barton, Keith, 49
Bass, Ellen, 141
"Battle of Seattle," 22
Bel Monte: After the Flood (documentary), 113
Bendell, Jem, 62, 67
Berlant, Lauren, 296–97
"Big Picture, The" (Bass), 142
binaries: hyper-separation and instrumentalizing reduction, 142; more-than-human world and, 148; SERF and, 203; undermining with games, 254, 257, 260
BIPOC (Black, Indigenous, and People of Color) students and communities: accountability paradox and, 212; climate injustice and, 134; food and, 102*table*; Got Green, 275; guilt and, 211, 215; police abolition and, 103*table*; "Principles of Environmental Justice" summit, 116–20; vocational exploration and, 300
Black communities: Black Joy movement, 223; Black radical thought, 76; climate injustice and, 134; climate trauma and, 134; hyper-separation and, 142
Black Lives Matter, 3, 181, 215
Bloodlands (Tailfeathers), 187
body mapping, 136–37. *See also* embodiment and somatics
Bogad, Larry, 189
Bogust, Ian, 226
Bordelon, Kayla, 193–94
Boykoff, Max, 165
brain processing, analytic and synergistic, 37
breathing, mindful, 39–40
brown, adrienne maree, 8, 167, 170, 178
Brulle, Robert J., 211
Brummel, Rachel F., 299–305
Bryan, Audrey, 9, 201–9, 244
Bryant, Andrew, 6, 22–30
Buddhism, 250–51
Burning Vision (Clements), 187
Butler, Octavia, 180–82

CAConrad, 150
calling, individual, 22, 29–30
"Calling All Grand Mothers" (Walker), 142
cancel culture, 4
Candy, Stuart, 157, 158
Canvas, 277

capabilities, emotional, 18–19
capitalism: bearing witness, 206–7; climate change art and, 203, 205; consumerism and consumption practices, 205, 206–7; decolonization and, 99; games and, 235; guilt and, 213–14; imagining beyond, 70, 172, 178; moving beyond, 76–77; naturalization of, 214. *See also* neoliberal ideology
carbon footprint concept, 215
carbon offset schemes, siting of, 101
cards, OLUP, 148–51
care, objects of, 195–96
careers and vocation: agency through vocational exploration, 299–304; "How will climate change affect my career?," 295–98
caring and empathy, 49–54
Carse, James P., 235
CCALL (Climate Communication, Advocacy and Literacy Lab) (University of Vermont), 279–84
Chalfin-Jacobs, Sophie, 68–70
Change-maker Personality Quiz, 47
"Characteristics of Life" (Dungy), 141
ChildAct, 287
Christian, Dan, 61–63
civic and community engagement (C&CE), 272–77, 274*table*
class modules. *See* lesson plans and class modules
classroom management policies, inclusive, 85–86
Clayton, Susan, 44, 195
Clements, Marie, 187
climate circles, 33–34
climate denial, 51, 163, 243
climate fiction reading groups, 285–90
climate generation, 49, 81–82, 116, 280, 284, 299
climate justice (overview), 3–7
Climate Justice and Civic Engagement Across the Curriculum, 272–77, 273*table*
Climate Justice Curriculum Repository, 272–74
Climate Therapy Alliance, 242
collaborative work and community: animal liberation and, 103*table*; CCALL and, 281–84; Climate Justice and Civic Engagement Across the Curriculum, 272–77; community engagement partnering activity, 54; Ecotopia versus Zombie Apocalypse game and, 256–62; in journalism, 191; more-than-human community, 250; relationality, collaborative, 229; R&R (resilience and resistance) activity and, 269–70; unexpected, 167; uniting, 27–28
colonized mind, 38
comedy narratives, 165
community. *See* collaborative work and community
confidentiality, in climate circles, 33
Contexture data set, 195
Conway, Erik, 215
Corbett, Julia, 166
co-regulation, 40
counselors, school, 27
COVID-19 pandemic: attendance policies and, 85–86; Butler's *Parable of the Sower* and, 182; educators and, 2–3; reading groups and, 290; Trees! initiative (Gainesville, FL) and, 247–51
Creane, Jessica, 196, 234–39
creativity: befriending the nervous system with, 40; climate fiction reading groups and, 290; creative activism as inspiration, 189; Ecotopia versus Zombie Apocalypse game and regenerative creativity, 257, 260; poetry, creative pieces with, 144
Crist, Meehan, 68
critical pedagogy, 90, 101
critical reflection, 44, 91–92
C-Roads platform, 226
cultural diversity, 84–85
cultural regeneration, 257–58
culture wars, 4
Cunsolo, Ashlee, 297
Curriculum for the Bioregion Initiative, 276, 277

DACA students, 86
DamNation (documentary; Knight and Rummel), 111
dams, 108–14
Dana, Deb, 40
Dante, 61–62
Davenport, Leslie: *Emotional Resiliency in the Era of Climate*, 267, 270; essay by, 31–41
Davis, Heather, 170
deBono, Edward, 257

decolonization: backgrounding and, 142; climate justice movement and, 99–100; of common sense, 249; fossil fuel infrastructure and, 76; Indigenous governance, 189; Land Back, 103*table*; leverage points and, 101, 103*table*; photovoice and, 89–91; synergistic thinking and, 32, 38
defenses, psychological, 32–33
deflection, 215
De Koven, Bernie, 224, 225
denialism, 51, 163, 243
dependent origination, 250–51
despair: action as salve for, 99; hope vs., 64, 66; "How will climate change affect my career?," 297; interconnection and, 256–57; pedagogy of discomfort and, 188; as roadblock, 256
Deszcz-Tryhubczak, Justyna, 287
Dickinson, Emily, 68
difficult knowledge: affect and, 49; art and, 205, 207; SERF and, 203; violence against Indigenous communities, 188
Dimaline, Cherie, 187
direct action. *See* action and activism
discomfort and difficult emotions: acknowledging, 91; as challenge, 6; Feel-Talk-Unite-Act process and, 25; newspaper assignment, fictional, 187–91; pedagogy of discomfort, 188; permission to feel, 123; photovoice and, 91; social ecology of responsibility, 201–7; trauma, dysregulation, and, 35–36; utility of fear, 193–98. *See also* anxiety
discussion, in-class. *See* group discussion; small groups
distribution, unequal, 18–19
diversity, cultural, 84–85
divestment as leverage point, 103*table*
Divine Comedy (Dante), 61–62
doom and gloom model: climate fiction and, 290; demoralization from, 222; despair and, 64; doomism, 32; doomsday scenarios, 259; imaginative hope and, 67–68; moving away from, 273*table*; problems with, 1, 6–7; resilience vs., 244; Romantic fatalism, 164–65; solutions focus vs., 273*table*; in survey, 130
Doucette, Sonya Remington, 272–78
"Do You Suffer from Eco-Despair? Seek Critical Thinking Treatment Right Away" (Ray), 44
dread, 130
drinking-water access, 52

drivers of climate change, hidden, 31, 32–36
due dates, inclusive policies on, 87
Duggan, Lisa, 69–70
Dungy, Camille T., 141
dysregulation, signs of, 35–36

ecophobia, 256, 287
Ecotopia versus Zombie Apocalypse game, 254–62
Education Ecologies Collective, 9
Eihei Dōgen, 251
Elders, 50, 51
Ellis, Neville, 297
embodiment and somatics: body mapping exercise, 136–37; Ecotopia versus Zombie Apocalypse game and, 260; Environmentalisms game and, 238; habituated responses to trauma, 149; individual stress responses, 134, 135*fig.*; learning your climate stress responses, 137–38; nervous system, trauma, and, 133–34; OLUP cards curriculum, 148–51; somatics, defined, 134; Spectrum Line of Climate Emotions and, 128–29; trauma and somatics, 134–35; Trees! initiative (Gainesville, FL), 247–51
emotional impact statements, 55–57
emotional intelligence, 32, 125
Emotional Resiliency in the Era of Climate Change (Davenport), 267, 270
emotions: climate justice, emotional, 17–21; connotation of movement and agitation, 196; cultural politics of, 198; denial as result of, 19; dynamic, 171; of educators, 2–3; Feel-Talk-Unite-Act process, 24–29, 25*fig.*; as hindrance under rationalism, 90; hope as action vs. feeling, 62; identity politics and, 4; importance of, 2; key ecological emotions and feelings, 127*fig.*; negative and positive, 25–26, 70, 222; resiliency and, 39–40; roller coaster of, 17; therapists vs. educators, 7; "ugly feelings," 112; voicing, in group discussions, 27, 123–30. *See also* affect; anxiety; despair; fear; grief; joy
empathy: as climate stress response, 135*fig.*; Environmentalisms game and, 238; historical, 49–54; hyper-empathy, 180, 181
endangered and extinct species, 53, 68–69
environmental impact statements (EIS), 55, 57
Environmentalisms game, 236–39

"Environmental Justice in the Anthropocene" course (Middlebury College), 74–77. *See also* "Principles of Environmental Justice"
Environmental Right Actions, 244
epistemic and affective dimensions of climate change, 201–2, 205–6
Erdrich, Louise, 171
Estes, Nick, 150
"everybody's fault" framework, 212
exceptionalism, existential, 69
exercises. *See* activities and exercises
"Existential Toolkit for Climate Justice Educators" workshop (Rachel Carson Center, Munich, 2020), 8

Facer, Keri, 157
faculty professional development, 272–77, 273*table*
fatalism, Romantic, 164
fear: as climate stress response, 135*fig.*; as defense mechanism, 188; Ecotopia versus Zombie Apocalypse game and, 259–60; fears and hopes list, 196–97; utility of, 193–98; worry vs., 195
Feel-Talk-Unite-Act process, 24–29, 25*fig.*
feminist-centered storytelling, 166
feminist theory, 90
figurations, 286
Finland, 123–30
First National People of Color Environmental Leadership Summit (1991), 116–20
Flight Behavior (Kingsolver), 193, 197
Florsheim, Morgan, 222
"For Those Who Would Govern" (Harjo), 142
fossil fuels: Big Oil propaganda, 215; decolonization and, 76; future careers and, 295; particulate matter (PM 2.5) pollution, 275; responsibility and, 203, 279–80
Frame, Alissa, 279–84
Franzen, Jonathan, 66
Freire, Paulo, 8, 190
Friederici, Peter, 162–69
frustration as primary emotion, 82
functional magnetic resonance imagery (fMRI), 36–37
futurists, 157, 171
futurity: climate fiction reading groups and, 285–89; Ecotopia versus Zombie Apocalypse game and, 257–61; "How will climate change affect my career?," 295–98; imaginative hope and, 67; imagining plural futures, 155–59; possible futures portfolios, 302–4; speculative futurism, 170–75. *See also* imagination and imaginaries

Gainesville, FL, 247–49, 252n2
games: about, 235–36; Ecotopia versus Zombie Apocalypse, 254–62; Environmentalisms, 236–39
gamification, 225
Gardiner, Stephen, 254, 259
Garrett, James, 49
Garvey, Michelle, 99–107
gaslighting, greenhouse, 18
Gatheru, Wanjiku "Wawa," 282–83
Geniusz, Mary Sissip, 150
George, Star, 175
Ghosh, Amitav, 166, 171
Glines Canyon Dam removal, 113*fig.*
Gokaslan, Hannah, 61–63
Good Grief Network, 166
Gore, Al, 280
Goto, Hiromi, 189
Graf, Charlotte, 279–84
grassroots praxis papers, 119
Gray, Summer, 177–83
Great Turning, 3, 179, 261
greening the self, 251
Green New Deal as leverage point, 103*table*
grief: art and productive grief, 207; eco-grief lesson plan, 42–47; guest speakers on, 27; hope and, 66; "How will climate change affect my career?," 296–97; interconnection and, 256; play and, 234–35; as roadblock, 256; stages of, 43
Griffis, Ryan, 148–50
Grossman, Zoltan, 229, 231
grounding exercise, 36
group discussion: Butler's *Parable of the Sower*, 181–82; careers and vocations, 296–97, 300–302; CCALL, 281–84; C&CE, 272*table*; Davenport's *Emotional Resiliency in the Era of Climate Change*, 270; difficult emotions and choice to participate, 34–36; eco-grief lesson plan, 42–45; emotional impact statements, 56; emotions, voicing, 27, 123–30; environmental racism, 52; films, 111–12; future as plural, 158; healthcare access, 85–86; hope

group discussion *(continued)*
resulting from, 47; "How will climate change affect my career?," 296–97; Indigenous knowledge, 50; Kimmerer's "Honorable Harvest," 231; Kingsolver's *Flight Behavior*, 197; oil industry climate denial, 51; photovoice, 89–92; poetry, 143–44; policy, 52; "Principles of Environmental Justice," 119–20; resilience-building, 242–44; Smith's "How Do Tribal Nations' Treaties Figure into Climate Change?," 231; Spectrum Line of Climate Emotions, 129; suffering, human and more-than-human, 53; "summit," 119–20; technology, 52; Wilson's "What Is an Indigenous Research Methodology?," 231–32. *See also* small groups
guest speakers, 27, 191
guilt: as climate stress response, 135*fig.*; as defense mechanism, 188; "everybody's fault" framework, 212; as key emotion, 126*fig.*; "roller coaster of emotions" and, 17; as slow violence, 215; sources of, 211; systemic drivers and accountability paradox, 210–15
Gulf Stream, Atlantic, 124
Gumbs, Alexis Pauline, 259

Hall, Shane D., 116–20
Haltinner, Kristin, 1, 194–96
Hamilton, Jo, 8, 129
Haraway, Donna, 167, 258–59, 286
Harjo, Joy, 142
Harpold, Terry, 247–53
Hauk, Marna, 254–63
Haus der Kulturen der Welt, 148, 151
"head, heart, hands" model, 117
healthcare access, 85–86
Heart and Mind, Structure, Behavior diagram, 46*fig.*
Heglar, Mary Annaise, 69, 171
Herrera, Juan Felipe, 141
hidden risks, 109
Hillman, Brenda, 113
Hiser, Krista, 195
historical trauma, 134
Holmes, Jessica, 267–71
"Honorable Feast, The" (Kimmerer), 230–31
hooks, bell, 8, 228
hope: *Active Hope* (Macy and Johnstone), 67, 182; as climate stress response, 135*fig.*; as commitment to dream, 157; fears and hopes list, 196–97; "How will climate change affect my career?," 297; Native/Indigenous rights and efficacy, 228–32; pedagogy of, 188, 190; "Politics of Hope" course (Middlebury College), 59–71; as result of class discussion, 47; Toivoa ja toimintaa (Hope and Action) project, 130
"How Do Tribal Nations' Treaties Figure into Climate Change?" (Smith), 230
Huizinga, Johan, 235
"Hydrology of California" (Hillman), 113
hyper-empathy, 180, 181
hyper-separation, 142

Idaho, 193–97
identity politics, 4, 13n8, 211
ideological denialism, 243
imagination and imaginaries: activity imagining place person, 53; assessment and, 191n11; climate change, radical imaginaries of, 167; colonization and, 170; Ecotopia versus Zombie Apocalypse game and, 256–62; as existential tool, 159; hope and, 67; photovoice and, 92; of plural futures, 155–59; storytelling, radical imaginaries through, 177–82; transformative climate imaginaries, 285–87. *See also* futurity
Imarisha, Walidah, 167, 172, 178
implicatedness: art and, 202; the implicated subject, 189–90, 207, 241–42; Social Ecology of Responsibility Framework (SERF), 203–5, 204*fig.*; systemic drivers and accountability paradox, 210–15
inclusion, 5, 83–87
Indigenous communities and knowledge: "academic-industrial complex" and, 147; climate injustice and, 82, 134; climate justice and, 18; climate justice methods and, 100; community resistance, 76–77; dams and, 110–11, 113; drinking-water access and, 52; dystopia stories, 166; felt history and, 114n5; guest speaker and group discussion, 50; hope and Native/Indigenous rights and efficacy, 228–32; Land Back, 103*table*; Line 3 pipeline (Minn.) and, 57; OLUP cards and, 150; plural futures and, 156–57; repatriation and decolonization, 189; Sámi people, 124–25; storytelling and, 102*table*; tragic narratives and, 164;

violence against, 187–88; vocation discussions and, 300
"information deficit" assumption, 1
infrastructure, "invisible," 108–9
injustice: BIPOC communities and, 82, 134; eco-anxiety silenced or ostracized, 124; emotional, 18; sources of environmental racism and injustice, 116–17
Inside the Sunrise Movement: How Climate Activists Put the Green New Deal on the Map (NBC News video), 45
instrumentalizing reduction, 142
intake questionnaires, 34–35
intergenerational and intragenerational aspects of climate justice, 275–76, 275*fig.*
Intergovernmental Panel on Climate Change (IPCC), 4, 40, 155–57
interviews, 51, 304
investment, green, 283

Jacob, Ruth, 295–98
Jemisin, N. K., 171
Jia Zhangke, 111
job interviews, mock, 304
Johansen, Jeffrey, 295–98
Johnson, Ayanna Elizabeth, 175
Johnstone, Chris: *Active Hope* (Macy and Johnstone), 67, 182; on the Great Turning, 261; "Three Stories of Our Time" (Macy and Johnstone), 179–80, 181
journaling: eco-grief lesson plan, 43–44; imagining future as plural, 158–59; for self-regulation, 40. *See also* writing assignments
journalism, fictional and critical, 187–91
joy and joyful climate work, 221–26, 234
Just Transition, 8

Kagawa, Fumiyo, 254, 256–57
Kanouse, Sarah, 147–51
Karazsia, Bryan T., 195
Karn, Sara, 49–54
Kaufman, Peter, 196
Kaur, Valarie, 1
Kaza, Stephanie, 251
Kelsey, Elin, 166, 297
Kennedy-Williams, Patrick, 47
Kermode, Frank, 165
Kimmerer, Robin Wall, 228–31, 290
Kingsolver, Barbara, 193, 196–97

Klein, Naomi, 187, 214
Knight, Ben, 111
knowledge, difficult. *See* difficult knowledge

Ladino, Jennifer, 110, 193–200
Land Back, 103*table*
Lee, Charles, 117–18, 120n7
Lee, Ingrid Fettell, 223
Le Guin, Ursula, 166, 214
Leiserowitz, A., 195–96
Lertzman, Renee, 1, 164
lesson plans and class modules: Climate Justice and Civic Engagement Across the Curriculum project, 272–77, 273*table*; eco-grief, 42–47; future imagination, 157–58; mega-dams, 108–14; Native/Indigenous rights and efficacy, 228–32; poetry, 141–45; radical imagination through storytelling, 177–82; R&R (resilience and resistance), 267–70; speculative futures, 170–75. *See also* activities and exercises; writing assignments
leverage points, 100–101, 103*table*
Levstik, Linda, 49
Lewis-Giggets, Tracey Michae'l, 223
Line 3 pipeline (Minn.), 57
Lipsitz, George, 118
logic vs. synergetic thinking, 37
Lorde, Audre, 141
love, 68
Lovero, Karissa, 47
Lynch, Matthew, 195

Machiorlatti, Jennifer, 178
Macy, Joanna: *Active Hope* (Macy and Johnstone), 67, 182; on connection, 261; on the Great Turning, 3, 179, 261; on self and other, 250–51; Spectrum Line and, 129; Spiral of the Work that Reconnects, 255–56, 255*fig.*; "Three Stories of Our Time" (Macy and Johnstone), 179–80, 181; the Work that Reconnects, 8
Mączyńska, Magdalena, 141–46
magazine page assignment, 173–75, 174*table*
Maher-Johnson, Louise, 142
Mann, Michael, 215
Manning, Christie M., 55–58
"Man on the TV Say" (Smith), 142
Marrow Thieves, The (Dimaline), 187

Marsh Billings Rockefeller National Park (Woodstock, VT), 236–39
Marvel, Kate, 225
masculinity, toxic, 198
McGonigal, Jane, 235–36
McGregor, Jean, 276
McIntosh, Peggy, 197
McKibben, Bill, 280
Mead, Michael, 22
Meadows, Donella, 100
meaning-making, collective, 92
media creation, 53
media literacy, 191n10
Meehan, Casey, 221–27
Meeker, Joseph, 164, 165
melancholia, environmental, 164
memory: affective dissonance and, 110; forgetting and amnesia, 69; misremembering of climate guilt, 212–15; somatics and, 138
Mental Health and Our Changing Climate (Clayton et al.), 44
Merchants of Doubt (Oreskes and Conway), 215
metacognitive reflection, 144, 182
methods activity, 100, 102*table*
Miéville, China, 287
military sites, 51–52
mindful breathing, 39–40
Mind & Life Institute, 9
Mind Map of Climate Emotions, 125–28, 127*fig.*, 128*table*, 129–30
mind map of positionality, 179, 180*fig.*
"Mississippi: An Anthropocene River" (Haus der Kulturen der Welt), 148
Mitchell, Sherri (Weh'na Ha'mu' Kwasset), 37–38
mock job interviews, 304
modules. *See* lesson plans and class modules
Moore, Kathleen Dean, 167
more-than-humans. *See* nonhumans/more-than-humans
Morrison, Scott, 18
Movement Generation, 8
Muñoz, José Esteban, 70
mutual aid as leverage point, 103*table*
Myers, Natasha, 249
mythos, 260

Næss, Arne, 250–51
narratives and story: apocalyptic stories, 171–72; comedy, 165; Ecotopia versus Zombie Apocalypse game and, 258–59; "everybody's fault" framework, 212; feminist-centered, 166; guilt narratives and accountability paradox, 211–15; imagining plural futures, 155–59; Indigenous dystopia stories, 166; as method, 102*table*; radical imagination through storytelling, 177–82; resilience-building and storytelling, 242–44; speculative futurism, 170–75; three pillars of storytelling, 178; tragic narratives and alternatives, 162–67
Native communities. *See* Indigenous communities
"Native Resilience" (TED Talk; Grossman), 229, 231
neoliberal ideology, 76, 212–14
nervous system, 40
neural dynamics, 37
newspaper assignment, fictional, 187–91
Ngai, Sianne, 112
Nietzsche, Friedrich, 62
Niger Delta, 240–44
NIMBY (Not In My Back Yard), 52, 101
Nixon, Rob, 68, 215, 280
nonhumans/more-than-humans: animal liberation as leverage point, 103*table*; empathy and care, historical, 50–54; Environmentalisms game and, 238; hope and, 68–69; sovereignty and ecological selfhood, 249–51; Trees! initiative (Gainesville, FL), 247–51
NOPE ("not on planet Earth") approach, 101
Norgaard, Kari, 1, 211
"Notes from a Climate Victory Garden" (Maher-Johnson), 142

Occupational Outlook Handbook, 303
O'Connor, Lydia, 279–84
Octavia's Brood (Imarisha and brown), 167
"Octavia Tried to Tell Us: Parable for Today's Pandemic" webinar, 182
"Ode to Dirt" (Olds), 142–45
Oesterby, Katrine, 155–61
oil industry: climate denial and, 51; Line 3 pipeline (Minn.), 57; Niger Delta and oil pollution, 243–44
Olds, Sharon, 142–45
On Being Project, 143
O*NET, 296, 303
optimism, "cruel," 59, 69, 296–97
oral histories, 51, 166

Oreskes, Naomi, 215
Osnes, Beth, 165
"Over the Levee, Under the Plow: An Experiential Curriculum" (OLUP), 148–51
oxytocin, 224
Oziewicz, Marek, 210–17

pain as connection, 256–57
paired activities: befriending the nervous system with creativity, 40; somatic, 136; Spectrum Line of Climate Emotions, 129
Palacios, Carolina, 188
Palahniuk, Chuck, 170–71
Parable of the Sower (Butler), 180–82
Paris Accord, 281
Parrish, Heather, 148–49
participation, 18–20
participatory action research (PAR) methods: about, 90; civic engagement, 272–77, 273*table*, 274*table*; methods and leverage points, 99–104, 102*table*, 103*table*; photovoice, 89–95; R&R (resilience and resistance), 267–70
particulate matter (PM 2.5) pollution, 275
Payne, Phillip, 150
Peltier, Autumn, 205
Penny, Laurie, 62, 69
Perry, Imani, 222
Petersen, B., 243
photography activity, 51
photovoice, 89–95, 94*fig.*
Pierson, Sidra, 60, 63–65
Pihkala, Panu, 8–9, 123–32, 164, 195
Pinker, Steven, 68
play: about, 223–25; grief and, 234–35; joyful climate work and, 221
pleasure activism, 8
Plumwood, Val, 142
poetry: alienation and, 141–42; CAConrad, 150; "Calling All Grand Mothers" (Walker), 142; "Characteristics of Life" (Dungy), 141; as creative meta-reflection, 182; *Divine Comedy* (Dante), 61–62; erasure poems activity, 143; "For Those Who Would Govern" (Harjo), 142; "Hydrology of California" (Hillman), 113; importance of, 141; "Man on the TV Say" (Smith), 142; "Notes from a Climate Victory Garden" (Maher-Johnson), 142; "Ode to Dirt" (Olds), 142–45; "The Big Picture" (Bass), 142

Poland, B., 242, 243
police abolition as leverage point, 103*table*
policies, environmental, 51–52
"Politics of Hope" course (Middlebury College), 59–71
polyphonic conversation, 238
polyvagal theory, 40
Porges, Stephen, 40
positionality: dams and, 110; defined, 179; importance of, 4; mind map of, 179, 180*fig.*; resilience-building and, 242; self-reflexivity and, 197
positive psychology, field of, 223
Practical Handbook for Climate Educators and Community, 8
Pratt, Jessica D., 42–48
Price, Heather U., 272–78
primary trauma, 134
"Principles of Environmental Justice" (First National People of Color Environmental Leadership Summit), 116–20
privilege: environmental privilege checklist, 197; existential exceptionalism and, 69; guilt and, 211; syllabus language and, 84
Process Model of Eco-Anxiety and Ecological Grief, 127*fig.*, 128*table*
Project-Based Learning, 81–88
Project Drawdown, 8
pseudo inefficacy effect, 6
psychological factors, 31, 32–36
Pyramids of Garbage (Shehab), 205, 206–7, 206*fig.*

questionnaire, reflexivity, 101, 104

racism, environmental, 52–53, 54. *See also* white supremacy
Ranganathan, Malini, 76
rationalism, 90
Ray, Sarah Jaquette: on benefits of change as focus, 194; on the climate generation, 81–82; on cultural politics of emotion, 198; "Do You Suffer from Eco-Despair? Seek Critical Thinking Treatment Right Away," 44; essay by, 1–14; on fatalism, 164; on guilt, 211; on hope, 67; on privilege, 150; on resilience and resistance, 70
reading groups, climate fiction, 285–90
Reavey, Kate, 228–33

reciprocity, 39, 228, 232, 249
recognition, 18–19
reflection: metacognitive, 144, 182; reflexivity questionnaire, 101, 104; self-reflexivity, 196–98
regenerative creativity, 257–61
Reis, Ashley, 81–88
relationality, 11, 229–32
renewable energy, 52
reparations as leverage point, 103*table*
repressing, 32
research: environmental activists research, 50; extinct and endangered species research, 53; human-caused disasters research, 51; military sites research, 51–52; Native/Indigenous rights and efficacy, 228–32; renewable energy and sustainable technologies research, 52; Wilson's "What Is an Indigenous Research Methodology?," 231–32. *See also* participatory action research (PAR) methods
resilience: Black Joy movement and, 223; CCALL and, 284; eco-grief and, 42–47; Ecotopia versus Zombie Apocalypse game and, 254, 260; emotional, 32, 39–40; environmental activists and, 50; hope and, 231; Indigenous power and, 230; joy and, 223; Moore's "presilience," 167; "Native Resilience" (TED Talk; Grossman), 229, 231; Niger Delta and resilience-building, 240–44; resilience hubs, 103*table*; resistance and, 70; somatics and, 133, 139; through play, 234–39. *See also* self-care
resistance: art as beginning of, 214; Black Joy movement and, 223; Indigenous, 76–77, 230; resilience and, 70; R&R (resilience and resistance) activity, 267–70. *See also* action and activism
resource exploitation, 50. *See also* fossil fuels
responsibility: ActNow and, 279–80; as climate stress response, 135*fig.*; Social Ecology of Responsibility Framework (SERF), 203–5, 204*fig.*; systemic drivers and accountability paradox, 210–15
restoration as leverage point, 103*table*
résumés, 301
Richards, Kimberly Skye, 187–92
Richardson, Michael, 296
Rights of Nature as leverage point, 103*table*
"Ripple Effect" podcast, 284

Rodriguez, Favianna, 118
Romantic fatalism, 164
Roosevelt, Franklin D., 110
Rosen, Michael, 224
Rosenthal, Debra J., 295–98
R&R (resilience and resistance) activity series, 267–70
Rummel, Travis, 111

safe space, 282–83
Sámi people, 124–25
Sarathchandra, Dilshani, 1, 194–96
Sawyer, R. Keith, 261
Scannell, Jillian, 279–84
scapegoating, 214
Science Education Resource Center (SERC), 274, 277
Scranton, Roy, 66, 68, 70, 75
secondary trauma, 134
Seidl, Amy, 279–84
Selby, David, 254, 256–57
self, sense of, 300–302
self-care: action and, 269; in climate circles, 33–34; distancing as, 128*fig.*; by exiting, 34–35; Florsheim on activism and, 222; hopefulness as act of, 65; more-than-self care, 250; in Process Model of Eco-Anxiety and Ecological Grief, 127*fig.*; resiliency and, 32, 242; self-support techniques, 138. *See also* resilience
selfhood, ecological, 250–51
self-regulation, emotional, 39–40
sense of self, developing, 300–302
sensorium, vegetalization of, 248–49
serotonin, 224
settler-colonial exploitation, 50, 148
Seymour, Nicole, 165
Shakespeare, William, 170
shared socioeconomic pathways (SSPs), 157–58
Shehab, Bahia, 205, 206–7
Simpson, Leanne, 147–48
skills audits, 300–302
slow violence, 215
small groups: affect of the apocalypse, 296; art essay, 119; befriending the nervous system with creativity, 40; climate fiction reading groups, 288–90; Climate Justice and Civic Engagement Across the Curriculum,

273*table*; eco-grief lesson plan, 45–47; emotional impact statements, 56; faculty institutes, 276; future as plural, 158; Olds's "Ode to Dirt," 143; R&R (resilience and resistance), 267–70; skills audits, 301; strengths and values inventory, 302; synergistic thinking, 38. *See also* group discussion
Smith, Anna, 229–31
Smith, N., 195–96
Smith, Patricia, 142
Sobel, David, 256, 257
Social Ecology of Responsibility Framework (SERF), 203–5, 204*fig*.
Solnit, Rebecca, 62, 64–65, 67, 69, 77
somatics. *See* embodiment and somatics
sovereignty, more-than-human, 249–51
Spectrum Line of Climate Emotions, 128–29
Speculate 2050 magazine page assignment, 173–75, 174*table*
speculative futurism, 170–75
Spiral of the Work that Reconnects (Macy), 255–56, 255*fig*.
splitting, 32–33
Stabinsky, Doreen, 155–61
Standard American English (SAE), 84
Stenson, Fred, 187
Stewart, Alan E., 195
Still Life (film; Jia), 111
Stockholm Declaration (1972), 51
Stoddard, Gianna, 196
stoicism, 166
story. *See* narratives and story
Story of Stuff Project, 47
strengths and values inventories, 301–2
stress: individual stress responses, 134, 135*fig*.; learning your climate stress responses activity, 137–38; nervous system and, 133; play and, 102*table*; post- and pre-traumatic stress disorder, 18, 240; resiliency and, 39. *See also* trauma, climate-related
Suarez, Daniel, 9, 59–73, 74–75
Suits, Bernard, 235
Sunrise Movement case study, 45
Supran, Geoffrey, 215
"Sustainability and Well-Being: A Happy Synergy" (Barrington-Leigh), 44
sustainable technologies, 52
Sutton-Smith, Brian, 224
syllabus language, inclusive, 84–87

synergistic/synergetic thinking, 36–39, 258–61
systemic drivers of climate change, 210–15
Sze, Julie, 116

Tactical Performance (Bogad), 189
Tailfeathers, Elle-Maija, 187
Táíwò, Olúfémi O., 8n13
Tallbear, Kim, 150
Taranta, Leif, 74–78
Teed, Corinne, 148, 150
Tellus Institute, 158
Terlesky, Annaliese, 60, 65–68
Three Gorges Dam, 111
"Three Stories of Our Time" (Macy and Johnstone), 179–80, 181
Thunberg, Greta, 194, 205, 280, 287
Tillerson, Rex, 163
time bank model, 87
Tippett, Krista, 143
Todd, Zoe, 170
toggling, 39–40
Toivoa ja toimintaa (Hope and Action) project, 130
Tools for Transformation project, 46*fig*., 47
trauma, climate-related: forced engagement with, 19–20; habituated bodily responses to, 149; nervous system and, 133; primary, secondary, and historic, 134; somatics and, 134; unequal distributions of, 19. *See also* stress
trauma-informed pedagogy: dysregulation, signs of, 35–36; importance of, 2; intake questionnaires, 34–35; somatic awareness exercises, 133–39; triggers and, 6
Trees! initiative (Gainesville, FL), 247–51
triggers and trigger warnings: about, 6, 34; for Butler's *Parable of the Sower*, 181; diversity of responses, 196; dysregulation, signs of, 35–36
Trott, Carlie D., 89–98
Turner, Robert, 277

Ulmer, Spring, 68, 69, 70
uncertainty: play and, 234, 260; Who can I be? and What can be done?, 299
UN Conference on the Human Environment (Stockholm, 1972), 51
undocumented students, 85–86
"Uninhabitable Earth, The" (Wallace-Wells), 56
United Kingdom, 285–90

United Nations Framework Convention on
    Climate Change (UNFCCC), 155
Un Lun Dun (Miéville), 287

values and beliefs, changed, 53
vegetalization of our sensorium, 248–49
Verlie, Blanche, 17–21
videos for change, 53
violence, slow, 215
vision papers, 118
Vizenor, Gerald, 167
vocation. *See* careers and vocation
Voskoboynik, Daniel, 157

Wakefield, Stephanie, 70
Walker, Alice, 142
Walker, Judith, 188
Wallace-Wells, David, 56, 63–64, 66
Wang, Susie, 195
Wardell, Emma, 279–84
Washington State, 272–77
Watts, Richard, 108–15
Western culture and perspective: analytic thinking in, 37; capabilities and, 18; emotion as hindrance under, 90; guilt complex and, 211; not imposing, 19; storytelling and, 166
"What Is an Indigenous Research Methodology?" (Wilson), 230–31
whiteness, 84, 150, 198
white supremacy, 2, 69, 99, 103
Who by Fire (Stenson), 187
whole-body sensing, 100, 102*table*
Whyte, Kyle, 166
wildfire smoke, 193, 275
Wilson, Isabella, 175
Wilson, Shawn, 228–31
Wirth, Jason, 251

Wise, Meghan, 8
Wong, Rita, 189
workshops: civic and community engagement (C&CE), 273*table*, 276; mock interviews, 304; sense self, developing, 300–302; telling your story, 182
World Trade Organization (WTO) protests, 1999, 22
World-Wide Teach-in for Climate Justice Education (Bard College), 9
worry: as climate stress response, 135*fig.*; definition of, 195; denial and, 19; as key emotion, 126*fig.*; objects of care and, 193–98
Wright, Emily (Em), 133–40
writing assignments: befriending the nervous system with creativity, 40; caring, 52–54; eco-grief lesson plan, 43–45; Ecotopia versus Zombie Apocalypse game, 254–62; emotional impact statements, 55–57; "Environmental Justice in the Anthropocene" course (Middlebury College), 74–77; "How will climate change affect my career?," 297–98; mega-dams, 112–13; Native/Indigenous rights and efficacy, 228–32; newspaper, fictional, 187–91; "Politics of Hope" course (Middlebury College), 59–71; possible futures portfolios, 302–4; "Principles of Environmental Justice" vision paper, art essay, and grassroots praxis paper, 116–20; radical imagination through storytelling, 177–82; skills audit, 300–301; speculative futures, 171–75, 174*table*; strengths and values inventory, 301–2; synergistic thinking, 38–39. *See also* journaling

Zapatistas, 156

Founded in 1893,
UNIVERSITY OF CALIFORNIA PRESS
publishes bold, progressive books and journals
on topics in the arts, humanities, social sciences,
and natural sciences—with a focus on social
justice issues—that inspire thought and action
among readers worldwide.

The UC PRESS FOUNDATION
raises funds to uphold the press's vital role
as an independent, nonprofit publisher, and
receives philanthropic support from a wide
range of individuals and institutions—and from
committed readers like you. To learn more, visit
ucpress.edu/supportus.